T0296251

CAMBRIDGE LIBRARY COLLECTION

Books of enduring scholarly value

Earth Sciences

In the nineteenth century, geology emerged as a distinct academic discipline. It pointed the way towards the theory of evolution, as scientists including Gideon Mantell, Adam Sedgwick, Charles Lyell and Roderick Murchison began to use the evidence of minerals, rock formations and fossils to demonstrate that the earth was older by millions of years than the conventional, Bible-based wisdom had supposed. They argued convincingly that the climate, flora and fauna of the distant past could be deduced from geological evidence. Volcanic activity, the formation of mountains, and the action of glaciers and rivers, tides and ocean currents also became better understood. This series includes landmark publications by pioneers of the modern earth sciences, who advanced the scientific understanding of our planet and the processes by which it is constantly re-shaped.

A System of Mineralogy

Robert Jameson (1774–1854) was a renowned geologist who held the chair of natural history at Edinburgh from 1804 until his death. A pupil of Gottlob Werner at Freiberg, he was in turn one of Charles Darwin's teachers. Originally a follower of Werner's influential theory of Neptunism to explain the formation of the earth's crust, and an opponent of Hutton and Playfair, he was later won over by the idea that the earth was formed by natural processes over geological time. He was a controversial writer, accused of bias towards those who shared his Wernerian sympathies such as Cuvier, while attacking Playfair, Hutton and Lyell. He built up an enormous collection of geological specimens, which provided the evidence for his *System of Mineralogy*, first published in 1808 and here reprinted from the second edition of 1816. Volume 2 continues 'earthy minerals' and covers saline and inflammable minerals, including coals.

A System of Mineralogy

VOLUME 2

ROBERT JAMESON

CAMBRIDGE
UNIVERSITY PRESS

CAMBRIDGE UNIVERSITY PRESS

Cambridge, New York, Melbourne, Madrid, Cape Town,
Singapore, São Paolo, Delhi, Tokyo, Mexico City

Published in the United States of America by Cambridge University Press, New York

www.cambridge.org
Information on this title: www.cambridge.org/9781108029742

This edition first published 1816
This digitally printed version 2011

ISBN 978-1-108-02974-2 Paperback

SYSTEM

OF

MINERALOGY.

A

SYSTEM

OF

MINERALOGY.

BY

ROBERT JAMESON,

REGIUS PROFESSOR OF NATURAL HISTORY, LECTURER ON MINERALOGY,
AND KEEPER OF THE MUSEUM IN THE UNIVERSITY OF EDINBURGH;
Fellow of the Royal and Antiquarian Societies of Edinburgh; President of the Wernerian Natural History Society; Honorary Member of the Royal Irish Academy, and of the Honourable Dublin Society; Fellow of the Linnean and Geological Societies of London, and of the Royal Geological Society of Cornwall;
Member of the Physical and Mineralogical Societies of Jena; of the Society of Natural History of Wetterau,
&c. &c. &c.

SECOND EDITION.

VOL. II.

EDINBURGH:

Printed by Neill & Company,

FOR ARCHIBALD CONSTABLE AND COMPANY, EDINBURGH; AND
LONGMAN, HURST, REES, ORME & BROWN, LONDON.

1816.

TABLE OF CONTENTS

OF

VOLUME SECOND.

SYSTEM OF ORYCTOGNOSY.

CONTINUATION OF CLASS I.—EARTHY MINERALS.

XVII. Hornblende Family.

XVIII. Chrysolite Family.

Basalt Family.

XX. Dolomite Family.

XXI. Limestone Family.

XXII. Apatite Family.

XXIII. Fluor Family.

XXIV. Gypsum Family.

XXV. Boracite Family.

XXVI. Baryte Family.

XXVII. Hallite Family.

CLASS II.—SALINE MINERALS.

FOSSIL SALTS.

Order I. Earthy Salts.

Genus 1. Salts of Alumina.

Genus 2. Salts of Magnesia.

Order II. Alkaline Salts.

Genus 1. Salts of Soda.

Genus 2. Salts of Potash.

Genus 3. Salts of Ammonia.

Order III.

Order III. Metallic Salts.

CLASS III.—INFLAMMABLE MINERALS.

I. Sulphur Family.

II. Bituminous Family.

III. Graphite

III. Graphite Family.

IV. Resin Family.

CONTENTS

OF

APPENDIX.

(*At the end of* Vol. II.)

NO. I. ADDITIONAL SPECIES.

NO. III. TABULAR VIEWS OF SYSTEMS OF MINERALOGY.

MINERAL SYSTEM.

CONTINUATION OF CLASS I.

EARTHY MINERALS.

XVII. HORNBLENDE FAMILY.

This Family contains the following species: Hornblende, Actynolite, Tremolite, Kyanite, Schiller-spar, Diallage, Bronzite, Anthophyllite, Hyperstene.

1. Hornblende.

Hornblende, *Werner*.

This species is divided into three subspecies, viz. Common Hornblende, Hornblende-Slate, and Basaltic Hornblende.

First Subspecies.

Common Hornblende.

Gemeiner Hornblende, *Werner.*

Corneus facie spatosa, striata, *Wall.* gen. 26. spec. 171.—Hornblende, *Kirw.* vol. i. p. 213.—Gemeiner Hornblende, *Estner.* b. ii. s. 699. *Id. Emm.* b. i. s. 322. & b. iii. s. 267.—Orniblenda commune, *Nap.* p. 276.—La Hornblende commune, *Broch.* t. i. p. 415.—Amphibole laminaire, *Hauy,* t. iii. p. 63. —Gemeiner Hornblende, *Reuss,* b. ii. s. 144. *Id. Lud.* b. i. s. 118. *Id. Suck.* 1r th. s. 118. *Id. Bert.* s. 185. *Id. Mohs,* b. i. s. 492. *Id. Hab.* s. 31. *Id. Leonhard,* Tabel. s. 24.— Amphibole schorlique commun, *Brong.* t. i. p. 452.—Gemeiner Hornblende, *Karsten,* Tabel. s. 38. *Id. Haus.* s. 91. —Amphibole lamellaire, *Hauy,* Tabl. p. 40.—Gemeiner Hornblende, *Steffens,* b. i. s. 304. *Id. Lenz,* b. i. s. 317. *Id. Oken,* b. i. s. 323.

External Characters.

Its colour passes from greenish-black into dark leek and olive green, and sometimes into liver-brown, and even into hair-brown.

It occurs massive, disseminated; and crystallised in the following figures:

1. Very oblique four-sided prism, in which the acuter lateral edges are sometimes truncated: when these truncations become larger, a six-sided prism is formed. The extremities of the prism are sometimes flatly acuminated by four planes, which are set either on the lateral edges or lateral planes: sometimes three of the acuminating planes increase so much, that the other three disappear,

appear, and thus there is formed an acumination of three planes; in other instances, two of the planes disappear, when the acumination is converted into a bevelment, which is either set on the lateral planes, or on the acuter or obtuser lateral edges; and sometimes three of the acuminating planes disappear, when the terminal planes of the prism appear set on obliquely *.

2. The preceding figure, with rounded lateral edges, forming a reed-like crystal.

The crystals are long and implanted, sometimes superimposed, and intersecting one another. They are deeply longitudinally streaked, and vary from middle-sized to very small.

Internally the lustre is shining and pearly.

The principal fracture is foliated, with a twofold oblique angular cleavage, in which the surfaces of the folia are longitudinally streaked: it is often also broad or narrow radiated, and either promiscuous or scopiform. The cross fracture is coarse-grained uneven.

The fragments are blunt-edged

The foliated varieties occur in concretions which are large, coarse, and fine, and generally long granular; the radiated varieties in wedge-shaped concretions.

The black coloured varieties are opaque, but the green generally translucent on the edges

It is intermediate between semi-hard and soft, but more inclining to the first.

It yields a mountain-green, inclining to greenish-grey coloured streak.

* According to Bournon, the primitive form of Hornblende is a rhomboidal tetrahedral prism, of 124° 30 and 55° 30, in which the terminal planes are inclined on the lateral edges 124° 30, so as to form with them angles of 105° and 75°.

It is rather brittle.

It is rather difficultly frangible.

Specific gravity, 3.202, 3.287, *Karsten*. 3.243, *Klaproth*.

Chemical Characters.

It melts before the blowpipe, with violent ebullition, into a greyish-black coloured glass.

Constituent Parts.

	Common Hornblende from Nora in Westmanland.
Silica, - - -	42.00
Alumina, - -	12.00
Lime, - -	11.00
Magnesia, - -	2.25
Oxide of Iron, -	30.00
Ferruginous Manganese,	0.25
Water, - - -	0.75
Trace of Potash.	
	98.25

Klaproth, Beit. b. v. s. 153.

Geognostic Situation.

It forms an essential ingredient in several mountain rocks; is sometimes accidentally intermixed with others; and it frequently occurs in beds of considerable magnitude. Thus, it forms an essential ingredient of syenite and primitive greenstone; also of transition syenite and greenstone; and of flœtz greenstone. It occurs accidentally in granite, gneiss, mica-slate, clay-slate, and porphyry; and beds of it, frequently associated with ores of different kinds, as magnetic-ironstone, and iron-pyrites, appear in gneiss, mica-slate, and clay-slate.

Geographic

Geographic Situation.

Europe.—It occurs very abundantly in Scotland, in greenstone, and syenite; and imbedded in limestone, gneiss, and mica-slate. It is found in similar rocks in England; and plentifully in the primitive and flœtz-trap rocks of Ireland. On the Continent, it occurs abundantly in Sweden; in Norway, as at Arendal, where it is associated with coccolite, felspar, quartz, granular limestone, titanite, and magnetic-ironstone; in Lower Saxony, as in the Hartz, where it forms a constituent part of transition greenstone; in Upper Saxony, as in the Erzgebirge; also in Hessia, Silesia, Franconia, Bavaria, Switzerland, Austria, Hungary, Transylvania, Italy, France, and Spain.

Asia.—It occurs very abundantly in many parts of Siberia, as Kolyvan, Irkutzk, Catharinenburgh, &c.

America.—In North America, it has been observed in primitive and flœtz, and also transition rocks, from Greenland, and the shores of Hadson's Bay to the Isthmus of Darien.

Observations.

1. This subspecies is characterised by its colour, frequent massive form, fracture, and lustre.

2. Hausmann, in his Entwurf, p. 92. mentions a subspecies of Hornblende, under the name *Talcaceous Hornblende*, of which he gives the following description in his Nordeutsche Beiträge:—" Its colour is intermediate between bronze and brass-yellow, and sometimes silver-white. It occurs in imbedded folia, like schillerstone, and also crystallised, in the following figures: 1. Oblique four-sided prism, the same as the primitive form of hornblende. 2. The preceding figure, bevelled on the extre-

 mities,

mities, the bevelling planes set on the obtuse lateral edges, 3. N° 2. in which the acute lateral edges are truncated. 4. N° 3. in which the edges formed by the meeting of the acuminating and bevelling planes are again truncated. The crystals are small and very small, generally single, seldom resting on one another. The bevelling and truncating planes are dull. Principal fracture foliated, with a double oblique angular cleavage; the cross fracture splintery or fibrous. The lustre of the principal fracture is splendent and metallic; of the splintery, dull; of the fibrous, pearly and glimmering. It is opaque, but translucent in thin folia. It is soft. Affords a greenish-grey streak. It is slightly flexible. It feels greasy. It occurs in primitive greenstone in the Hartz. Is nearly allied to Schillerstone."

Second Subspecies.

Hornblende-Slate.

Hornblende Schiefer, *Werner.*

Corneus rigidus non nitens, apparenter lamellis parallelis; Corneus fissilis, *Wall.*—Schistose Hornblende, *Kirw.* vol. i. p. 222. —La Hornblende schisteuse, *Broch.* t. i. p. 428.—Schiefriger Hornblende, *Reuss,* b. ii. s. 151.— Hornblende-Schiefer, *Lud.* b. i. s. 120. *Id. Suck.* 1r th. s. 238. *Id. Bert.* s. 187. *Id. Hab.* s. 32. *Id. Leonhard,* Tabel. s. 25.—Amphibole hornblende shisteux, *Brong.* t. i. p. 453.—Schiefrige Hornblende, *Karst.* Tabel. s. 38.—Hornblende-Schiefer, *Steffens,* b. i. s. 310.—Schiefrige Hornblende, *Lenz,* b. i. s. 321.—Hornblende-Schiefer, *Oken,* b. i s. 323.

External Characters.

Its colour is intermediate between greenish-black and blackish-green.

It

It occurs massive.

Internally it is glistening, passing into shining and pearly.

The fracture is slaty in the large, but promiscuous radiated in the small.

The fragments are thick and tabular.

It yields a greenish-grey coloured streak.

It is semi-hard, passing into soft.

It is rather difficultly frangible.

In other characters it agrees with the foregoing.

It is not always pure, being frequently intermixed with mica and felspar.

Geognostic Situation.

It occurs in beds, in granite, gneiss, mica-slate, sometimes also in clay-slate, and frequently along with beds of primitive limestone. It occasionally accompanies metalliferous beds, that contain magnetic-ironstone, chlorite, and other minerals. It is frequently intermixed with mica; and sometimes with quartz or iron-pyrites.

Geographic Situation.

Europe.—In Scotland, it occurs in gneiss, in the districts of Braemar and Aberdeen, in Aberdeenshire; in Banffshire, as near Portsoy; in Argyleshire, as in the islands of Coll, Tiree, &c.; in Inverness-shire, as in the islands Rona, Lewis, &c.; and in many other parts in Scotland; and also in England and Ireland, as will be mentioned in the third volume of this work. On the Continent, it occurs in Norway, Sweden, Saxon Erzgebirge, Lusatia, Bohemia, Silesia, Franconia, Bavaria, Moravia, Switzerland, Stiria, the Tyrol, Hungary, France, and Spain.

Asia.—It occurs abundantly in many places in Siberia, as Nertschinsk, Kolywan, and Catharinenburgh.

Third Subspecies.

Pasaltic Hornblende.

Basaltische Hornblende, *Werner.*

Schorl opaque rhomboidal, *Romé de Lisle,* t. ii. p. 379.—Basaltische Hornblende, *Wid.* s. 417.—Basaltine, *Kirw.* vol i. p. 219 —Basaltische Hornblende, *Estner,* b ii s. 719. *Id. Emm.* b. ii. s. 330. *Id.* b. iii. s. 269.—Orniblenda basaltica, *Nap.* p. 281.—Amphibole, *Lam.* t. ii. p. 330.—Amphibole crystallizée, *Hauy,* t. iii. p. 58.—Basaltische Hornblende, *Reuss,* b. ii. s. 159. *Id. Lud.* b. i. s 120. *Id Suck.* 1r th. s. 242. *Id. Bert.* s. 188. *Id. Mohs,* b. i. s. 500. *Id. Hab.* s. 32. *Id. Leonhard,* Tabel. s. 25.—Amphibole schorlique basaltique, *Brong.* t i. p. 452 —Basaltische Hornblende, *Haus.* s. 91. *Id. Karsten,* Tabel. s. 38. *Id. Steffens,* b. i. s. 311. *Id. Lenz,* b. i. s. 322. *Id. Oken,* b. i. s. 324.

External Characters.

Its colours are velvet-black or brownish-black.

It occurs crystallised, in the following figures:

1. Unequiangular six-sided prism, flatly acuminated with three planes, set on the alternate lateral edges of the prism *, fig. 109.
2. The preceding prism, flatly acuminated on one extremity by four planes, which are set on the four opposite lateral planes, and on the other extremity bevelled, the bevelling planes set on the two opposite lateral edges †, fig. 110.

3. The

* Amphibole dodecaedre, Hauy.

† Amphibole equi-different, Hauy.

3. The six-sided prism flatly acuminated at one extremity by three planes, which are set on the alternate lateral edges; on the other bevelled, the bevelling planes set on the opposite lateral edges *.
4. Six-sided prism, in which two opposite lateral planes are broader than the others, and doubly acuminated on the extremities; first, with four planes, which are set on those edges which one of the broader lateral planes always forms with an adjacent smaller one; and again acuminated with four planes, which are set on the first, under very obtuse angles †.

The crystals are small and middle sized, seldom large, and are imbedded, and all around crystallised. Their surfaces are smooth.

The lustre of the principal fracture is splendent and vitreous, approaching to pearly; that of the cross fracture is glistening.

The principal fracture is perfect and straight foliated, with a double oblique angular cleavage; the cross fracture is small-grained uneven.

The fragments are indeterminate angular, and sometimes indistinctly rhomboidal.

It is always opaque.

It is semi-hard, inclining to soft.

It is rather brittle.

It is more easily frangible than the preceding subspecies.

Specific gravity, 3.158, 3.199, *Karsten.*

Chemical

* Amphibole ondecimal, Hauy.

† Amphibole surcomposé, Hauy.

Chemical Characters.

Before the blowpipe, it melts into a black glass, but is rather more refractory than common hornblende.

Constituent Parts.

	Basaltic Hornblende from Fulda.
Silica, - - -	47.00
Alumina, - -	26.00
Lime, - - -	8.00
Magnesia, - -	2.00
Oxide of Iron, -	15.00
Water, - - -	0.50
	98.50

Klaproth, Beit. b. v. s. 154.

Geognostic Situation.

It occurs imbedded in basalt, along with olivine and augite; also in wacke and trap-tuff; in small quantity in some kinds of porphyry, and frequently in lava.

Geographic Situation.

Europe.—It occurs in the basalt of Arthur's Seat, and other similar hills around Edinburgh; also in the basalt of Fifeshire, and that of the islands of Mull, Canna, Eig, and Skye. It is also an inmate of the basaltic rocks in England and Ireland. Upon the Continent, it is very widely and abundantly distributed: thus, it occurs in the flœtz-trap rocks of Fulda, the Saxon Erzgebirge, Bohemia, Silesia, Bavaria, Austria, Stiria, Hungary, Transylvania, Italy, France, and Spain.

America.—It is frequent in the basaltic rocks of Mexico.

Observations.

Observations.

1. It is distinguished from the other subspecies of *Hornblende* by its colour, crystallization, lustre, and easy frangibility. It was confounded with Schorl, until Werner pointed out its characters and place in the system.

2. It decomposes more slowly than basalt: hence we frequently find unaltered crystals dispersed through the clay formed by the decomposition of basaltic rocks

3. Beyer describes a mineral under the name *Kohlenhornblende*, (Coal Hornblende), which appears to be nearly allied to hornblende; hence it deserves to be noticed in this part of the system. He describes it in the following terms:—Its colour is velvet-black, passing into brownish-black. It occurs massive and disseminated. The principal fracture is imperfect foliated, almost slaty, sometimes straight, sometimes curved, and inclining to fibrous; the cross fracture is small-grained uneven. The lustre of the principal fracture is shining and glistening, and pearly; the cross fracture glimmering, or dull. It is opaque. It affords a dark greenish grey-coloured streak. It is soft. It emits a clayey smell when breathed on. It occurs imbedded in pitchstone-porphyry, between Zwickau and Planitz *.

3. Actynolite.

---

* Vid. Beyer, in Crell's Chem. Annal. 2. 11. s. 381. Lenz, Tabel. 33. Leonhard, Tasch. b. i. s. 267.

2. Actynolite.

Strahlstein, *Werner*.

This species is divided into four subspecies, viz. Asbestous Actynolite, Common Actynolite, Glassy Actynolite, and Granular Actynolite.

First Subspecies.

Asbestous Actynolite.

Asbestartiger Strahlstein, *Werner*.

Asbestartiger Strahlstein, *Wid.* s. 479.—Amianthinite, *Kirw.* vol. i. p. 164.—Asbestartiger Strahlstein, *Emm.* b. i. s. 416. —Asbestoid, *Lam.* t. ii. p. 371.—Actinote aciculaire, *Hauy,* t. iii. p. 75.—La Rayonante asbestiforme, *Broch.* t. i. p. 504. —Asbestartiger Strahlstein, *Reuss,* b. ii. 1. s. 174. *Id. Lud.* b. i. s. 140. *Id. Suck.* 1r th. s. 252. *Id. Bert.* s. 156. *Id. Mohs,* b. i. s. 581. *Id. Hab.* s. 61. *Id. Leonhard,* Tabel. s. 30.—Amphibole actinote aciculaire, *Brong.* t. i. p. 455.—Asbestartiger Strahlstein, *Karst.* Tabel. s. 40. *Id. Haus.* s. 99.—Asbestinite, *Kid,* vol. i. p. 116.—Amphibole, *Hauy,* Tabl. p. 40.—Asbestartiger Strahlstein, *Steffens,* b. i. s. 281. *Id. Lenz,* b. ii. s. 683.—Strahlige Hornblende, *Oken,* b. i. s. 322.

External Characters.

Its colour is greenish-grey, which passes on the one side through mountain-green into a kind of sky-blue, on the other through olive-green into yellowish-brown, and liver-brown.

It

It occurs massive, and in capillary crystals: the crystals are sometimes elastic-flexible.

Internally the lustre is glistening and pearly.

The fracture is scopiform fibrous, sometimes passing into radiated.

The fragments are splintery and wedge-shaped.

It occurs in distinct concretions, which are wedge-shaped, and promiscuously aggregated.

It is opaque, or slightly translucent on the edges.

It is soft.

It is rather sectile.

It is rather difficultly frangible.

Specific gravity, 2.579, *Kirwan*. 2.809, *Karsten*.

Chemical Characters.

It melts with difficulty before the blowpipe, into a black or dark green coloured glass.

Constituent Parts.

Silica, - -	47.0
Lime, - -	11.3
Magnesia, - -	7.3
Oxide of Iron, -	20.0
Oxide of Manganese, -	10.0
Loss, - - -	4.4
	100.0

Vauquelin, in Hauy, t. iv. p. 335.

This is an analysis of the variety of Asbestous Actynolite, named *Byssolite* by Saussure.

Geognostic Situation.

It occurs in beds in gneiss, mica-slate, and granular limestone, along with magnetic ironstone, iron-glance, iron-

iron-pyrites, copper-pyrites, variegated copper-ore, malachite, galena, blende, common actynolite, amethyst, garnet, and asbestus.

Geographic Situation.

Europe.—In Norway, it occurs at Arendal, Kongsberg, and Röraas; in Sweden, at Sala, and other places; in the Hartz, the Saxon Erzgebirge, Bohemia, Franconia, Silesia, Switzerland, Hungary, Italy, and France.

America.—Greenland; and the metalliferous mountains of Zacatecas in Mexico.

Observations.

1. It is distinguished from the other subspecies by its colour-suite, capillary crystallizations, pearly lustre, fibrous fracture, and inferior degree of hardness. It is distinguished from *Asbestus* by its lustre, fracture, distinct concretions, and weight.

2. Those varieties of asbestous actynolite which occur in very thin scopiformly aggregated acicular elastic-flexible crystals, have been considered as forming a distinct species, and named *Byssolite* by Saussure, *Amianthoid* by Hauy, and *Asbestoid* by some French mineralogists.

Second

Second Subspecies.

Common Actynolite.

Gemeiner Strahlstein, *Werner.*

Basaltes radiis minimis, fibrosis, nitidis, compositus; Basaltes fibrosus, *Wall.* gen. 22. spec. 152.—Gemeiner Strahlstein, *Wid.* s. 480.—Schorlaceous Actynolite, and Common Asbestoid, *Kirw.* vol. i. p. 166. & 168.—Gemeiner Strahlstein, *Estner,* b. ii. s. 887. *Id. Emm.* b. i. s. 418.—Stralite commune, *Nap.* p. 323.—Zillerthite, *Lam.* t. ii. p. 357.—Actinote etale, *Hauy,* t. iii. p. 75.—La Rayonnante commune, *Broch.* t. i. p. 507. —Gemeiner Strahlstein, *Reuss,* b. i. s. 176. *Id. Lud.* b. i. s. 140. *Id. Suck.* 1r th. s. 140. *Id. Bert.* s. 185. *Id. Mohs,* b. i. s. 583. *Id. Hab.* s. 59. *Id. Leonhard,* Tabel. s. 31.—Amphibole Actinote hexaedre, *Brong.* t. i. p. 454.—Gemeiner Strahlstein, *Karst.* Tabel. s. 40. *Id. Haus.* s. 99.—Actynolite, *Kid,* vol. i. p. 116.—Amphibole comprimé, *Hauy,* Tabl. p. 40.—Gemeiner Strahlstein, *Steffens,* b. i. s. 284. *Id. Lenz,* b. ii. s. 681. *Id. Oken,* b. i. s. 322.

External Characters.

Its principal colour is leek-green: it occurs also olive-green, grass-green, and mountain-green.

It occurs massive, and disseminated.

Internally it is shining, inclining to glistening.

The fracture is small, scopiform, and promiscuous radiated, with a double oblique angular cleavage.

It frequently occurs in thick columnar concretions.

It is generally translucent on the edges, sometimes translucent.

It

It is semi-hard.

It is rather brittle.

It is rather difficultly frangible.

Specific gravity, 2.994, 3.293, *Kirwan.* 3.450, *Brisson.*

Chemical Characters.

Before the blowpipe, it melts into a greenish-grey or blackish glass.

Constituent Parts.

Silica, - -	64.00
Magnesia, - -	20.00
Alumina, - -	2.70
Lime, - - -	9.30
Iron, - - -	4.00
	100 *Bergmann.*

Geognostic Situation.

It occurs in beds in gneiss, mica-slate, and talc-slate, sometimes alone, sometimes accompanied with ores of different kinds, as galena, magnetic ironstone, copper-pyrites, and blende. Small and irregular veins occasionally occur in transition trap, and minute portions in flœtz-trap rocks.

Geographic Situation.

It occurs at Eilan Reach in Glenelg, in Inverness-shire; in the parish of Sleat, in the isle of Skye; different places in the isle of Lewis. In Cornwall, as in the neighbourhood of Redruth *. On the Continent, it is not

* Greenough.

not uncommon in Saxony, Bohemia, Silesia, Sweden, and Norway.

Observations.

1. This is the most common subspecies of Actynolite. It never occurs regularly crystallised; the crystallised varieties of actynolite formerly included under this subspecies being now referred by Werner to the Glassy Actynolite.

2. It has been frequently confounded with Epidote or Pistacite; but these minerals are distinguished from each other by the characters stated in vol. i. p. 97.

Third Subspecies.

Glassy Actynolite.

Glasartiger Strahlstein, *Werner.*

Glasartiger Strahlstein, *Wid.* s. 438.—Glassy Actynolite, *Kirw.* vol. i. p. 168.—Glasartiger Strahlstein, *Estner,* b. ii. s. 893. *Id. Emm.* b. i. s. 422.—Stralite vetrosa, *Nap.* p. 326.—La Rayonnante vitreux, *Broch.* t. i. p. 510.—Glasartiger Strahlstein, *Reuss,* b. i. s. 182. *Id. Lud.* b. i. s. 141. *Id. Bert.* s. 155. *Id. Mohs,* b. i. s. 386. *Id. Leonhard,* Tabel. s. 31. —Amphibole actinote fibreux, *Brong.* t. i. p. 455.—Glasartiger Strahlstein, & Muschlicher Strahlstein, *Karst.* Tabel. s. 40.—Glasartiger Strahlstein, *Haus.* s. 99 —Amphibole etalé et fibreux, (in part), *Hauy,* Tabl. p. 40.—Glasartiger Strahlstein, *Steffens,* b. i. s. 286. *Id. Lenz,* b. ii. s. 685. *Id. Oken,* b. i. s. 322.

External Characters.

Its colour is mountain-green, which passes into grass-green, and leek-green, also into greenish-white.

It occurs massive; and crystallised in the following figures:

Very oblique rhomboidal four-sided prism, in which either the acute or the obtuse edges are truncated, or both are truncated in the same crystal. Some crystals appear to be truncated on the terminal edges and angles *.

The crystals are middle-sized, also small, and very small; seldom single, generally many resting on one another, or they are scopiformly aggregated.

The external and internal lustre is vitreous, slightly inclining to pearly.

The fracture is narrow and straight radiated, sometimes passing into fibrous, generally scopiformly diverging, seldom parallel, and principally in the varieties in which the fracture inclines to fibrous.

The fragments are splintery and wedge-shaped.

It occurs in thick prismatic concretions, which inclose smaller ones of the same kind.

It is translucent.

It is brittle.

It is uncommonly easily frangible.

It is traversed by rents.

It is semi-hard: it scratches glass, but is scratched by quartz.

Specific gravity, 3.175, *Karsten*. 3.4050, *Hausmann*.

Chemical

* According to Count de Bournon, the primitive form of actynolite differs from that of hornblende, although they are said by Hauy to be identical. He says the primitive crystal of actynolite is a rhomboidal tetrahedral prism, of about 130° and 50°, in which the terminal planes are inclined to the axis, so as to form with the edges 50° of the prism angles of 94° and 85°.—Vid. Bournon's *Catalogue Min.* p. 86.

Chemical Characters.

Before the blowpipe, it melts difficultly into an opaque green-coloured glass.

Constituent Parts.

	Glassy Actynolite from Zillerthal in the Tyrol.
Silica, - -	50.00
Magnesia, - -	19.25
Alumina, - -	0.75
Lime, - - -	9.75
Potash, - -	0.50
Oxide of Iron, -	11.00
Oxide of Manganese,	0.50
Oxide of Chrome, -	3.00
Carbonic Acid, and Water,	5.00
Loss, - -	0.25

Laugier, Annales du Mus. t. v. p. 79.

Geognostic and Geographic Situations.

Europe.—It occurs in primitive rocks in the isle of Skye: in veins, along with rock-crystal, axinite, and epidote, at Bourg d'Oisans in Dauphiny; in beds of indurated talc, with limestone and common talc, on St Gothard; also in a similar repository in the Zillerthal, in the Tyrol; and in Sweden.

Asia.—It appears to be associated with talc at Bialoyarsk, in the Uralian Mountains.

Observations.

1. It is distinguished from the preceding subspecies by its vitreous lustre, crystallizations, and parallel cross rents.

2. Hauy is of opinion, that Actynolite and Hornblende belong to the same species, because his observations shew an identity in their primitive forms. Count de Bournon, on the contrary, proves that their primitive forms are not the same; hence, he infers that they are distinct species. These two minerals are further distinguished by their colour, fracture, crystallizations: the crystallizations of actynolite being few in number, and simple; whereas those of hornblende are more numerous, and complex; and, lastly, they differ remarkably in geognostic situation.

3. The fibrous varieties of glassy actynolite have been confounded with Amianthus; but they are distinguished from it by lustre, cross rents, and the rough feel of their powder.

Fourth Subspecies.

Granular Actynolite.

Körniger Strahlstein, *Werner.*

Körniger Strahlstein, *Karsten,* Tabel. n. 42. s. 91. *Id. Steffens,* b. i. s. 289. *Id. Lenz,* b. ii. s. 688. *Id. Oken,* b. i. s. 322.

External Characters.

Its colour is grass-green.

It occurs massive.

Internally its lustre is shining and vitreous.

The principal fracture is imperfect foliated, with a twofold cleavage: the cross fracture is splintery.

It occurs in large, coarse, and small granular distinct concretions.

It

It is faintly translucent.
It is semi-hard, approaching to hard.
It is rather brittle.
It is rather easily frangible.

Geognostic and Geographic Situations.

It occurs, along with precious garnet and quartz, in the Saualpe, and Tainach, in Stiria.

Observations.

The Smaragdite of Saussure, (the Green Diallage of Hauy), has been confounded with this subspecies of Actynolite; but it is distinguished from it by its pearly lustre, single cleavage, uneven cross fracture, inferior hardness, and sectility.

3. Tremolite *.

Tremolith, *Werner.*

This species is divided into three subspecies, viz. Asbestous Tremolite, Common Tremolite, and Glassy Tremolite.

* The name is derived from *Tremola*, a valley in the Alps, where it is said to have been first found. It would, however, appear, that it was first discovered in Transylvania by M. Von Fichtel, and described by him, under the name *Saulen* and *Stern-spath*: it was afterwards, in the year 1788, found in the valley of Tremola.

First Subspecies.

Asbestous Tremolite.

Asbestartiger Tremolith, *Werner*.

Asbestartiger Tremolith, *Emm.* b. i. s. 425. *Id. Estner*, b. ii. s. 893.—Grammatite, *Hauy*, t. iii. p. 227.—La Tremolith asbestiforme, *Broch.* t. i. p. 514.—Asbestartiger Tremolith, *Reuss*, b. i. s. 136. *Id. Lud.* b. i. s. 142. *Id. Suck.* 1r th. s. 272. *Id. Bert.* s. 166. *Id. Mohs*, b. i. s. 589. *Id. Leonhard*, Tabel. s. 31.—Grammatite, *Brong.* t. i. p. 475. *Id. Lucas*, p. 77. *Id. Brard*, p. 188.—Asbestartiger Tremolith, *Karsten*, Tabel. s. 44.—Amphibole blanc et soyeux, *Hauy*, Tabl. p. 41.—Asbestartiger Tremolith, *Steffens*, b. i. s. 290. *Id. Lenz*, b. i. s. 689.—Asbestartiger Grammatite, *Oken*, b. i. s. 327.

External Characters.

Its most common colour is greyish-white; it is found also yellowish-white, and greenish-white.

It occurs massive.

Internally it is shining, approaching to glistening, and is pearly *.

The fracture is fibrous, and is either scopiform or stellular fibrous.

The fragments are wedge-shaped, or splintery.

It occurs in distinct concretions, which are thick, wedge-shaped, and promiscuously aggregated.

It is translucent on the edges.

It is rather easily frangible.

It

* It has a lower lustre than any of the other subspecies.

It is soft, approaching to very soft.
It is rather sectile.
Specific gravity, 2.683, *Karsten*.

Physical Characters.

When struck gently, or rubbed in the dark, it emits a pale reddish coloured light; when pounded, and thrown on coals, a greenish coloured light. It phosphoresces more than any of the other subspecies.

Chemical Characters.

Before the blowpipe it melts into a white opaque mass.

Geognostic Situation.

It occurs most frequently in granular foliated limestone, which is either primitive or transition, or in dolomite: sometimes in chlorite; and more rarely in flœtz-trap rocks.

Geographic Situation.

It occurs, along with actynolite, in Glenelg in Inverness-shire; in dolomite in Aberdeenshire and Icolmkill; and in basalt in the Castle Rock of Edinburgh. In Norway, it is an inmate of transition limestone; in Bohemia, it is imbedded in limestone, along with calcareous-spar, slate-spar, brown-spar, fluor-spar, and quartz; at Dognatska in Hungary, with galena, copper-pyrites, iron-pyrites, compact and foliated magnetic ironstone, and garnet; Switzerland in dolomite; in granular limestone, along with augite, on Mount Vesuvius.

Second Subspecies.

Common Tremolite.

Gemeiner Tremolith, *Werner.*

Gemeiner Tremolith, *Estner,* b. ii. s. 901. *Id. Emm.* b. i. s. 426. —Grammatite, *Hauy,* t. iii. p. 227.—La Tremolithe commune, *Broch.* t. i. p. 515.—Gemeiner Tremolith, *Reuss,* b. i. s. 188. *Id. Lud.* b. i. s. 142. *Id. Suck.* 1r th. s. 274. *Id. Bert.* s. 164. *Id. Mohs,* b. i. s. 590.—Grammatite, *Lucas,* p. 77.—Tremolith, *Hab.* s. 61. *Id. Leonhard,* Tabel. s. 31.—Grammatite, *Brong.* t. i. p. 475. *Id. Brard,* p. 188.—Gemeiner Tremolith, *Karsten,* Tabel. s. 44.—Gemeiner Grammatite, *Haus.* s. 97.—Amphibole grammatite, *Hauy,* Tabl. p. 40.—Gemeiner Tremolith, *Steffens,* b. i. s. 291. *Id. Lenz,* b. ii. s. 691. —Gemeiner Grammatite, *Oken,* b. i. s. 327.

External Characters.

Its colours are greyish, greenish, yellowish and reddish white of these the most frequent is the greyish-white; also dark smoke-grey.

It occurs massive; and crystallised in the following figures:

1. Very oblique four-sided prism, truncated on the acute lateral edges.
2. Same prism, truncated on the obtuse lateral edges.
3. Same prism, truncated on all the lateral edges.
4. Same prism, bevelled on the obtuse lateral edges. When these bevelling planes increase so much that the original ones disappear, there is formed
5. An extremely oblique four-sided prism.
6. Very oblique four-sided prism, very flatly bevelled on

on the extremities, the bevelling planes set on the acute lateral edges *.

7. The preceding figure †, truncated on the acute lateral edges.
8. N° 6. truncated on all the lateral edges ‡.
9. N° 6., in which all the lateral edges are rounded off ||.

The lateral planes are longitudinally streaked.

The crystals are middle-sized or small; sometimes singly imbedded, sometimes superimposed, or promiscuously aggregated.

The lustre is shining, and intermediate between vitreous and pearly ¶.

The fracture is broad radiated, with a double oblique angular cleavage, which gives to the fracture-surface a longitudinally streaked appearance: the cross fracture is uneven §.

The

* Grammatite di-tetraedre, Hauy.

† Grammatite bis-unitaire, Hauy.

‡ Grammatite tri-unitaire, Hauy.

|| Grammatite cylindroide, Hauy.

¶ It has a higher degree of lustre than any of the other subspecies.

§ This mineral splits easily, not only in the direction of the planes of the prism, but also in that of its diagonals, particularly the longest diagonal. When we break across one of these prisms, we observe on the fracture-surface a line in the direction of the longer diagonal, which is so strongly marked, that at first sight we are apt to consider it as pointing out these as hemitrope or twin-crystals. The name *Grammatite*, formerly given to this mineral by Hauy, is derived from the character just stated. It is also worthy of remark, that in the fracture of tremolite, even in crystals, there is a tendency to the fibrous structure: the stroke of a hammer, or even the simple pressure of the finger in some cases, will separate folia or radia into fibres as delicate as those of amianthus, and which are somewhat elastic-flexible. All these characters are foreign to hornblende and actynolite, and may therefore be used as characters for distinguishing tremolite from these species.—Vide Bournon's *Catalogue Min.* p. 87, 88. where all the particulars here stated are more fully enumerated.

The fragments are indeterminate angular, seldom slightly rhomboidal.

It is translucent, or semi-transparent.

It occurs in large and coarse granular concretions, which sometimes approach to the diverging wedge-shaped columnar variety.

It scratches glass; and if we draw it very hard over the surface of quartz, it slightly scratches it.

It is rather brittle.

It is easily frangible.

Its powder is rough to the feel.

Specific gravity, 2.9257, 3.2, *Hauy*. 2.832, *Karsten*, 3.000, *Wid.*

Chemical Characters.

Before the blowpipe, it loses its colour and transparency, melts with great difficulty, often only on the edges, and with considerable ebullition, into an opaque glass.

Constituent Parts.

Silica,	27.0	Silica,	35.5	Silica,	52.0	Silica, -	50
Magnesia,	18.5	Magnesia,	16.5	Magnesia,	12.0	Magnesia,	25
Lime,	21.0	Lime,	16.5	Lime,	20.0	Lime, -	18
Alumina,	6.0	Carbonic acid		Carbon. acid,	12.0	Carbonic acid	
Carbon. acid,	26.0	and Water,	23.0	A trace of iron.		and Water,	5
Chenevix.		*Bucholz.*		*Lowiz.*		*Laugier.*	

Geognostic Situation.

Like the asbestous subspecies, it occurs principally in granular limestone, or dolomite, and in metalliferous beds. These beds contain, besides the tremolite, quartz, calcareous-spar, garnet, blende, galena, copper-pyrites, and vitreous

vitreous copper-ore, or copper-glance. It sometimes occurs in indurated talc, along with rhomb-spar; or in common talc, with calcareous-spar and rutile. It occurs rarely in serpentine and granite.

Geographic Situation.

Europe.—In Scotland, it is found in Aberdeenshire, and Inverness-shire; also in the island of Unst, one of the Zetland group of islands. On the Continent, it occurs at Kongsberg in Norway, along with ores of silver; in the island of Senjen in Nordland, in thin beds, resting on limestone, and covered with a bed of massive garnet; in Sweden, Hessia, Bohemia, Silesia, Moravia, Switzerland, the Tyrol, Carinthia, Carniola, Hungary, Transylvania, Italy, and France.

Asia.—It is found on the borders of the Lake Baikal.

America.—It occurs in Pennsylvania.

Africa.—Egypt.

Third Subspecies.

Glassy Tremolite.

Glasartiger Tremolith, *Werner.*

Glasartiger Tremolith, *Estner.* b. ii. s. 907. *Id. Emm.* b. i. s. 429.—Grammatite, *Hauy,* t. iii. p. 227.—La Tremolithe vitreuse, *Broch.* t. i. p. 516.—Glasartiger Tremolith, *Reuss,* b. i. th. 1. s. 193. *Id. Lud.* b. i. s. 145. *Id. Suck.* 1r th. s. 277. *Id. Bert.* s. 165. *Id. Mohs,* b. i. s. 392.—Grammatite, *Lucas,* p. 77.—Glasartiger Tremolith, *Leonhard,* Tabel. s. 31.—Tremolith, *Hab.* s. 61.—Grammatite, *Brong.* t. i. p. 475. *Id. Brard,* p. 188.—Glasartiger Tremolith, *Karsten,* Tabel. s. 44. *Id. Haus.* 97.—Amphibole Grammatite, *Hauy,* Tabl. p. 40.—Glasartiger Tremolith, *Steffens,* b. i. s. 294. *Id. Lenz,* b. ii. s. 694.—Glasartiger Grammatite, *Oken,* b. i. s. 327.

External Characters.

Its colours are greyish, greenish, yellowish, and reddish-white.

It occurs massive, and crystallised in acicular crystals.

Its lustre is shining, but in a lower degree than the preceding subspecies, and intermediate between vitreous and pearly.

The fracture is straight, long, narrow, and scopiform radiated, with frequent parallel cross rents.

The fragments are splintery.

It occurs in distinct concretions, which are thick wedge-shaped columnar, including thin columnar. The latter are promiscuously arranged.

It is translucent.

It

It is intermediate between soft and semi-hard.

It is very easily frangible.

It is rather brittle.

Specific gravity, 2.863, *Karsten.*

Physical Character.

It is phosphorescent in a low degree, but not by rubbing.

Chemical Character.

It is said to be infusible before the blowpipe.

Constituent Parts.

		Tremolite from St Gothard.					
Silica,	65.00	Silica,	35.5	Silica,	28.4	Silica,	41.00
Magnesia,	10.33	Lime,	26.5	Lime,	30.6	Lime,	15.00
Lime,	18.00	Magnesia,	16.5	Magnesia,	18.0	Magnesia,	15.25
Iron,	0.16	Water, and		Water, and		Water, and	
Carbon. acid		Carbon. acid,	23.0	Carbon. acid,	23.0	Carbon. acid,	23.00
and Water,	6.05		101.5		100.0	Loss, -	5.75
							100.0
Klaproth.		*Laugier*, Annales du Museum, 34 cahier, t. vi. p. 232.					

Geognostic Situation.

It is the same as that of the preceding subspecies, occurring principally along with granular limestone.

Geographic Situation.

Europe.—It occurs at Arendal in Norway; in Bavaria, Salzburg, the Tyrol, Switzerland, and Hungary.

Asia.—In the island of Ceylon; in the Uralian mountains.

Observations.

Observations.

1. Asbestous Tremolite might be confounded with *Amianthus*; but the following distinctive characters will enable us to discriminate them with facility:—The fracture of amianthus is parallel and curved fibrous, but that of asbestous tremolite is scopiform fibrous: amianthus does not occur in distinct concretions, as is the case with asbestous tremolite; the powder of amianthus is soft, whereas that of asbestous tremolite is dry and rough to the feel; and, lastly, amianthus is not phosphoric by friction, whereas asbestous tremolite becomes phosphoric by friction.

2. Hauy now considers tremolite but as a variety of actynolite, and actynolite as a variety of hornblende; consequently he arranges these three minerals under one head, as a single species, under the name *Amphibole*. Bournon having proved that these minerals do not agree in their primitive forms, and Werner having shewn their difference in other characters, we are still inclined to consider them as distinct species.

3. On a general view, Tremolite is characterised by its white colours: Actynolite by its light green colours; and Hornblende by its dark green colours.

4. Kyanite,

4. Kyanite, or Cyanite.

Kyanit, *Werner*.

Sappare, *Saussure*, Voyages, § 1900. & Jour. de Phys. 1789, p. 213.—Cyanite, *Wid.* s. 475. *Id. Kirwan*, vol. i. p. 209. *Id. Estner*, b. ii. s. 690. *Id. Emm.* b. i. s. 412. *Id. Nap.* p. 328. *Id. Lam.* t. ii. p. 256.—Disthene, *Hauy*, t. iii. p. 220. —La Cyanite, *Broch.* t. i. p. 501.—Cyanit, *Reuss*, b. ii. 2. s. 61. *Id. Lud.* b. i. s. 139. *Id. Suck.* 1r th. s. 463. *Id. Bert.* s. 285. *Id. Mohs*, b. i. s. 575. *Id. Hab.* s. 34.—Disthene, *Lucas*, p. 76.—Cyanit, *Leonhard*, Tabel. s. 30.—Disthene, *Brong.* t. i. p. 423. *Id. Brard*, p. 186.—Cyanit, *Karsten*, Tabel. s. 48.—Kyanit, *Haus.* s. 102.—Cyanite, *Kid*, vol. i. p. 182.—Disthene, *Hauy*, Tabl. p. 54.—Kyanit, *Steffens*, b. i. s. 299.—Cyanit, *Lenz*, b. ii. s. 696.—Talkschorl, *Oken*, b. i. s. 303.

External Characters.

Its colours are milk-white, passing into bluish-grey, Berlin-blue, and sky-blue, which latter borders on celandine-green. The white varieties are often marked with blue-coloured flame delineations.

It occurs massive, disseminated; and crystallised in the following figures:

1. Oblique four sided prism, with two opposite broad and two opposite narrow planes *.
2. Oblique four-sided prism, truncated on the two opposite acute lateral edges †, fig. 111.

3. Preceding

* Forme primitive, Hauy.

† Disthene perihexaedre, Hauy.

3. Preceding figure, in which all the lateral edges are truncated.
4. Twin-crystal: it may be considered as two flat four-sided prisms joined together by their broader lateral planes *.

The narrow lateral planes are longitudinally streaked, and glistening: the broad are smooth and splendent.

The crystals are middle-sized, small, and very small; are singly imbedded, or intersect one another.

The lustre is splendent and pearly.

The fracture is broad and slightly curved radiated, sometimes inclining to foliated, and is scopiformly or promiscuously radiated. The fracture of the crystals is foliated, with a threefold cleavage: of these cleavages, that parallel with the broad lateral planes is the most distinct; the others are much less so. The cross fracture is uneven.

The fragments are splintery, or imperfectly rhomboi dal.

It occurs in distinct concretions: the massive varieties in long granular, the radiated in wedge-shaped prismatic concretions.

The massive varieties are translucent; the crystals in general transparent.

It scratches glass: the broad planes can be scratched with steel; the narrow planes do not yield to it.

It is rather brittle.

It is easily frangible.

Specific gravity, 3.470, *Karsten.* 3.517, *Saussure.* 3.680, *Klaproth.*

Physical

* Disthene double, Hauy.

Physical Characters.

When pure, it is idio-electric. Some crystals, by rubbing, acquire negative electricity, even on perfectly smooth planes, others positive electricity: hence the name *Disthene* given by Hauy to this mineral, on account of its double electrical powers.

Chemical Character.

It is infusible before the blowpipe.

Constituent Parts.

Silica, -	29.2	Silica, -	38.50	Silica, -	43.00
Alumina,	55.0	Alumina,	55.50	Alumina,	55.50
Magnesia,	2.0	Lime, -	0.50	Iron, -	0.50
Lime, -	0.25	Iron, -	2.75	Trace of Potash.	
Iron, -	6.65	Water, -	0.75		——
Water, -	4.9		——		99.00
	——			*Klaproth*, Beit. b. v. s. 10.	
Saussure the Son, Voyages dans les Alpes, N° 1900.		*Laugier*, Annales du Mus. t. v. 25 cahier, p. 17.			

Geognostic Situation.

It has been hitherto found only in primitive mountains, where it occurs in whitestone, mica-slate, and talc-slate, accompanied with several other minerals.

Geographic Situation.

Europe.—It occurs in primitive rocks, near Banchory in Aberdeenshire; and in mica-slate in Mainland, the largest of the Zetland islands. At Airolo, on St Gothard, it is found in a beautiful silver-white mica-slate, associated with felspar, garnet, grenatite, and quartz; in the Saualp in Carinthia, with quartz, calcareous-spar, garnet, and common actynolite; in the Zillerthal in the Tyrol, with quartz and hornblende; and imbedded in whitestone at Waldenberg, in the Saxon Erzgebirge; also in France, Transylvania, Hungary, and Spain.

Asia.—In the Uralian Mountains; also in India.

America.—Near Baltimore; and at Maniquarez in South America.

Uses.

In India it is cut and polished, and sold as an inferior kind of sapphire.

Observations.

1. It is distinguished from *Actynolite* by its cleavage, and its infusibility; from *blue coloured Quartz*, and *Sapphire*, by its inferior hardness; from *Mica* by its superior hardness, its infusibility, and its being common flexible, whereas mica is elastic-flexible: from *Tremolite*, by colour, figure, and infusibility.

2. It was first described as a kind of schorl, under the names *violet schorl*, *blue schorl-spar*, *pseudo-schorl*; afterwards as belonging to the mica or talc species, under the names *blue mica* and *blue talc*. Some observers arranged it with felspar, and named it *skye-blue foliated felspar*; and by others it was denominated *foliated beryl*. It was Werner who first correctly pointed out its characters. Although it appears to be allied to talc, actynolite, and tremolite, yet the affinity is so distant, that at present it stands almost isolated in the system. If increasing the number of divisions in the system were not an evil, it might be an improvement to place it as a member of a distinct family, which would be named the Kyanite Family.

3. We

3. We sometimes meet with crystals, formed partly of kyanite, partly of grenatite, and the two substances at their junction are intermixed,—a proof of their cotemporaneous formation. It is also worthy of remark, as noticed by Steffens, that kyanite and grenatite agree very nearly in chemical composition.

4. Professor Nau, in the first volume of the Annals of the Society of Wetterau, describes, in the following terms, a mineral under the name *Fibrous Cyanite:* "Colour reddish-white, passing into flesh-red, and pale peach-blossom red; also yellowish, greenish, and bluish-grey; massive; dull, glistening, and silky; diverging, seldom parallel fibrous, which sometimes passes into perfect foliated; fragments splintery; opaque, or very feebly translucent on the thinnest edges; soft; white-coloured streak; difficultly frangible; 3.100. It occurs in gneiss, along with schorl, and titanitic iron-one, near Aschaffenburg."

5. Schlottheim, in the Magazine of the Society of the Friends of Natural History in Berlin, gives an account of a fossil from India, which he conjectures to be nearly allied to Kyanite, and names it *Sapparite.* The following is his description of it:

" *Sapparite* —Colour pale Berlin-blue, but when held in particular directions, shews a silver-white splendent opalescence.

It appears to be crystallised in rectangular four-sided prisms.

The longitudinal fracture is foliated; the cross fracture uneven, or imperfect conchoidal.

It is translucent.

It is semi-hard, inclining to soft.

It affords a pale greyish-white dull streak.

Geographic Situation.—It was brought from Pegu or Ceylon, imbedded in a druse of spinel crystals.

Observations.—It is distinguished from *Kyanite* by inferior hardness, opalescence, and fracture *.".

5. Schiller-Spar.

Schillerstein, *Werner.*

Schillerspath, ou Spath chatoyant, *Broch.* t. i. p. 421.—Schillerende Hornblende, *Reuss,* b. ii. 1. s. 153.—Schillerstein, *Lud.* b. i. s. 134. *Id. Suck.* 1r th. s. 134. *Id. Bert.* s. 532. *Id. Mohs,* b. i. s. 557. *Id. Hab.* s. 30. *Id. Leonhard,* Tabel. s. 28.—Diallage chatoyante, *Brong.* t. i. p. 442.—Smaragdit, *Karsten,* Tabel. s. 40.—Schillerende Hornblende, *Haus.* Nordeutsche Beit. b. i. s. 1.—Diallage metalloide, *Hauy,* Tabl. p. 47.—Schillerstein, *Steffens,* b. i. s. 371. *Id. Lenz,* b. ii. s. 661.

External Characters.

Its colours are celandine-green, leek-green, olive-green, which latter passes into pinchbeck-brown, and brass-yellow; also reddish-brown, silver-white, and greenish-black, inclining to emerald-green.

It occurs in folia, which are sometimes indeterminate angular, sometimes rounded or hexangular, and imbedded in parallel interrupted portions †.

The

* Schlottheim, Magaz. d. Gesellsch. Nat. Freunde zu Berlin, 1. 4. p. 303.

† The primitive form of Schillerstone, according to Hausmann, is the same as that of Hornblende, with this difference, that the folia parallel with the basis of the figure, are very distinct and large, whereas the others are indistinct, and of small dimensions. According to the more correct observations of Bournon, the primitive form is a rectangular four-sided prism, in which the bases, which are rectangles, are set on the broadest lateral planes, in such a manner as to form with them angles of 95° and 85°. The prism is

The lustre of the principal fracture is splendent and metallic; that of the cross fracture dull, where splintery, pearly glimmering where fibrous.

The principal fracture is foliated; the cross fracture splintery, sometimes passing into fibrous.

It is opaque, but translucent in thin folia.

The streak is greenish-grey, and dull.

It is sectile.

It is scratched by common hornblende.

It is very slightly common flexible.

It feels meagre.

Geognostic and Geographic Situations.

It occurs imbedded in serpentine in Fetlar and Unst in Zetland, and at Portsoy in Banffshire; in the greenstone rocks of Fifeshire; in the porphyritic rock of the Calton Hill *, and in similar rocks near Dunbarton; in serpentine between Ballantrae and Girvan in Ayrshire †. In Cornwall it occurs in serpentine aud hornblende-slate. At Basta in the Hartz, it is found in primitive greenstone, which rests on granite, associated with compact felspar, pinchbeck-brown mica, amianthus, (which appears passing into fibrous schillerstone), mountain-leather, precious serpentine, steatite, copper-pyrites, and iron-pyrites. Also disseminated in the serpentine of Zöblitz, of Gastein in Salzburg, and of the Pinzgau in the Tyrol.

is divisible in the direction of all the planes, but most easily parallel to that of the terminal planes. These terminal planes, and the cleavage, which is parallel with them, have a splendent metallic lustre, whilst the other planes of the primitive form, and the cleavages, are dull, somewhat resembling the surface of steatite and serpentine, with which substances the schillerstone seen in these directions has been frequently confounded.

* Bournon. † Allan.

6. Diallage.

Smaragdit, *Saussure,* Voyage, v. p. 198. § 1313.; p. 269. § 1362. —Feldspath vert, *R. de Lisle,* t. ii. p. 544.—Schorl feuilleté verdatre en grand lames, *De Born,* Catal. t. i. p. 380.—Emeraudite, Smaragdite, *Daubenton,* Tabl. p. 15.—Diallage verte, *Hauy,* t. iii. p. 126.—Schmaragdit, *Leonhard,* Tabel. s. 29. —Diallage verte, *Brong.* t. i. p. 442.—Schmaragdit, *Karsten,* Tabel. s. 40. *Id. Haus.* s. 92.—Diallage verte, *Hauy,* Tabl. p. 46.—Diallage, *Steffens,* b. i. s. 326.—Smaragdit, *Lenz,* b. ii. s. 658.—Diallage, *Oken,* b. i. s. 330.

External Characters.

Its colours are grass-green, apple-green, and hair-brown?

It occurs massive and disseminated.

Its lustre is shining and glistening, and pearly.

The fracture is foliated; twofold nearly rectangular cleavage; only one of the cleavages distinct.

The fragments are indeterminate angular.

It is translucent on the edges.

It is hard, inclining to semi-hard.

It is brittle.

It is difficultly frangible.

Specific gravity, 3.140, *Saussure.* 3.0, *Kopp.*

Chemical Characters.

It is said to melt easily before the blowpipe.

Constituent

Constituent Parts.

Silica, - - -	50.00
Magnesia, - -	6.00
Alumina, - -	11.00
Lime, - - -	13.00
Chrome, - - -	7.05
Iron, - - -	6.30
Copper, - - -	1.50

Vauquelin.

Geognostic and Geographic Situations.

Europe.—It occurs in the island of Corsica, along with saussurite; and with the same mineral on Mont Rosa in Switzerland; along with kyanite and precious garnet, in the Saualp in Carinthia; and in primitive rocks in Transylvania.

Asia.—In India, along with quartz and rutile.

America.—In Labrador, associated with saussurite.

Uses.

The mixture of diallage and saussurite, named *Gabbro* by the Italians, *Euphotide* by the French, and by artists *Verde di Corsica duro*, when cut and polished, has a beautiful appearance: hence it is much prized, and is cut into tables, and a variety of ornamental articles, such as snuff-boxes, ring-stones, &c.

Observations.

1. It is so nearly allied to Granular Actynolite, that it has been frequently confounded with it; but its pearly lustre, single cleavage, (at least the second is very indis-

 tinct,

tinct, and meets the other nearly at right angles,) uneven cross fracture, greater softness and sectility, distinguish it from that mineral. It is distinguished from *Green Felspar* by its inferior hardness; it scratches glass with difficulty, but felspar scratches it with ease: further by its fracture; in felspar, both cleavages are distinct, in diallage only one: from *Hornblende*, by its nearly rectangular cleavage, and having but one perfect cleavage.

2. It was formerly considered by some as a felspar, by others as a schorl, or hornblende, or of the nature of emerald. Hauy remarks, that as the minerals with which this substance had been confounded, have at least two distinct cleavages, whereas it has but one, he chose for it a name which would recall this difference: hence the origin of the name *Diallage*, which signifies difference.

7. Bronzite.

Bronzit, *Karsten*.

Blättriger Anthophyllith, *Werner*.

Bronzit, *Leonhard*, Tabel, s. 29.—Diallage metalloide, *Brong.* t. i. p. 443.—Bronzit, *Karst.* Tabel. s. 40 —Diallage metalloide, *Hauy*, Tabl. p. 47.—Bronzit, *Steffens*, b. i. s. 325. *Id. Lenz*, b. ii. s. 663. *Id. Oken*, b. i. s. 330.

External Characters.

Its colour is intermediate between yellowish-brown and pinchbeck-brown, sometimes approaching to brass-yellow.

It occurs massive, and coarsely disseminated.

Its lustre is shining, and semi-metallic.

Its

The fracture is foliated, with a very distinct single cleavage: the fracture-surface is streaked

The fragments are blunt-edged.

It occurs in distinct concretions, which are coarse granular.

It is opaque in the mass, but translucent in thin folia.

It affords a white-coloured streak.

It is semi-hard.

It is very brittle.

It is very easily frangible.

Specific gravity, 3.200, *Klaproth.*

Chemical Characters.

It is infusible before the blowpipe.

Constituent Parts.

Silica, - -	60.00
Magnesia, - -	27.50
Iron, - - -	10.50
Water, - - -	0.50
	98.50

Klaproth, Beit. b. v. s. 34.

Geognostic and Geographic Situations.

Europe.—It occurs in a syenitic rock in Glen Tilt in Perthshire; and in large masses in beds of serpentine, near Kraubat in Upper Stiria.

America.—In the island of Cuba.

Observations.

1. It differs from *Smaragdite,* with which it has been confounded, not only in external characters, but also in chemical composition: Bronzite contains neither alumina nor

nor lime, but smaragdite both; and the quantity of magnesia in bronzite is four times that in smaragdite. Bronzite is infusible before the blowpipe, but diallage melts into a slag.

2. Its single cleavage distinguishes it from *Anthophyllite*, in which there is a distinct double cleavage, and the surfaces of these cleavages are smooth and shining.

8. Anthophyllite.

Anthophyllith, *Schumacher*.

Anthophyllith, *Schumacher*, Verzeichniss, s. 96. *Id. Leonhard*, Tabel. s. 42. *Id. Brong*. t. i. p. 444. *Id. Karsten*, Tabel, s. 32. *Id. Haus*. s. 92. *Id. Hauy*, Tabl. p. 58. *Id. Steffens*, b. i. s. 324. *Id. Lenz*, b. i. s. 527.

External Characters.

Its colour is intermediate between dark yellowish-grey and clove-brown.

It occurs massive, and crystallized in reed-like crystals, which appear to be four-sided prisms. The surface of the crystals is longitudinally streaked.

The lustre is shining and glistening, and pearly, approaching to semi-metallic.

The fracture is radiated: the surface of the rays streaked; double cleavage, parallel with the sides of a rectangular prism: other less distinct cleavages are to be seen parallel with the diagonal of the prism. The radiated fracture is sometimes scopiform, sometimes promiscuous.

It sometimes occurs in wedge-shaped and long granular distinct concretions.

It is translucent on the edges, or translucent.

It

It is hard, but in a low degree: it scratches fluor-spar with ease, but glass with difficulty.

It is rather easily frangible.

Specific gravity, 3.285, *Hauy*. 3.118, *Schumacher*. 3.156, *John*.

Chemical Characters.

It becomes dark greenish-black before the blowpipe, but is infusible.

Constituent Parts.

Silica, - - -	56.00
Alumina, - -	13.30
Magnesia, - -	14.00
Lime, - -	3.33
Iron, - -	6.00
Oxide of Manganese,	3.00
Water, - - -	1.43

John, Chem. Untersuchungen, 1. s. 200, 201.

Geognostic and Geographic Situations.

It occurs at Kongsberg in Norway, along with hornblende; and it is said to have been discovered in granite, near Microw in Mecklenburg *.

Observations.

It is named *Anthophyllite*, from the resemblance of its colour to that of the Anthophyllum. This name was given to it by Schumacher, the naturalist who first described it.

9. Hyperstene.

* Leonhard, Taschenbuch, t. iv. s. 256.

9. Hyperstene.

Hyperstene, *Hauy.*

Labrador Hornblende, *Kirw.* vol. i. p. 221.—Diallage metalloide, *Hauy,* t. iii. p. 127.—Labradorische Hornblende, *Leonhard,* Tabel. s. 25.—Hyperstene, *Brong.* t. i. p. 444. *Id. Karst.* Tabel. s. 40. *Id. Haus.* s. 98. *Id. Hauy,* Tabl. p. 44. *Id. Steffens,* b. i. s. 322. *Id. Lenz,* b. ii. s. 664. *Id. Oken,* b. i. s. 328.

External Characters.

Its colours are greyish-black, greenish-black, and brownish-black, which latter passes into chesnut-brown.

Internally it reflects a colour intermediate between copper-red, pinchbeck-brown, and gold-yellow.

It has hitherto been found only in rolled pieces.

Internally its lustre is shining and semi-metallic.

Its fracture is perfect foliated, with a twofold cleavage, in which the folia meet under angles of 80° and 100°.

The fragments are rather oblique rhomboidal.

It occurs in large granular and lamellar concretions.

It is opaque.

It becomes greenish-white in the streak.

It is so hard as to scratch common hornblende.

Specific gravity, 3.390, *Klaproth.* 3 376, *Karsten.*

Chemical Character.

It is infusible before the blowpipe.

Constituent

Constituent Parts.

Silica, - -	54.25
Magnesia, - -	14.00
Alumina, - -	2.25
Lime, - - -	1.50
Oxide of Iron, -	24.50
Water, - -	1.00
Oxide of Manganese, a trace.	
	97.50

Klaproth, Beit. b. v. s. 40.

Geognostic and Geographic Situations.

It has hitherto been found only on the coast of Labrador, where it occurs as a constituent part of a rock composed of Labrador-felspar, and sometimes also of common hornblende, and magnetic ironstone.

Observations.

This mineral, although nearly allied to Anthophyllite, differs from it in being harder, heavier, its cleavage less distinct, and its lustre more metallic.

XVIII. CHRYSOLITE

XVIII. CHRYSOLITE FAMILY.

This Family contains the following species: Sahlite, Augite, Diopside, Chrysolite, Olivine, and Lievrite.

1. Sahlite.

Sahlit, *Werner.*

Sahlit, *D'Andrada,* Scherer's Journal, b. iv. 19. s. 81. *Id. Schumacher,* Verzeichniss, s. 32. *Id. Hauy,* t. iv. p. 379. *Id. Broch.* t. ii. p. 518. *Id. Reuss,* b. ii. 1. s. 474. *Id. Lud.* b. ii. s. 158.—Malacolith, *Suck.* 1r th. s. 186. *Id. Bert.* s. 162. *Id. Mohs,* b. i. s. 488.—Malacolithe, *Lucas,* p. 201.—Sahlit, *Leonhard,* Tabel. s. 31.—Malacolithe, *Brong.* t. i. p. 445. *Id. Brard,* p. 414.—Sahlit, *Karsten,* Tabel. s. 44.—Salait, *Haus.* s. 98.—Pyroxene laminaire gris-verdatre, *Hauy,* Tabl. p. 42. —Malacolith, *Steffens,* b. i. s. 354.—Sahlit, *Lenz,* b. ii. s. 700. —Schaliger Pyroxene, *Oken,* b. i. s. 333.

External Characters.

Its colour is greenish-grey, passing into a variety intermediate between mountain-green and asparagus-green.

It occurs massive, and crystallised in rectangular four-sided prisms, in which the terminal planes are set obliquely on the lateral planes *. A great many varieties of

* According to Bournon, the primitive form of Sahlite is a rectangular four-sided prism, having rectangular bases, which are inclined on the two opposite sides of the prism, so as to form angles of 106° 15′, and 73° 45′. This measurement, which was made with the greatest care, differs from that of Hauy, who affirms that the primitive form is the same as that of augite.

of form, proceeding from this figure, are described by Count de Bournon, in his valuable *Catalogue Mineralogique.*

The crystals are occasionally superimposed, and are middle-sized and small.

Internally the lustre of the principal fracture is shining, splendent and vitreous; that of the cross fracture dull.

The fracture is foliated, with a fivefold cleavage: one of the cleavages is parallel with the terminal planes; two with the lateral planes; and two with the diagonals of the prism: the prism splits easily in the direction of all the planes, but most easily in the direction of the terminal planes. The cross fracture is uneven.

The fragments are sometimes rhomboidal.

It occurs in straight lamellar concretions, in which the the surfaces of those parallel with the most distinct cleavage are shining and pearly; also in coarse granular concretions.

It is translucent.

It is semi-hard; not so hard as augite, which readily scratches it.

It is rather brittle.

It is easily frangible.

Specific gravity, 3.223, *Hauy.* 3.236, *D'Andrada.* 3.473, *Dr Wollaston.*

Chemical Characters.

It melts with great difficulty before the blowpipe.

Constituent

Constituent Parts.

Silica, - -	53.00
Magnesia, - -	19.00
Alumina, - -	3.00
Lime, - -	20.00
Iron, and Manganese,	4.00

Vauquelin, in Hauy, t. iv. p. 302.

Geognostic Situation.

It occurs in beds in primitive mountains.

Geographic Situation.

Europe.—It occurs in the island of Unst in Zetland: in granular limestone in the island of Tiree, one of the Hebrides: in the silver mines of Sala, in Westmannland in Sweden, associated with asbestous actynolite, calcareous-spar, iron-pyrites, and galena; near Arendal in Norway, along with magnetic ironstone, common hornblende, calcareous-spar, and seldom with felspar and black mica.

Asia.—At Odon-Tschelong, near the river Amour in Siberia, along with beryl, mica, and calcareous-spar.

America.—On the banks of Lake Champlain.

Observations.

1. Werner, Bournon, and other mineralogists, consider it as a distinct species; whereas Hauy, and many French mineralogists, view it but as a variety of *Augite*. The following are some of the characters in which they differ, and which shew that they are distinct species. Both species have green colours, but those of sahlite are pale, whereas those of augite are dark: the primitive form of sahlite is not the same with that of augite; in sahlite, the

the prism splits easily in the direction of all the cleavages, but most easily in the direction of the terminal planes; in augite, the splitting is effected with difficulty in every direction, but with the greatest difficulty in the direction of the terminal planes of the prism: the cross fracture of sahlite is uneven and dull; that of augite, although uneven, generally inclines to conchoidal, and has a considerable degree of lustre: the distinct concretions of sahlite have a shining surface, which is not the case with those of augite: lastly, it contains more lime and magnesia, and less iron, than augite.

2. It was first observed and described by M. D'Andrada of Lisbon, who named it *Sahlite*, from Sala, the place where he first found it.

2. Augite *.

Augit, *Werner*.

This species is divided into two subspecies, viz. Common Augite, and Coccolite.

* Augite is the name applied to a shining stone by Pliny, and is derived from the Greek word *αὐγή*, *lustre*.

First Subspecies.

Common Augite.

Gemeiner Augit, *Steffens.*

Schorl des Volcans, *Daubenton*, Tabl. p. 11.—Augite, *Estner*, b. ii. s. 129. *Id. Emm.* b. iii. s. 241.—Volcanite, *Lam.* t. ii. p. 327.—L Augite, *Broch.* t. i. p. 179.—Pyroxene, *Hauy*, t. iii. p. 80.—Augite, *Reuss*, b. ii. s. 138. *Id. Lud.* b. i. s. 62. *Id. Suck.* 1r th. s. 188. *Id. Bert.* s. 158. *Id. Mohs*, b. i. s. 49.—Pyroxene, *Lucas*, p. 57.—Augit, *Hab.* s. 29. *Id. Leonhard*, Tabel. s. 2.—Pyroxene Augite, *Brong.* t. i. p. 447. *Id. Brard*, p. 148.—Augit, *Karsten*, Tabel. s. 40. *Id. Haus.* s. 98.—Pyroxene, *Hauy*, Tabl. p. 41.—Gemeiner Augit, *Steffens*, b. i. s. 340.—Augite, *Hoff.* b. i. s. 448. *Id. Lenz*, b. i. s. 208, 209.—Pyroxen, *Oken*, b. i. s. 335.

External Characters.

Its principal colour is blackish-green, which sometimes passes on the one hand into leek-green, olive-green, and liver-brown, on the other into greenish-black, and velvet-black.

It occurs massive, in roundish grains; and crystallised in the following figures:

1. Unequiangular six-sided prism, in which the terminal planes are set obliquely on the lateral planes *, fig. 112 †.

2. The

* According to Hauy, the primitive form is an oblique rhomboidal prism, the alternate angles of which are 92° 18′, and 87° 42′.

† Pyroxene perihexaedre, Hauy.

2. The preceding figure, truncated on the acute lateral edges, forming an eight-sided prism.
3. N° 2. in which the four edges between the broader and smaller lateral planes are truncated, thus forming a twelve-sided prism *.
4. N° 1. flatly bevelled on the extremities, the bevelling planes set obliquely on the acuter lateral edges, and the proper edge of the bevelment oblique †, fig. 113.
5. The preceding figure, truncated on the acute lateral edges ‡, fig. 114.
6. The preceding figure, in which the proper edge of the bevelment is truncated ||.
7. N° 5. in which the acute angle of the bevelment is horizontally truncated by a slightly curved face §, fig. 115.
8. N° 5. in which the obtuse terminal edges of two and two opposite planes, that meet under acute angles, are truncated ¶, fig. 116.
9. Eight-sided prism, acuminated with four planes, which are set on the alternate lateral planes, and two of the acuminating edges truncated **.
10. The preceding figure, in which the apex of the acumination, and also the four planes on which the acumination are set, and all the edges which

 are

* **Pyroxene equivalent, Annales du Museum, t. x. p. 13. fig. 3.**

† **Pyroxene bis-unitaire, Hauy.**

‡ **Pyroxene tri-unitaire, Hauy**

|| **Pyroxene sexogonal, Hauy.**

§ **Pyroxene soustractif, Hauy.**

¶ **Pyroxene dioctaedre, Hauy.**

** **Pyroxene octoduodecimal, Hauy.**

are not altered in the preceding figure, semi-truncated *.

11. Twin-crystal, in which the crystals are joined together by their broader lateral planes †, fig. 117.

12. Twin-crystal, in which the crystals intersect each other ‡, fig. 118.

The crystals are all around crystallised, also imbedded, and superimposed: are middle-sized, and small, seldom large.

Internally it alternates from specular splendent to glistening, and is resinous.

The fracture varies considerably, being sometimes fine grained uneven, sometimes foliated, from imperfect to highly perfect foliated: this latter occurs principally in crystals, where it forms the longitudinal fracture. The cross fracture is conchoidal. In the foliated fracture, a twofold slightly oblique angular cleavage is to be observed. Besides these, a concealed foliated fracture occurs.

The massive variety occurs in distinct concretions, which are coarse and small granular, and very much grown together.

It extends on the one side from translucent nearly to opaque; on the other nearly to transparent, as is particularly the case with the foliated and conchoidal.

It scratches glass with difficulty.

It is brittle.

It is rather easily frangible.

Specific gravity, 3.226, *Hauy*. 3.286, *Karsten*.

Chemical

---

* Pyroxene trioctonal, Hauy.

† Pyroxene hemitrope, Hauy.

‡ Pyroxene hemitrope, Hauy.

Chemical Characters.

It melts before the blowpipe, but with difficulty, and only in small pieces.

Constituent Parts.

From Etna.		From Frascati.		From the Rhongebirge.		From Norway	
Silica,	52.50	Silica,	48.00	Silica,	52.00	Silica,	50.25
Magnesia,	10.00	Magnesia,	8.75	Magnesia,	12.75	Magnesia,	7.00
Alumina,	3.30	Alumina,	5.00	Alumina,	5.75	Alumina,	3.50
Lime,	13.20	Lime,	24.00	Lime,	14.00	Lime,	25.06
Iron,	14.66	Iron,	12.00	Iron,	12.25	Iron,	10.50
Manganese,	2.00	Manganese,	1.00	Manganese,	0.25	Manganese,	2.25
Loss,	4.00	Trace of Potash.		Water,	0.25	Water,	0.50
						Trace of Chrome.	
Vauquelin.		*Klaproth.*		*Klaproth.*		*Simon.*	

Geognostic Situation.

It occurs in basalt, either alone, or with olivine; also in some flœtz greenstones. It occurs massive, aggregated with other minerals, or in drusy cavities in the beds of primitive trap at Arendal in Norway; and the minerals with which it is associated in these beds, are garnet, hornblende, epidote, felspar, mica, actynolite, asparagus-stone, calcareous-spar, and magnetic ironstone. A rock entirely composed of augite, and named *Augite-rock*, occurs in beds of considerable magnitude in primitive limestone in the Pyrenees. Masses of ejected granular foliated limestone, containing drusy cavities lined with augite, hornblende, meionite, vesuvian, felspar, and nepheline, are found in the neighbourhood of Vesuvius. These masses are by Mohs conjectured to belong to a particular flœtz formation; but Von Buch supposes that they are derived from beds in mica-slate or gneiss. It is found abundantly in the lavas of Vesuvius, Etna, Stromboli, and other volcanoes.

Geographic Situation.

Europe.—In Scotland, it is imbedded in the basalt of Arthur's Seat, and in the basalt, and other flœtz-trap rocks around Edinburgh; in the basalt hills and cliffs of Fifeshire, and other flœtz-trap districts in the mainland of Scotland; also in the Orkney islands, and those of Arran, Mull, Eigg, Canna, Rume, and Skye. In Ireland, it is not unfrequent in the flœtz-trap rocks of the northern districts of that island. On the Continent, it is very widely distributed: thus, it occurs at Arendal in Norway, in Hessia, Saxon Erzgebirge, Bohemia, Silesia, the Tyrol, Hungary, Transylvania, Italy, and France

Africa.—In the basalt of the islands of St Helena, Teneriff, and Isle de Bourbon.

America.—In the flœtz-trap rocks of Mexico and the United States.

Observations.

1. Distinctive Characters.—*a.* Between Augite and *Basaltic Hornblende.* Basaltic hornblende has always a velvet-black colour, augite most generally a green colour; the lustre of basaltic hornblende is vitreous, that of augite resinous; basaltic hornblende is softer than augite; and a simple chemical distinction may be added, basaltic hornblende is easily fusible before the blowpipe, augite difficultly —*b.* Between Augite and *Schorl.* The crystals of schorl have their lateral planes longitudinally streaked, those of augite are smooth; the lustre of schorl is vitreous, that of augite resinous; it is harder than augite;

gite; it becomes electrical by heating, but augite does not; and schorl is more easily fusible than augite.—*c.* Between cross twin-crystals of Augite, and cross crystals of *Grenatite.* In grenatite, the crystals cross each other under angles of 60° and 90°; in augite, they cross each other less determinately.—*d.* Between Augite and *Olivine.* In olivine, the colours are lighter, the crystallizations are different, and the hardness and specific gravity less considerable than in augite.—*e.* Between Augite and *Epidote.* In epidote, the green colours are lighter than those of augite: further, its crystallizations, and cleavage, are not the same; and in general, it is a more transparent mineral.

2. It is by Werner divided into four subspecies, viz. Granular, Foliated, Conchoidal, and Common; and the following are a few of the characters used by him in their description:

(1.) *Granular Augite.*—Colour is greenish black. Occurs massive; and crystallised in the following figures: 1. Broad six-sided prism, with two opposite acute lateral edges; and generally flatly bevelled on the extremities. 2. Six-sided prism, acuminated with four planes, which are set on the obtuse lateral edges. 3. Six-sided prism, with convex terminal planes. 4. N° 1. truncated on the acute lateral edges. When these increase, so as to be of equal magnitude with the lateral planes, there is formed, 5. A broad and nearly equiangular eight-sided prism. 6. When the two broader lateral planes of N° 1. disappear, a four-sided prism is formed. Crystals seldom perfectly sharp-edged; generally superimposed, and form druses. Internally glistening, and resinous. Fracture imperfect foliated, or uneven. Occurs in large and small angulo-granular distinct

 concretions.

concretions. It is opaque. This subspecies has hitherto been found only at Arendal in Norway.

(2.) *Foliated Augite.*—Colour passes from velvet-black through greenish-black, into blackish-green, and sometimes approaches dark leek-green. Occurs only crystallised, and its crystallizations are nearly the same with those of the granular subspecies. Internally shining, approaching to splendent, and the lustre intermediate between resinous and vitreous. Principal fracture perfect foliated, with a double slightly oblique angular cleavage, parallel with the smaller lateral planes of the six-sided prism; also a third cleavage, parallel with the truncations on the acuter edges of the six-sided prism. Cross fracture conchoidal. It is sometimes opaque, sometimes translucent on the edges: the crystals which occur in the basalt of Bohemia, in general belong to this subspecies; also those from Frascati, Etna, and Vesuvius.

(3.) *Conchoidal Augite*.—Colour greenish-black, passing into blackish-green; also into a very rare dark olive-green, sometimes even into liver-brown. Occurs in imbedded grains. Lustre splendent, and intermediate between resinous and vitreous. Fracture imperfect, but flat conchoidal. Translucent on the edges, or translucent. It occurs only in the flœtz-trap formation, and is the rarest of the four subspecies. It occurs in the basalt of Fulda, and near Cassel in Hessia.

(4.) *Common Augite.*—Colour blackish-green, and velvet-black. Occurs in large and small imbedded grains. Internally intermediate between shining and glistening, and lustre resinous. Fracture coarse and small grained uneven. Translucent on the edges, seldom translucent. Occurs in the flœtz-trap formation.

3. Steffens, in his Handbuch, describes a species under the name *Keraphyllite*, which Karsten and Werner refer

refer to the foliated augite, and Hauy to hornblende. As far as I can judge from the accounts published by mineralogists, it appears to be very nearly allied to augite; yet the following description shews that it cannot well be arranged either with augite or hornblende. "Colour generally dark greenish-black, seldom velvet-black. Occurs massive and disseminated. Internally specular-splendent on the principal fracture, and the lustre intermediate between resinous and vitreous. Fracture perfect and straight foliated, with double oblique angular cleavage, under an angle of 55° ½′; cross fracture conchoidal. Fragments very sharp-edged. Equally hard with augite. Specific gravity, 3.085, *Klaproth*. It melts with difficulty into an uniform olive-green, opaque, and shining slag. Silica, 52.52. Alumina, 7.25. Magnesia, 12.50. Lime, 9.00. Potash, 0.50. Oxide of iron, 16.25." *h laproth*, Beit. B. iv. s. 189. It is found in the Saualp in Carinthia, in a bed in primitive rocks, along with quartz, cyanite, garnet and zoïsite. It is distinguished from *Augite* by its splendent lustre, perfect conchoidal cross fracture, more perfect foliated fracture, different cleavage, and inferior weight: from *Hornblende* by its stronger lustre, and kind of lustre, perfect conchoidal cross fracture, greater hardness, and the action of the blowpipe on it *.

4. Karsten describes a mineral, under the name Slaggy Augite, (*Schlackiger Augite*), in Klaproth's Beiträge, vol. iv. p. 190. It is by some referred to the Conchoidal Augite of Werner, but which differs from augite in specific

* Vid. Karsten, in Klap. Beit. b. iii. s. 185.; Karsten, Tabel, s. 40.; Hauy, Annales du Mus. t. xiv. p. 290.; Leonhard, Taschenbuch, b. iv. s. 132.

cific gravity, fusibility, and composition. The following is the account given of it by Karsten and Klaproth:

Slaggy Augite.—Its colour is deep black, approaching in some places to dark leek-green. Occurs massive, and coarsely disseminated. Internally shining and resinous. Fracture small and perfect conchoidal. Fragments very sharp-edged. Opaque. Hard. Specific gravity, 2.666, *Karsten.* It intumesces before the blowpipe. Silica, 55.00. Alumina, 16.50. Magnesia, 1.75. Lime, 10.00. Oxide of Iron, 13.75. Trace of Oxide of Manganese. Wa. 1.50.—*Klaproth.* It is found near Guiliana in Sicily, in a bed of limestone.

5. Augite was known to the older mineralogists, under the name *Volcanic Schorl:* it was afterwards arranged with basaltic hornblende, where it remained, until Werner gave it its present name and place in the system. Hauy names it *Pyroxene,* which implies that it is a stranger to fire; because, although it is found in lava, it is not a production of volcanic fire.

Second

Second Subspecies.

Coccolite *.

Kokkolith, *Werner.*

Körniger Augit, *Karsten.*

Coccolith, *D'Andrada,* Scherer's Journal, b. iv. 19. s. 30 — *Schumacher,* Verzeichn. s. 30. *Id. Broch.* t. ii. p. 504. *Id. Hauy,* t. iv. p. 355. *Id. Reuss,* b. i. s. 86. *Id. Lud.* b. ii. s. 134. *Id. Suck.* 1r th. s. 184. *Id. Bert.* s. 159. *Id. Mohs,* b. i. s. 55. *Id. Lucas,* p. 194. *Id. Leonhard,* Tabel. s. 2.—Pyroxene Coccolithe, *Brong.* t. i. p. 447.—Pyroxene granuleux, *Brard,* p. 141.—Körniger Augit, *Karst.* Tabel. s. 40. *Id. Haus.* s. 98.—Pyroxene granuliforme, *Hauy,* Tabel. p. 42. Kokkolith, *Steffens,* b. i. s. 347. *Id. Hoff.* b. i. s. 443.—Körniger Augit, *Lenz,* b. i. s. 208.—Coccolith, *Oken,* b. i. s. 336.

External Characters.

Its principal colour is leek-green, which passes on the one side into pistachio-green, blackish-green, even into olive-green, and oil-green, and on the other into mountain-green.

It occurs massive; and crystallised in the following figures:

1. Six-sided prism, with two opposite acute lateral edges, and bevelled on the extremities; the bevelling planes set on the acute lateral edges. Sometimes two additional planes occur, when the bevelment passes into a four-planed acumination;

---

* Coccolite, from the Greek word κοκκος, *granum,* and λιθος, on account of the granular concretions that characterise this subspecies.

nation; and in other varieties two of the opposite lateral planes disappear, when the planes that meet under acute angles form a

2. Four-sided prism.

The crystals are generally blunt, or rounded on the angles and edges, even appear with convex lateral faces, and hence pass into longish grains.

The crystals are generally middle-sized, seldom small, and occur either singly imbedded, as is the case with the grains, or in druses.

Externally the crystals are sometimes smooth, sometimes rough; the first is shining and glistening, the other strongly glimmering.

Internally it is shining, sometimes approaching to glistening, and the lustre is vitreous, inclining to resinous.

The principal fracture is foliated, generally rather imperfect, with a twofold slightly oblique cleavage; the cross fracture is uneven.

The fragments are more or less sharp-edged.

It occurs in distinct concretions, which are coarse or small, seldom fine, angulo-granular, and are so loosely connected together, as frequently to be separable by the simple pressure of the fingers. Sometimes it occurs in longish granular concretions.

The surfaces of the concretions are sometimes slightly rough and strongly glimmering, sometimes smooth and glistening.

It is translucent, or translucent on the edges.

It is hard in a low degree: it scratches glass.

It is brittle.

It is very easily frangible.

Specific gravity, 3.316, *D'Andrada*. 3.303, *Karsten*. 3.15, 3.06, *Schumacher*. 3.373, *Hauy*.

Chemical

Chemical Character.

It is very difficultly fusible before the blowpipe.

Constituent Parts.

Silica, - -	50.0
Lime, - - -	24.0
Magnesia, - -	10.0
Alumina, - -	1.5
Oxide of Iron, -	7.0
Oxide of Manganese, -	3.0
Loss, - -	4.5
	100 *Vauquelin.*

Geognostic Situation.

It occurs in mineral beds subordinate to the primitive trap formation, where it is associated with granular limestone, garnet, and magnetic ironstone.

Geographic Situation.

It occurs at Arendal in Norway; in the iron mines of Hellsta and Assebro in Sudermannland; and in many places in Nericke, in Sweden. The mountain-green variety is found at Barkas in Finland. It is mentioned as occurring in the Harzeburg Forest in the Hartz, in Lower Saxony; and also in Spain.

Observations.

1. *Distinctive Characters.—a.* Between Coccolite and *Common Garnet:* The internal lustre of common garnet is resinous, that of coccolite vitreous; the fracture of common garnet is uneven, that of coccolite is foliated; common garnet scratches quartz, coccolite only glass: common

common garnet has a specific gravity of 3.75, coccolite of 3.33.—*b.* Between Coccolite and *Common Augite:* The colour-suites of the two minerals are different; the lustre of common augite is resinous, that of coccolite is vitreous; in common augite the distinct concretions are grown together, whereas they are so loosely aggregated in coccolite, as frequently to be separable by the mere pressure of the fingers; coccolite is rather softer than augite; and the concretions in coccolite are frequently enveloped in an extremely delicate crust, which is not the case with common augite.

2. It was first described by D'Andrada, under its present name.

Diopside.

Diopsid, *Werner.*

Cristaux gris-verdatres, transparens; formes tres prononcés, du Depart. du Po; *Alalite* de *Bonvoisin,* Journal de Physique, Mai 1806, p. 409 &c.—Varieté du *Diopside,* Journal des Mines, n. 115. p. 65. &c.—Cristaux gris-verdatres, ou blancs gris, tres, offrant la forme primitive peu prononcé, de Depart. du Po; *Mussite* de *Bonvoisin,* Journal de Physique, ib.—Variété du Diopside, Journal des Mines, ib.—*Pyroxene;* also Pyroxene cylindroide, comprimé, et fibro-granulaire, *Hauy,* Tabl. p. 41, 42 —Diopsid, *Steffens,* b. i. s. 349. *Id. Hoff.* b. i. s. 467. *Id. Lenz,* b. i. s. 212.—Strahliger Pyroxen, *Oken,* b. i. s. 335.

External Characters.

Its colour is greenish-white, slightly inclining to grey; also pale mountain-green, and emerald-green.

It

It occurs massive, disseminated; and crystallised in the following figures:

1. Low, very slightly oblique four-sided prism, sometimes equilateral, sometimes broad. This is the fundamental or primitive form *.
2. The preceding figure truncated on the acute lateral edges, bevelled on the obtuse edges, and the edge of the bevelment truncated: also rather acutely acuminated by four planes, two of the large acuminating planes set on the truncating planes of the acute lateral edges of the prism, the two smaller on the truncating planes of the bevelment. The two edges of the larger truncating planes, and the truncating planes of the acute edges of the prism, and the apex of the acumination, truncated. The broader lateral planes of the prism, and the truncating planes of the acumination, belong to the primitive form †.
3. Eight-sided prism, with alternate broader and smaller lateral planes, acuminated with four planes, the acuminating planes set on the smaller lateral planes, and this acumination again acuminated with four planes, set obliquely on the planes of the lower acumination, and of which two adjacent planes are large, and two small. The summit of the second acumination, and also the angles of the truncated summit, and the lateral edges of the larger planes, are truncated. The smaller lateral planes of the prism, and the acuminating planes of the upper acumination, belong to the primitive form ‡.

The

* Diopside primitive, Hauy; Mussite, Bonvoisin.

† Diopside didodecaedre, Hauy; Alalite, Bonvoisin.

‡ Diopside octovigesimal, Hauy.

The broader lateral planes are deeply longitudinally streaked; but the smaller lateral planes, and the acuminating planes, are smooth.

The crystals are middle-sized and small; occur resting on one another, intersecting one another, and collected into scopiform groups.

Externally it is shining and glistening, and pearly; internally it is shining and vitreous.

The fracture is foliated, with a fourfold cleavage: of these two are parallel with the truncations on the lateral edges, a third with the smaller lateral planes, and the fourth with the broader lateral planes; some varieties incline to radiated. The cross fracture is imperfect and small conchoidal.

The fragments are splintery.

It occurs in thin and very thin lamellar distinct concretions, which sometimes approach to columnar.

It is translucent.

It scratches fluor-spar, but hardly glass.

It is easily frangible.

Specific gravity, 3.310, *Hauy.*

Chemical Character.

It melts with difficulty before the blowpipe.

Constituent Parts.

Silica, - - -	57.50
Magnesia, - -	18 25
Lime, - - -	16.50
Iron and Manganese, -	6.00

Laugier.

Geognostic

Geognostic and Geographic Situations.

It is found in the hill of Ciarmetta in Piedmont; also in the Alp of La Mussa, near the town of Ala, in veins, along with epidote or pistacite and hyacinth-red garnets; and in the same district, in a vein traversing serpentine, along with prehnite, calcareous-spar, and iron-glance, or specular iron-ore. It is said also to occur at St Nicolas, in the Upper Valais.

Observations.

1. This mineral was discovered by Dr Bonvoisin, who formed it into two species, named *Alalite*, and *Mussite*: the white-coloured and massive varieties, and those crystallised in the form N° 2. he refers to alalite; the green, with scopiformly aggregated crystals, and radiated fracture, he considered as mussite. Hauy having ascertained that mussite and alalite have the same primitive form, ranged them in the system as one species, under the name *Diopside*; afterwards, he ascertained that the primitive form of diopside did not differ from that of augite, and consequently, in conformity with his system, abolished the diopside species, and arranged it with augite. Although the primitive form of diopside and augite be alike, we are still inclined with Werner to consider them as distinct species. Diopside differs from *Augite* in colour, crystallization, fracture, its streaked surface, thin and straight lamellar concretions, transparency, hardness, and geognostic situation *.

 4. Chrysolite.

* The only specimens I have ever seen of this mineral, are those in my possession, sent me by Mr Hewland of London.

4. Chrysolite.

Krisolith, *Werner*.

Yellowish-green Topaz. Chrysolith, *Cronstedt*, § 46. 5. p. 54.—Gemma pellueidissima, duritia sexta, colore viridi, subflavo, in igne fugaci; Chrysolithus, *Wall.* gen. 18. spec. 119. p. 255. —Krysolith, *Wid.* s. 264.—Chrysolite, *Kirw.* vol. i. p. 262. —Krysolith, *Estner*, b. ii. s. 122. *Id. Emm.* b. i. s. 26.—Chrysolito nobile, *Nap.* p. 127.—Peridot, *Lam.* t. ii. p. 250. —La Chrysolithe, *Broch.* t. i. p. 170.—Peridot, *Hauy*, t. iii. p. 198.—Chrysolith, *Reuss*, b. i. s. 49. *Id. Lud.* b. i. s. 60, *Id. Suck.* 1r th. s. 540. *Id. Bert.* s. 138. *Id. Mohs*, b. i. s. 42. *Id. Leonhard*, Tabel. s. 1.—Peridot, *Lucas*, p. 74.—Peridot Chrysolithe, *Brong.* t. i. p. 440.—Peridot, *Brard*, p. 179.—Chrysolith, *Haus.* s. 98. *Id. Karsten*, Tabel. s. 40. —Chrysolite, *Kid*, vol. i. p. 120.—Peridot, *Hauy*, Tabl. p 52. —Chrysolith, *Steffens*, b. i. s. 365. *Id. Lenz*, b. i. s. 203, —Harter Peridot, *Oken*, b. i. s. 333.

External Characters.

Its colour is pistachio-green, which sometimes approaches to olive-green, seldom to asparagus-green, and pale grass-green. Very rarely we observe in the same specimen, besides the green, also in a particular direction a pale cherry-red, inclining to broccoli-brown colour.

It occurs in angular pieces, (that appear to be original), which are often notched, and have a scaly splintery surface; also crystallised, in the following figures:

1. Broad rectangular four-sided prism, in which the lateral edges are truncated. The broader lateral faces are generally cylindric convex. The prism is acuminated with six planes: of these planes,

planes, two are set on the broader lateral planes, and the other four on the truncating planes of the lateral edges. The apex of the prism is truncated *, fig. 119. †.

2. The preceding figure, but acuminated with eight planes, which are set on the lateral and truncating planes of the prism, and the apex of the prism deeply truncated, fig. 120. ‡.
3. The preceding figure, in which the edge between the truncating plane of the apex of the acumination, and the acuminating plane which rests on the smaller lateral plane, is truncated, fig. 121. ||.
4. Very oblique four-sided prism, in which the acuter edges are bevelled, and the edges of the bevelment truncated; and acuminated on the extremities with four planes, which are set on the lateral edges, and the apex of the acumination slightly truncated. This figure is formed from N° 1. when the truncations on the lateral edges increase, until the broader lateral planes disappear. Sometimes the bevelling and acuminating planes are very small, but the truncating planes of the bevelment large, so that the crystal appears like a four-sided table of heavy-spar, bevelled on the terminal planes, or it assumes a reed-like aspect, §, fig. 122.

* The primitive form, according to Hauy, is a four-sided prism, with rectangular bases. The cleavage of the smaller lateral planes pretty distinct, the others indistinct.

† Peridot triunitaire, Hauy.

‡ Peridot monostique, Hauy.

|| Peridot subdistique, Hauy.

§ Peridot continu, Hauy.

5. Less oblique four-sided prism, in which the obtuse edges are truncated, the acute bevelled, and the bevelment truncated; acuminated on the extremities with eight planes, of which four are set obliquely on the lateral planes of the prism, four straight on the truncated lateral edges. The apex of the acumination truncated *, fig. 123.
6. N° 3. in which the lateral edges, in place of being truncated, are bevelled †, fig. 124.

Some crystals are very thin and reed-like, or table-shaped.

The crystals are middle-sized and small, and appear to have been imbedded.

The broader lateral planes in all the varieties, with exception of N° 4., are deeply longitudinally streaked; but the smaller lateral planes are often smooth, and the acuminating planes always smooth.

Internally the lustre is splendent and vitreous,

The fracture is perfect flat conchoidal.

The fragments are very sharp-edged.

It is transparent, and refracts double, particularly when viewed through the broader acuminating planes, and the obliquely opposite broader lateral planes of N° 1.

It is hard, but in a low degree; scratches glass and felspar, but is scratched by quartz and tourmaline.

It is brittle.

It is easily frangible.

Specific gravity, 3.340, 3.420, *Werner*. 3.428, *Hauy*. 3.301, 3.472, *Karsten*. 3.343, *Lowry*.

Chemical

* Peridot doublant, Hauy.

† Peridot quadruplant, Hauy.

Chemical Characters.

Its colours change, but it does not melt, without addition, before the blowpipe; but with borax, it melts into a transparent green glass.

Constituent Parts.

Silica, -	39.00	38.00	38.00
Magnesia,	43.50	39.50	50.50
Iron, -	19.00	19.00	9.50
	101.50	96.50	100
	Klaproth, Beit. b. i. s. 110.	Id. *Klap.* s. 107.	*Vauquelin.*

Geognostic Situation.

This mineral has hitherto been found only in a loose state: some mineralogists conjecture that it occurs in veins in serpentine, or greenstone; and also in flœtz-trap rocks.

Geographic Situation.

It is brought to Europe from the shores of the Red Sea; and it is alleged to have been found in trap rocks in Bohemia, and the Isle de Bourbon.

Uses.

This mineral has an agreeable colour and lustre; hence it is employed as a precious stone in different kinds of jewelling, particularly for ring-stones, when it is set in a gold foil. It is the softest of the precious stones; hence jewels of it wear soon, if not carefully worn and kept. It is said that ring-stones of chrysolite, by wearing, soon become dull on the surface, but that the lustre may be restored by immersing them in vegetable oils

Observations.

1. This mineral is characterised by its pistachio-green colour, the other varieties occurring rarely; the fine splintery or scaly surface of the angular pieces; its crystallizations, internal lustre, fracture, inferior hardness, and weight.

2. *Distinctive Characters.—a.* Between Chrysolite and *Vesuvian.* If the vesuvian be in rolled pieces, it can be distinguished from chrysolite, by its wanting the fine scaly or splintery surface which characterises that mineral: if in crystals, by their being very slightly longitudinally streaked, having a fine-grained uneven fracture, and resinous internal lustre; whereas the crystals of chrysolite are deeply longitudinally streaked, the fracture is conchoidal, and the lustre vitreous. A simple chemical distinctive character may be mentioned: vesuvian is fusible before the blowpipe, chrysolite is infusible.—*b.* Between Chrysolite and yellowish brown and olive-green *Tourmaline.* Tourmaline becomes strongly electric by heating, but the chrysolite only by rubbing; tourmaline is harder than chrysolite, as it scratches glass more readily.—*c.* Between Chrysolite and *Asparagus-stone* of Werner. Asparagus-stone does not scratch glass so easily as chrysolite, and refracts single, whereas chrysolite refracts double.

3. Werner is of opinion, that the stone described by the ancients, under the name *Yellow Chrysolite,* is not the true chrysolite, but our topaz. The celebrated traveller Bruce, mentions an Emerald Island in the Red Sea; but says that the substance he there met with, was scarcely harder than glass. Dr Kid remarks, " May not this have been a chrysolite, and this island the Topaz Island mentioned by Pliny ?" Romé de Lisle and Born, de-

scribe

scribe the asparagus-stone of Werner under the name Chrysolite; and other writers have confounded it with Chrysoberyl, and oil-green Beryl. Hauy considers it as a variety of olivine, but Werner places it in the system as a distinct species, in which he is followed by Mohs, Steffens, and other mineralogists.

5. Olivine.

Olivin, *Werner*.

Olivin, *Werner*, Bergm. Journ. 3. 2. s. 56.—Chrysolit en grains irreguliers, *De Born*. t. i. p. 70.—Olivin, *Wid*. s. 261. *Id. Kirw*. vol. i. p. 263. *Id. Emm*. b. i. s. 35.—Crysolito commune, *Nap*. p. 131.—Olivine, *Lam*. t. ii. p. 278. *Id. Broch*. t. i. p. 175.—Peridot granuliforme, *Hauy*, t. iii. p. 205.—Olivin, *Reuss*, b. ii. s. 49. *Id. Lud*. b. i. s. 61. *Id. Suck*. 1r th. s. 556. *Id. Bert*. s. 151. *Id. Mohs*, b. i. s. 45.—Chrysolith Olivin, *Hab*. s. 56.—Peridot granuliforme, *Lucas*, p. 74.—Olivin, *Leonhard*, Tabel. s. 2.—Peridot Olivine, *Brong*. t. i. p. 441.—Peridot granuliforme, *Brard*, p. 179.—Olivin, *Karst*. Tabel. s. 40. *Id. Kid*, vol. i. p. 122.—Peridot granuliforme, & lamelliforme, *Hauy*, Tabl. p. 52.—Olivin, *Steffens*, b. i. s. 363. *Id. Lenz*, b. i. s. 206.—Weicher Peridot, *Oken*, b. i. s. 334.

External Characters.

Its colour is olive-green, which passes on the one side into asparagus-green, on the other into oil-green, and into a colour intermediate between ochre and cream yellow, and into pale yellowish-brown.

It occurs massive, in roundish pieces, in grains, and rarely crystallised, in broad rectangular four-sided prisms, which are imbedded, and so easily broken, that it is difficult to ascertain their form.

Internally the lustre is shining and glistening, and is intermediate between resinous and vitreous.

The fracture is small-grained uneven, sometimes passing into imperfect small conchoidal. The crystals have an imperfect foliated fracture, with a rectangular twofold cleavage, the folia being parallel with the lateral planes of the prism.

The fragments are indeterminate angular, and rather sharp-edged.

The massive varieties occur in small and angulo-granular concretions.

It is translucent, passing into semi-transparent, seldom transparent.

It is hard, but in a lower degree than chrysolite.

It is brittle.

It is easily frangible.

Specific gravity, 3.225, *Werner*. 3.265, *Klaproth*.

Chemical Characters.

It is infusible before the blowpipe, without addition: with borax, it melts into a dark-green bead. It loses its colour in nitrous acid, the acid dissolving the iron, which is its colouring ingredient.

Constituent Parts.

	Olivine of Unkel.	Olivine of Karlsberg.
Silica, - -	50.0	52.00
Magnesia, -	38.50	37.75
Lime, -	0.25	0.12
Oxide of Iron,	12.00	10.75
	100.75	100.62
	Klaproth, Beit, b. v. s. 118.	Id. *Klaproth*, s. 121.

Geognostic

Geognostic Situation.

It occurs imbedded in basalt, greenstone, porphyry, and lava, and generally accompanied with augite.

Geographic Situation.

Europe.—It occurs in the flœtz-trap rocks of the Lothians, and other districts in Scotland; and in those of the Hebrides. Sparingly in the flœtz-trap rocks in the north of Ireland *. It is found in Iceland; and on the Continent, in Bohemia, Saxony, Stiria, Austria, Hungary, France, Italy, Spain, &c.

Africa.—Teneriffe; St Helena; Isle de Bourbon.

America.—Greenland; and the Cordilleras of South America.

Observations.

1. *a.* This mineral was first established as a distinct species by Werner: before his time, it had been confounded with Chrysolite. Hauy is still of opinion, that it is but a variety of chrysolite; but the following comparison of the characters of the two minerals, shew that they may be viewed as different species:—The colours of olivine do not agree with those of chrysolite: the most common external shape of olivine is granular, whereas that of chrysolite is angular and notched; the rolled pieces of olivine have not a scaly and splintery surface, as is the case with chrysolite: olivine is seldom crystallised, and when it is so, the crystal-suite is inconsiderable, whereas chrysolite is frequently crystallised, and its forms very various; the lateral planes of the crystals of olivine

* Greenough.

olivine are smooth, those of chrysolite streaked, shining, and glistening; the lustre of olivine is intermediate between resinous and vitreous, but that of chrysolite is splendent and vitreous; the fracture of olivine is imperfect conchoidal, or imperfect foliated, whereas that of chrysolite is perfect conchoidal; olivine often occurs in distinct concretions, chrysolite never; olivine is translucent, passing into semi-transparent, chrysolite is transparent; olivine is very easily frangible, chrysolite easily frangible; olivine is softer than chrysolite; and the specific gravity of olivine is 3.225 to 3.265, that of chrysolite 3.34 to 3.4.

b. Olivine is nearly allied to Augite: this alliance is not so much a consequence of agreement in external characters, as rather a similarity in geognostic relations. Both species occur in the same species of rock, and the one seldom without the other; and large masses and grains of olivine sometimes contain small angular grains of augite, which take, as it were, the place of single distinct concretions,—a fact which shews their mutual affinity. It is distinguished from *Augite* by its paler colours, external shape, kind of lustre, fracture, superior transparency, and its inferior hardness and weight.

c. It is distinguished from *Common Green Garnet*, by its greater transparency, inferior hardness, and weight, and geognostic situation.

2. Leonhard, Reuss, and others, divide this species into two subspecies, named Conchoidal and Foliated; but Karsten has shewn that the foliated subspecies is augite.

3. It is named *Olivine*, from its olive-green colour.

4. It frequently decays, or falls into an earth, which much resembles iron-ochre. When it begins to exhibit

on

on the surface iridescent colours, it is a proof of its having already begun to decay.

5. A yellow substance, nearly allied to olivine, occurs in the Siberian meteoric iron.

7. Lievrite.

Lievrit, *Werner.*

Yenite, *Lelievre,* Journal des Mines, N. 121. p. 65. *Id. Hauy,* Tabl. p. 42. & 182.—Ilvait, *Steffens,* b. i. s. 356.—Lievrit, *Hoff.* b. ii. s. 376.—Yenit, *Lenz,* b. i. s. 215.

External Characters.

Its colour is intermediate between dark greyish-black, and iron black, but sometimes passes through raven-black into blackish-green.

It occurs massive; and crystallised in the following figures:

1. Oblique four-sided prism, acuminated on the extremities with four planes, which are set on the lateral planes *.
2. Four-sided prism, which is almost rectangular, bevelled on the extremities, and the bevelling planes set on the obtuse edges.
3. The preceding figure, in which the angles of the bevelment are bevelled.
4. The preceding figure, in which the angles of the second bevelment are truncated, and the obtuse lateral edges of the prism bevelled.

The

* Yenite quadrioctonal, Hauy.

The crystals vary from acicular to the thickness of half an inch: they are frequently scopiformly aggregated, sometimes superimposed, and sometimes imbedded.

The lateral planes of the crystals are longitudinally streaked.

The lustre of the principal fracture is shining; that of the cross fracture glistening, and is semi-metallic.

The principal fracture is small, and scopiform diverging radiated; the cross fracture uneven.

The fragments are indeterminate angular, rather sharp edged.

It occurs in thin and straight prismatic distinct concretions.

It is opaque.

It is hard in a low degree: it scratches glass with ease, and gives a few sparks with steel, but is scratched by adularia.

It does not change its colour in the streak.

It is easily frangible.

Specific gravity, 3.825, 4.061. *Lelievre.*

Chemical Characters.

It is attacked by the three mineral acids, but does not gelatinate with them. When exposed to heat, it becomes magnetic; its colour is changed from black into dark reddish-brown, and it loses about 2 *per cent.* of weight. Before the blowpipe, it melts easily, and without intumescence, into an opaque black grain, which has a dull metallic aspect, and is attracted by the magnet, but does not possess polarity. It dissolves in glass of borax, with a slight ebullition.

Constituent

Constituent Parts.

Silica,	28.0	Silica, -	29	Silica,	30.0	Silica,	30.0
Alumina,	0.6	Lime, -	12	Lime,	12.5	Alumina,	1.0
Lime,	12.0	Oxide of Iron		Oxide of Iron		Lime,	14.8
Oxide of Iron,	55.0	and Oxide of		and Oxide of		Oxide of Iron,	49.0
Oxide of Man-		Manganese,	57	Manganese,	57.5	Oxide of Man-	
ganese,	3.0		98		100.0	ganese,	2.0
	98.6		*Vauquelin.*		*Vauquelin.*		96.8
	Descotils.						*Vauquelin.*

Geognostic and Geographic Situations.

It occurs in primitive limestone, along with epidote, quartz, garnet, magnetic ironstone, and crystallised arsenic-pyrites, at Rio la Marine, and Cape Calamite, in the island of Elba. It is said also to occur in Siberia.

Observations.

1. Colour, crystallization, kind of lustre, fracture, distinct concretions, opacity, hardness, and considerable weight, distinguish this mineral from all others with which it might be confounded.

2. Werner places it in the system between Schorl and Epidote: in the systems of Hauy, Steffens, and Lenz, it follows Augite; which latter arrangement is here followed. The quantity of iron it contains, is remarkable; and if not accidental, shews that Lievrite probably belongs to a family different from any in the Wernerian system.

3. The French naturalist who first directed the attention of mineralogists to this mineral, named it *Yenite*, in honour of the battle of Jena! Hoffmann, in Saxony, affects to be ignorant of the origin of this name. Lenz is

is not so ill informed: he asks, What has mineralogy to do with the battle of Jena? Steffens pities the weakness and absurdity of those with whom such an association could arise; and this name, originating in folly, has been banished from the system, by Werner changing Yenite into *Lievrite.*

XIX. BASALT

XIX. BASALT FAMILY.

This Family contains the following species: Basalt, Wacke, Clinkstone, and Iron-Clay.

1. Basalt.

Basalt, *Werner*.

Basaltes figura columnari, lateribus inordinatis, crystallisatus, *Wall.* gen. 22. spec. 150.—Basalt, *Wid.* s. 423.—Figurate Trap, *Kirw.* vol. i. p. 231.—Basalt, *Estner*, b. ii. s. 726. *Id. Emm.* b. i. s. 339.—Basalto, *Nap.* p. 284.—Le Basalte, *Broch.* t. i. p. 430.—Lava lithoides basaltique, *Hauy*, t. iv. p. 474. —Basalt, *Reuss*, b. ii. 2. s. 125. *Id. Lud.* b. i. s. 121. *Id. Suck*, 1r th. s. 513. *Id. Bert.* s. 221. *Id. Mohs*, b. i. s. 502. *Id. Hab.* s. 17. *Id. Leonhard*, Tabel. s. 25. *Id. Brong.* t. i. p. 455. *Id. Karsten*, Tabel. s. 36. *Id. Kid*, vol. ii. App. p. 6. *Id. Steffens*, b. i. s. 333. *Id. Lenz*, b. ii. s. 607. *Id. Oken*, b. i. s. 364.

External Characters.

Its most frequent colour is greyish-black, of different degrees of intensity; from this it passes into ash-grey, or greenish-black.

It occurs massive, in rolled pieces, and sometimes vesicular.

Internally it is dull, sometimes feebly glimmering.

The fracture of the coarser varieties is large or small grained uneven; of the more crystalline varieties even inclining to large and flat conchoidal, and seldom to splintery.

The

The fragments are indeterminate angular, and rather sharp-edged.

It occurs in distinct concretions. These are generally columnar, varying from a few inches to several fathoms, even to upwards of 100 feet in length: the number of sides varies from three to nine, and of these the nine-sided are the rarest; they are straight, bent, and either parallel or diverging; sometimes they are articulated, and the joints have concave and convex faces. In mountains, these concretions are collected into large groups, and many of these groups or colossal concretions form a hill or mountain. Sometimes it occurs in tabular, sometimes in globular concretions; these latter are frequently composed of concentric lamellar concretions, or of columnar concretions radiating from the centre. Some varieties are composed of large, coarse, and fine granular concretions.

It is opaque, or feebly translucent on the edges.

It yields a light grey-coloured streak.

It is semi-hard, bordering on hard.

It is rather brittle.

It is difficultly frangible.

Specific gravity, 3.065, *Klaproth*. 3.082, *Karsten*.

Chemical Characters.

Before the blowpipe it melts easily, without addition, into an opaque black-coloured glass. According to Dr Kennedy, the basalt of the Castle Rock of Edinburgh softens at 45° of Wedgwood; that of Staffa at 38°; and I obtained similar results with the basalts of Arran, as is mentioned in my mineralogical description of that island *.

Constituent

* Vid. Mineralogical Travels, vol. i.

Constituent Parts.

Basalt of the Hassenberg.				Basalt of Staffa.	
Silica, -	44.50	Silica, -	50	Silica, -	48.0
Alumina,	16.75	Alumina,	15	Alumina,	16.0
Lime, -	9.50	Lime, -	8	Lime, -	9.0
Magnesia,	2.25	Magnesia, -	2	Soda, -	4.0
Soda, -	2.60	Oxide of Iron,	25	Muriatic Acid,	1.0
Oxide of Iron,	20.00		——	Oxide of Iron,	16.0
Oxide of Manganese,	0.12		100	Moisture, & Volatile Matter,	5.00
Water, -	2.00		*Bergmann.*		——
	——				99.9
	97.72			*Kennedy*, Edinburgh Phil. Trans. vol. v. p. 89.	
Klaproth, Beit. t. iii. s. 253.					

Geognostic Situation.

It occurs in beds, veins, and imbedded masses, in several of the flœtz formations, as the first or old red sandstone formation, the coal formation, and the newest flœtz-trap of Werner. It is said to be a production of transition districts; and it occurs in beds, imbedded masses, and veins, in primitive rocks, as granite and mica-slate.

Geographic Situation.

Europe.—It occurs in considerable abundance in the three great divisions of Scotland, viz. the northern, middle, and southern *, but more abundantly in the middle and southern, than in the northern divisions. It is a rare rock

* The *northern* division of Scotland is the country to the north of the chain of lakes which extends from Loch Ness to the West Sea: the *middle* division the country between the Frith of Forth and the chain of lakes; and the *southern* division, the country between the Frith of Forth and the borders of England.

rock in the Orkney Islands; and I do not know that it has been observed in the Zetland Islands. In Ireland, it occurs abundantly in the northern counties; but it is not a frequent rock in England. It abounds in the island of Iceland; and is a member of the series of trap rocks of which the Feroe Islands are composed. It occurs rarely in Norway, and there only in beds, along with transition rocks; and it is scarcely more abundant in Sweden. It is a frequent rock in several parts in Germany, as the Siebengebirge, the Rhongebirge, the vicinity of Frankfort on the Mayne, the Bohemian Mittelgebirge, the Erzgebirge, &c. In France, it occurs frequently in the interesting country of Auvergne; and the Euganean mountains, and other districts in Italy, afford very interesting displays of this rock. It abounds in several districts in Spain; and forms considerable hills in Portugal, particularly around Lisbon.

Asia.—At Hadie Andjor in Yemen; on the banks of the Amour in Dauria; and in Kamschatka.

Africa.—Islands of Teneriffe and St Helena.

Polynesia.—Otaheité, Easter Island, Kergeulen's Island, and the Sandwich Islands.

America.—Abundantly in Mexico; also in different parts in the United States.

Uses.

As it is very compact, and unites readily with mortar, it is considered as a good building-stone; and it is observed, that if pulverised basalt be added to the mortar, its binding quality is thereby increased. It is sometimes hollowed into water-troughs: it is used as a paving-stone; and it is sometimes cut by jewellers, and used as a touchstone. The most fusible varieties are occasionally used as a flux with calcareous ironstones; and they are some-

times

times melted, and blown into glass bottles, which have a dark green colour, and are said to be harder and more durable than those made of common green glass. Although it is harder, more brittle, and less obedient to the chisel, and its colours not so pleasing and durable as those of marble, yet the ancients, who were acquainted with its greater indestructibility, executed several fine works in it. Pliny describes several fine pieces of sculpture said to have been executed in this stone; and the famous statue of Minerva, at Thebes, is described by travellers as being of basalt. Antiques of basalt are always in a much better state of preservation than those of marble. Even those dug out of the earth, are never covered with that tufaceous crust we find investing those made of marble; they still retain the original polish, and the finest touches of the chisel are preserved unimpaired. Many of the antique basalts preserved in collections, are evidently greenstone, syenite, or hornblende-rock.

Observations.

1. The popular name of this mineral in Scotland, is *Whinstone;* but all whinstones are not basalt; for the name is applied to greenstone, clinkstone, wacke, trap-tuff, porphyry, grey-wacke, and other rocks.

2. Humboldt, Voigt, Faujas St Fond, and Hauy, maintain its volcanic origin;—Werner, Mohs, Steffens, Klaproth, and Karsten, its neptunian origin:—Von Buch, Dolomieu, Daubuisson, and others, its double origin, sometimes volcanic, sometimes neptunian;—and Hutton, Playfair, Hall, &c. its plutonic formation. The theory of its formation will be considered in the geognostic part of this work.

 2. Wacke.

2. Wacke.

Wacke, *Werner*.

Wacke, *Karsten*, Magaz. f. Helvet Naturkunde, 3. s. 234. *Id. Hoffmann*, Bergm. Journ. 1. 2. s. 507.—Wacken, *Kirw*. vol. i. p. 223.—Wacke, *Estner*, b. ii. s. 737. *Id. Emm*. b. i. s. 335. *Id. Nap*. s. 228.—La Wakke, *Broch*. t. i. p. 434. ; t. ii. p. 606. *Id Reuss*, b. ii. 2. s. 119. *Id. Lud*. b. i. s. 122. *Id. Suck*. 1r th. s. 511. *Id. Bert*. s. 220. *Id. Mohs*, b. i. s. 506. *Id. Hab*. s. 17. *Id. Leonhard*, Tabel. s. 25. *Id. Karsten*, Tabel. s. 38. *Id. Kid*, vol. ii. Appen. p. 19. *Id. Steffens*, b. i. s. 336. *Id. Lenz*, b. ii. s. 610. *Id. Oken*, b. i. s. 364.

External Characters.

Its colour is greenish-grey, of various degrees of intensity: from light greenish-grey, it passes into yellowish-grey, which is sometimes mixed with brown, mountain-green, and olive-green. Those varieties that incline to basalt, approach in colour to greyish-black.

It occurs massive, and vesicular: when the vesicles are filled, it is said to be amygdaloidal.

Internally it is dull.

The fracture is large and flat conchoidal in the most characteristic specimens; but small-grained uneven in those which are less so.

The fragments are blunt-edged.

It is opaque.

It is more or less shining in the streak.

It is soft.

It is sectile.

It is rather easily frangible.

It

It feels rather greasy.

Specific gravity, 2.617, 2.887, *Kirwan*. 2.790, *Karsten*.

Chemical Character.

Before the blowpipe, it melts into a greenish slag.

Constituent Parts.

Silica, - - -	63.00
Alumina, - -	13.00
Lime, - - -	7.00
Iron, - - -	17.00

Withering.

The above analysis is of amygdaloid, not of pure wacke: so that an analysis of that substance, in its unmixed and characteristic form, is still wanting.

Geognostic Situation.

It occurs in transition mountains, in the form of amygdaloid; in flœtz mountains, either in the pure state, or in the amygdaloidal form, in beds, mountain-masses, and veins, generally associated with basalt.

Geographic Situation.

It is a frequent rock, either in the pure or amygaloidal form, in the different districts in Scotland where basalt occurs; and on the Continent of Europe, it also occurs in basaltic countries.

Observations.

1. Its colour, streak, and inferior hardness, distinguish it from *Basalt*.

2. It was first established as a distinct species by Werner.

3. A transition is to be observed from wacke into Basalt; also from wacke into Greenstone,—a fact which shews the near alliance of these minerals.

3. Clinkstone.

Klingstein, *Werner*.

Phonolith, *Daubuisson*.

Hornslate, *Kirw.* vol. i. p. 307.—Porphirschiefer, *Estner,* b. ii. s. 747. *Id. Emm.* b. iii. s. 344.—Pierre sonnante, *Broch.* t. i. p. 437.—Klingstein, *Klap.* Beit. b. iii. s. 229. *Id. Reuss,* b. ii. s. 340. *Id. Lud.* b. i. s. 123. *Id. Suck.* 1r th. s. 364. *Id. Bert.* s. 222. *Id. Mohs,* b. i. s. 509. *Id. Hab.* s. 16. *Id. Leonhard,* Tabel. s. 26.—Feldspath compacte sonoré, *Lucas,* p. 266.—Klingstein, *Karsten,* Tabel. s. 38.—Clinkstone, *Kid,* vol. ii. App. p. 18.—Klingstein, *Steffens,* b. i. s. 338. *Id. Lenz,* b. ii. s. 613. *Id. Oken,* b. i. s. 363.

External Characters.

Its most frequent colour is greenish-grey, which sometimes passes into yellowish-grey, and ash-grey; and occasionally olive-green.

It occurs massive.

The lustre of the principal fracture is glistening and pearly; that of the cross fracture is faintly glimmering, almost dull.

The principal fracture is slaty, generally thick, and often curved slaty, with a scaly aspect; the cross fracture is splintery, passing into even.

The

The fragments are indeterminate angular, and often slaty.

It occurs in columnar, globular, and tabular distinct concretions.

It is strongly translucent on the edges.

It is intermediate between hard and semi-hard.

It is rather easily frangible.

It is brittle.

In thin plates, it emits, when struck, a ringing sound.

Specific gravity, 2.575, *Klaproth.*

Chemical Characters.

It melts before the blowpipe into a grey-coloured glass, but with more difficulty than basalt.

Constituent Parts.

Silica, - -	57.25
Alumina, - -	23.50
Lime, - - -	2.75
Natron, - -	8.10
Oxide of Iron, -	3.25
Oxide of Manganese,	0.25
Water, - -	3.00
	98.10

Klaproth, Beit. b. iii. s. 243.

Geognostic Situation.

This mineral occurs principally along with rocks of the flœtz-trap series. When it forms part of a hill or range of hills, with wacke, basalt, and greenstone, we generally find it resting on the basalt or wacke, and covered by the greenstone.

Geographic Situation.

Europe.—The Bass rock at the mouth of the Frith of Forth, North Berwick Law, Traprain Law, and the Girleton Hills, all in East Lothian, are principally composed of clinkstone, and afford many beautiful and highly characteristic varieties of this mineral. It occurs in the island of Arran, isle of Lamlash, and other parts of Scotland, as will be mentioned in the Geognosy. On the Continent of Europe, it is found in many districts where basalt abounds, as in the Bohemian Mittelgebirge; also in Bavaria, Suabia, Lusatia, Hessia, France, Italy, Hungary, &c.

Africa.—Along with basalt, in the island of Teneriffe.

America.—Along with trap rocks, both in North and South America.

Observations.

1. Charpentier was the person who first directed the particular attention of mineralogists to this substance: in his Mineralogical Description of the Electorate of Saxony, he gives a very interesting account of it, under the name *Hornslate*, (Hornschiefer) *. Werner afterwards examined it with more minute attention, and introduced it into the oryctognostic system as a distinct species, under the name *Clinkstone*.

2. It is distinguished from *Basalt* by colour, lustre, fracture, and transparency.

3. It passes on the one hand into Basalt, and on the other into Felspar.

4. Iron-Clay.

* Older mineralogists were of opinion, that clinkstone was the same mineral as that described by Wallerius under the name *Corneus fissilis*: hence they gave it the name Hornslate; but in this they erred, as Wallerius's mineral appears to be hornblende-slate. Born, Ferber, and others, include under their hornslate, also some varieties of clay-slate, and of mica-slate.

4. Iron-Clay.

Eisenthon, *Werner.*

Eisenthon, *Karsten,* Tabel. s. 38. *Id. Steffens,* b. i. s. 340. *Id. Lenz,* b. ii. s. 614.—Eisenwacke, *Oken,* b. i. s. 366.

External Characters.

Its colours are reddish-brown, and brownish-red, passing into a tint between yellowish-brown and laurel-berry-brown.

It occurs massive, and vesicular, as the basis of some varieties of amygdaloid.

Internally it is dull or glimmering.

The fracture is small and fine-grained uneven.

The fragments are blunt-edged.

It is semi-hard, passing into soft.

It is opaque.

It is rather brittle.

It is easily frangible.

It is rather heavy, but in a middling degree.

Geognostic and Geographic Situations.

It occurs in beds, and in the form of amygdaloid, along with flœtz-trap rocks, in Fifeshire, island of Skye, and other parts of Scotland; and in the north of Ireland, as at the Giants Causeway, it is associated with rocks of the same nature. It also occurs in Iceland, Feroe Islands, and in the County of Glatz in Silesia.

Observations.

This species, which was first established by Werner, appears to be intermediate between wacke and basalt, and is distinguished from both by colour, fracture, hardness, (which is less than that of basalt, but more than of wacke,) and easy frangibility.

XX. DOLO-

XX. DOLOMITE FAMILY.

This Family contains the following species: Dolomite, Brown-Spar, Miemite, and Gurhofite.

1. Dolomite.

Dolomit, *Werner.*

This species is divided into four subspecies, viz. Common Dolomite, Dolomite-Spar or Rhomb-Spar, Columnar Dolomite, and Compact Dolomite or Magnesian Limestone.

First Subspecies.

Common Dolomite.

Gemeiner Dolomit, *Lenz.*

External Characters.

Its colours are snow, greyish, yellowish, and bluish white; greenish and bluish grey, and pale leek-green.

It occurs massive, in rolled pieces, and in loose grains.

Internally it is sometimes shining, sometimes glimmering.

The fracture in the large is imperfect curved slaty; in the small, common or scaly foliated, sometimes passing into splintery, and almost into uneven.

The fragments are rather blunt-edged.

It occurs in fine granular concretions, which are sometimes so loosely aggregated, that the mass is flexible.

It

It is faintly translucent, or only translucent on the edges.

It is semi-hard, sometimes approaching to soft.

It is brittle.

It feels rough and meagre.

It is easily frangible.

Specific gravity, Dolomite of Alps of Carinthia, 2.835, *Klaproth*.

Chemical and Physical Characters.

It effervesces very feebly with acids,—a character that distinguishes it from granular limestone, which effervesces briskly.

It in general phosphoresces when placed on heated iron, or when rubbed in the dark; and this property is much stronger in some varieties than in others.

Constituent Parts.

	St Gothard.	Apennines.	Carinthia.	Antique.
Carbonate of Magnesia,	46.50	35.00	48.00	48.00
Carbonate of Lime,	52.00	65.00	52.00	51.50
Oxide of Manganese,	0.25			
Oxide of Iron, - -	0.50	-	0.20	
Loss, - - -	0.75			
	100	100	100.20	99.50
	Klaproth, Beit. b. iv. s. 209.	*Klaproth*, Id. s. 215.	*Klaproth*, Id. s. 219.	*Klaproth*. Id. s. 222.

Iona.	
Carbonic Acid, - -	48.00
Lime, - -	31.12
Magnesia, - -	17.06
Insoluble Matter, - -	4.00

Tennant, Phil. Trans. for 1799.

Geognostic

Geognostic Situation.

It occurs in primitive and transition mountains.

Geographic Situation.

Europe.—Beds of it, containing tremolite, occur in the island of Iona. In the mountain-group of St Gothard, it occurs in beds, often of great thickness, containing imbedded crystals of tremolite, grains of quartz, and scales of mica and talc. In the Apennines, it occurs in imbedded portions, in a dark ash-grey small splintery limestone: in Carinthia, it forms whole ranges of mountains: in Bareuth, it occurs in beds along with granular foliated limestone: at Sala in Sweden, it is mixed with mica, talc, and quartz: on the mountain of Maladetta in Spain: a beautiful white variety, used by ancient sculptors, is found in the isle of Tenedos: in veins, traversing granite, in the valley of Sesia in Italy; and it is found loose on Monte Somma.

America.—Province of New-York, with tremolite *.

Asia.—Bengal †, with imbedded tremolite; also in Siberia.

Uses.

It appears to have been used by ancient sculptors in their finest works.

Observations.

1. It is named Dolomite, in honour of the celebrated French geologist Dolomieu.

2. The only mineral with which it is likely to be confounded, is granular foliated limestone; but a simple chemical test at once distinguishes them:—a drop of mineral acid causes a violent effervescence, when poured on granular

* Dr Bruce. † Sir John Murray.

granular foliated limestone, but a very feeble one with dolomite.

3. The flexible variety of dolomite was first noticed in the Borghese Palace in Rome, by Ferber: it was afterwards found on the mountain of Campo Longo, in the St Gothard group, by Fleuriau de Bellvue. It was sold at a very high price, until the publication of Fleuriau de Bellvue's experiments, by which it appeared, that the other varieties of dolomite, and also common granular limestone, could be rendered flexible, by exposing them in thin and long slabs, for six hours, to a heat of 200° of Reaumur.

Second Subspecies.

Dolomite-Spar, or Rhomb-Spar.

Rautenspath, *Werner.*

Bitterspath, *Wid.* s. 518.—Crystallized Muricalcite, *Kirw.* vol. i. p. 92.—Bitterspath, *Emm.* b. iii. s. 353.—Spato Magnesiano, *Nap.* p. 358.—Bitterspath, *Lam.* t. ii. p. 347.—Chaux carbonaté magnesiée, *Hauy,* t. ii. p. 187.—Le Spath magnesien, ou le Bitterspath, *Broch.* t. i. p. 560.—Bitterspath, *Reuss,* b. ii. 2. s. 330.—Rautenspath, *Lud.* b. i. s. 154.—Gemeiner Bitterspath, *Suck.* 1r th. s. 634.—Rautenspath, *Bert.* s. 113. *Id. Mohs,* b. ii. s. 96. 98.—Bitterspath, *Hab.* s. 83.—Gemeiner Bitterspath, *Leonhard,* Tabel. s. 35.—Chaux carbonatée lente, Picrite, *Brong.* t. i. p. 230.—Rhomboedrischer Dolomit, *Karsten,* Tabel. s. 50.—Gemeiner Bitterspath, *Haus.* s. 128.—Chaux carbonatée magnesifere, *Brard,* p. 38.—Rhomb-spar, *Kid,* vol. i. p. 57.—Chaux carbonaté magnesifere primitive, *Hauy,* Tabl. p. 5.—Rautenspath, *Lenz,* b. ii. s. 710. *Id. Oken,* b. i. s. 393.

External Characters.

Its colours are greyish-white, yellowish-white, and yellowish-grey, which latter passes into pea-yellow.

It

It occurs massive, disseminated, and crystallised in rhombs, which are sometimes rounded, or truncated on the edges.

The crystals are middle-sized and small; the surface is sometimes smooth, sometimes rough, and either shining or glimmering.

Internally the lustre is splendent, between vitreous and pearly *.

The principal fracture is foliated, with a threefold oblique angular cleavage: the alternate angles of which measure 106° 15′ and 73° 45′; the cross fracture imperfect conchoidal †.

The fragments are rhomboidal.

It is semi-hard; harder than calcareous-spar, or brown-spar.

It is easily frangible.

It is brittle.

Specific gravity, 2.880, 3.000. 2.8901, *Murray* *.

Chemical Characters.

Before the blowpipe it is infusible, without addition: even when pounded, it effervesces but feebly; and dissolves slowly in muriatic acid.

Constituent

---

* The lustre in general is stronger than that of calcareous-spar.—*Bournon.*

† Dr Wollaston.

‡ Newton-Stewart, Galloway.

Constituent Parts.

Hall, in the Tyrol.	
Carbonate of Lime,	68.00
Carbonate of Magnesia,	25.50
Carbonate of Iron,	1.00
Water,	2.00
Clay intermixed,	2.00
	98.50

Klaproth, Beit. b. iv. s. 238.

Taberg, in Wermeland.	
Carbonate of Lime,	73.00
Carbonate of Magnesia,	25.00
Oxide of Iron, mixed with Manganese,	2.25
	100.25

Klaproth, b. i. s. 306.

Near Newton-Stewart in Galloway.	
Carbonate of Lime,	56.60
Carbonate of Magnesia,	42.00
	98.60

Or by another result:

Carbonate of Lime,	56.2
Carbonate of Magnesia,	43.5
	98.9

With a trace of Manganese and Iron.—*Murray* *.

Lime,	28.00
Magnesia,	25.05
Carbonic Acid,	48.00
Oxide of Manganese,	1.05
With a trace of Iron.	

Bucholz.

Geognostic Situation.

It occurs imbedded in chlorite-slate, talc-slate, limestone, and serpentine, occasionally associated with asbestus and tremolite; in the salt formation, where it is imbedded in anhydrite; in drusy cavities in compact dolomite, and in metalliferous veins.

Geographic

* The above analysis was communicated to me by my friend Dr Murray.

Geographic Situation.

Europe.—It occurs imbedded in chlorite-slate on the banks of Loch Lomond; in a vein in transition rocks, along with galena, blende, copper-pyrites, and calcareous-spar, near Newton-Stewart in Galloway; in compact dolomite in the Isle of Mann and the north of England; in chlorite-slate and talc in the Upper Palatinate; in the mountain of Chalance in Dauphiny, along with asbestus, talc, and chlorite; also at Brienz in Switzerland; in the mountains of Salzburg; in granular limestone, in the silver mines of Sala, and in the Taberg in Wermeland in Sweden.

America.—In Greenland, imbedded in common and indurated talc; and in Mexico, along with amethyst, common quartz, and felspar.

Observations.

1. This mineral was formerly named *Bitter-Spar*, from the magnesia contained in it, which is denominated Bitter Salt by the Germans, because obtained easily from sulphate of magnesia or Epsom salt. It was named *Muricalcite* by Kirwan, from the magnesia and lime contained in it: magnesia having been called muriatic earth, as being the base of one of the salts contained in sea-water. Werner named it *Rhomb-Spar*, from its form; and it is here named *Dolomite-Spar*, from its relation to Dolomite, and its sparry structure.

2. It is distinguished from *Calcareous-spar* by the shape of its rhomboid, superior hardness, specific gravity, and dissolving slowly in the mineral acids.

Third

Third Subspecies.

Columnar Dolomite.

Stänglicher Dolomit, *Klaproth.*

Stänglicher Dolomit, *Klaproth,* Mag. der Gesellsch. Naturf. Freunde, b. v. s. 402.

External Characters.

Its colour is pale greyish-white.

It occurs massive, generally in pieces about two inches long, which are covered with an isabella-yellow small botryoidal crust, and interwoven with greyish-white straight fibres of asbestus.

The lustre is vitreous, inclining to pearly.

The longitudinal fracture is narrow radiated, with delicate cross rents; the cross fracture is uneven, but specular in the cross rents.

It occurs in distinct concretions, which are thin, long, and straight prismatic.

It breaks into acicular-shaped fragments.

It is feebly translucent.

It is brittle.

Specific gravity, 2.765.

Constituent Parts.

From the Mine Tschistagowskoy.

Lime, - - -	28.20
Magnesia, - -	19.74
Oxide of Iron, -	0.50
Carbonic Acid, -	39.25
Water, - -	11.31
Loss, - -	1.00

Klaproth, Magaz. der Gesellch. Naturf. Freunde, t. v. s. 402. & 403.

Geognostic and Geographic Situations.

It occurs in serpentine, in the mine Tschistagowskoy, on the river Mjafs, in the Government of Orenburg in Russia.

Observation.

It was at one time considered to be a variety of Strontianite; but in external characters, it is much more nearly allied to Tremolite.

Fourth Subspecies.

Compact Dolomite, or Magnesian Limestone.

This subspecies is subdivided into two kinds, viz. Common Compact Dolomite, and Flexible Compact Dolomite.

First Kind.

Common Compact Dolomite.

Magnesian Limestone, *Tennant.*

Tennant, Transactions of Royal Society of London for 1799.—*Thomson,* Annals of Philosophy for December 1814.

External Characters.

Its colours are yellowish-grey, yellowish-brown, and a colour intermediate between chesnut-brown and yellowish-brown; seldom bluish-grey.

It occurs massive.

Internally it is glistening or glimmering, and the lustre is between pearly and vitreous.

The

The fracture is minute foliated, often combined with splintery, and even conchoidal.

The fragments are rather blunt edged.

It occurs in minute granular concretions.

It is translucent.

It is semi-hard ; it is harder than calcareous-spar.

It is brittle.

Specific gravity of the crystals, 2.823, *Tennant.* 2.777, 2.820, *Berger.* 2.791, *Thomson.*

Chemical Characters.

It dissolves slowly, and with but feeble effervescence, in nitrous acid. When deprived by heat of its carbonic acid, it is much longer of re-absorbing it from the atmosphere than common limestone.

Constituent Parts.

Yorkshire.	Building Hill, near Sunderland.	Humbleton Hill, near Sunderland.
Lime, 29.5 to 31.07	Carbonate of Lime, 56.80	Carbonate of Lime, 51.50
Magnesia, 20.3 to 22.05	Carbonate of Magnesia, - 40.84	Carbonate of Magnesia, - 44.84
Carbonic Acid, 47.2	Carbonate of Iron, 0.36	Insoluble matter, 1.60
Alumina & Iron, 0.8 to 1.24	Insoluble matter, 2.00	Loss, - - 2.06
	100.00	100.00
Tennant, Phil. Tr. for 1799.	*Thomson*, Annals of Philosophy, vol. iv. p. 416.	*Thomson*, ib. p. 417.

Geognostic Situation.

In the north of England it occurs in beds of considerable thickness, and great extent, and appears to rest on the Newcastle coal-formation [*]; but in the Isle of Man, it occurs in a limestone which rests on grey-wacke, and

 contains

* Thomson, Annals of Philosophy, vol. iv. p. 416.

contains imbedded portions of quartz, dolomite-spar, and sparry iron-ore *.

Geographic Situation.

It occurs in Nottinghamshire, Derbyshire, Northamptonshire, Leicestershire, Northumberland, and Durham †: also in Ireland, at Portumna in Galway, Ballyshannon in Donnegal, Castle Island near Killarney ‡. It has been observed among the limestone rocks near Erbefeld and Gemarek, in Westphalia ‖. It also occurs in veins, as in those of Derbyshire, where it is associated with galena §.

Uses.

Like common limestone, it is burnt and made into mortar, but it remains much longer caustic than quicklime from common limestone; and this is the cause of a very important difference between magnesian and common limestone, with regard to their employment in agriculture: Lime, from magnesian limestone, is termed *hot*, and when spread upon land in the same proportion as is generally practised with common quicklime, greatly impairs the fertility of the soil; and when used in a greater quantity, is said by Mr Tennant to prevent all vegetation ¶.

Second

* Berger, Geological Society Transactions, vol. ii. p. 44.

† Greenough.

‡ Greenough.

‖ Bournon, Traité de Mineralogie, t. i. p. 268.

§ Bournon, Traité, ib.

¶ In regard to this limestone, Dr Thomson has the following remarks: "This magnesian limestone has been long burnt in prodigious quantity in the

Second Kind.

Flexible Compact Dolomite.

Flexible Limestone, *Thomson.*

External Characters.

Its colour is yellowish-grey, passing into cream-yellow.

It occurs massive.

It is dull.

The fracture is earthy in the small, but slaty in the large.

It is opaque.

It yields readily to the knife, but with difficulty to the nail.

In thin plates it is remarkably flexible.

Specific gravity, 2.544, *Thomson.* This is probably below the truth, as the stone is porous.

Chemical Characters.

It dissolves in acids as readily as common carbonate of lime.

 Constituent

the neighbourhood of Sunderland, and sent coastwise, both to the north and to the south. It goes in great abundance to Aberdeenshire. As no complaints have ever been made of its being injurious, when employed as a manure, it would be curious to know whether this circumstance be owing to the soil on which it is put, or to the small quantity of it used, in consequence of its price, occasioned by its long carriage; for it appears, from Mr Tennant's statement, that at Ferrybridge, the farmers are aware that it does not answer as a manure so well as pure carbonate of lime."—*Annals of Philosophy*, vol. iv. p. 418.

Constituent Parts.

Carbonate of Lime,	62.00
Carbonate of Magnesia,	35.96
Insoluble matter, -	1.60
Loss, - -	0.44
	100.00

Thomson, Annals of Phil. vol. iv. p. 418.

Geographic Situation.

It occurs about three miles from Tinmouth Castle.

Observations.

This curious mineral was discovered by Mr Nicol, Lecturer on Natural Philosophy. To that gentleman I am indebted for the following particulars in regard to it. He finds, that its flexibility is considerably influenced by the quantity of water contained in it. When saturated with water, it is remarkably flexible; as the evaporation goes on, it becomes more and more rigid, until the water be reduced to a certain limit, when the flexibility becomes scarcely distinguishable. From this point, however, the flexibility gradually increases, as the moisture diminishes; and as soon as the water is completely exhaled, it becomes nearly as flexible as it was when saturated with that fluid.

2. Miemite.

2. Miemite.

Miemite, *Klaproth.*

This species is divided into two subspecies, viz. Granular Miemite, and Prismatic Miemite.

First Subspecies.

Granular Miemite.

Magnesian Spar of *Thompson.*—Miemite, *Klaproth,* Beit. b. iii. s. 292.—Chaux carbonatée lente, Miemite, *Brong.* t. i. p. 230. Körniger Bitterspath, *Haus.* s. 128. *Id. Leonhard,* Tabel. s. 36.—Chaux carbonatée magnesifere lenticulaire, *Hauy,* Tabl. p. 6.—Miemit, *Lenz,* b. ii. s. 716.—Halbgeformter Bitterspath, *Oken,* b. i. s. 393.

External Characters.

Its colour is pale asparagus-green, which passes into greenish-white.

It occurs massive, and crystallised in flat double three-sided pyramids, in which the lateral planes of the one are set on the lateral edges of the other.

The crystals are middle-sized, or small; are either attached by their lateral edges, or intersect each other; and their surface is drusy.

Internally it is splendent and pearly.

The fracture is curved foliated.

The fragments are rather blunt-edged.

It occurs in distinct concretions, which are large and coarse, and are angulo-granular.

It is translucent.

It is semi-hard.

It is brittle.

Specific gravity, 2.885.

Chemical Characters.

It dissolves slowly, and with little effervescence, in nitrous acid; but more rapidly, and with increased effervescence, when the acid is heated.

Constituent Parts.

Carbonate of Lime, - -	53.00
Carbonate of Magnesia, -	42.50
Carbonate of Iron, with a little Manganese, - - - - -	3.00
	98.50

Klaproth, Beit. b. iii. s. 296.

Geognostic and Geographic Situations.

It is found at Miemo in Tuscany, imbedded in gypsum; and Mr Gieseké met with it in kidneys, along with wavellite, arragonite, and calcedony, in decomposed wacke, at Kannioak, in Omenaksfiord in Greenland.

Observations.

This mineral was first observed by the late Dr Thompson of Naples, who sent specimens of it to Klaproth for analysis. It is named *Miemite*, after the place where it was discovered.

Second

Second Subspecies.

Prismatic Miemite.

Stänglicher Bitterspath, *Klaproth.*

Strahliger Kalkstein, *Von Schlottheim,* Hoff's Magaz. fur die Gesammte Mineralogie, b. i. s. 156.—Stänglicher Bitterspath, *Klaproth,* b. iii. s. 297. *Id. Leonhard,* Tabel. s. 36. *Id. Haus.* s. 128. *Id. Lenz,* b. ii. s. 712. *Id. Oken,* b. i. s. 393.

External Characters.

Its colour is asparagus-green.

It occurs in low, nearly rectangular three-sided pyramids, which are deeply truncated on all the edges.

The crystals are small, and very small, and sometimes they form only drusy crusts. The tetrahedrons adhere by their bases, and are sometimes reniformly aggregated.

The lateral planes of the tetrahedrons are granular, and only glistening; but the truncating planes are smooth, and splendent and pearly.

Internally it is shining and vitreous.

The fracture passes from concealed foliated to splintery.

The fragments are rather blunt-edged.

It occurs in prismatic distinct concretions.

It is strongly translucent.

It is semi-hard in rather a high degree.

It affords a greyish-white coloured streak.

Specific gravity, 2.885, *Karsten.*

Chemical Characters.

It dissolves slowly, and with but feeble effervescence, in nitrous acid.

Constituent Parts.

Lime, - -	33.00
Magnesia, - -	14.50
Oxide of Iron, - -	2.50
Carbonic acid, -	47.25
Water and Loss, -	2.75
	100

Klaproth, Beit. b. iii. s. 303.

Geognostic and Geographic Situations.

It occurs in cobalt veins that traverse the first sandstone formation at Glücksbrunn in Gotha.

3. Brown-Spar, or Pearl-Spar.

Braunspath, *Werner.*

This species is divided into three subspecies, viz. Foliated Brown-Spar, Fibrous Brown-Spar, and Columnar Brown-Spar.

First

First Subspecies.

Foliated Brown-Spar.

Blättriger Braunspath, *Werner.*

Spath perlé, *Romé de Lisle,* t. i. p. 605.—Braunspath, *Wid.* s. 515.—Sidero-calcite, *Kirw.* vol. i. p. 105.—Braunspath, *Estner,* b. ii. s. 999. *Id. Emm.* b. i. s. 79.—Brunispato, *Nap.* p. 356.—Le Spath brunissant, ou le Braunspath, *Broch.* t. i. p. 563.—Gemeiner Braunspath, *Reuss,* b. i. s. 50. *Id. Lud.* b. i. s. 153. *Id. Suck.* 1r th. s. 630. *Id. Bert.* s. 118. *Id. Mohs,* b. ii. s. 108.—Spathiger Braunkalk, *Hab.* s. 82.—Chaux carbonatée manganesifere, *Lucas,* p. 8.—Spathiger Braunkalk, *Leonhard,* Tabel. s. 35.—Chaux carbonatée brunissante, *Brong.* t. i. p. 237.—Gemeiner Braunspath, *Karst.* Tabel. s. 50.—Chaux carbonatée ferro-manganesienne, *Bournon,* Traité, t. i. p. 277.—Pearl-Spar, *Kid,* vol. i. p. 56.—Chaux carbonatée ferro-manganesifere, *Hauy,* Tabl. p. 5.—Gemeiner Braunkalk, *Lenz,* b. ii. s. 717.—Gemeiner Braunspath, *Oken,* b. i. s. 394.

External Characters.

Its colours are milk white, yellowish-white, reddish-white, and greyish-white; yellowish-grey, and pearl-grey; flesh-red, rose-red, and brownish-red; clove brown, reddish-brown, liver brown, chesnut-brown, and brownish olive green; also cream-yellow, and ochre-yellow; and brownish-black, and pitch-black. Frequently several colours occur in the same mass; and it is often variegated with a yellow, pinchbeck-brown, and bronze-like tarnish.

It occurs massive, disseminated, globular, stalactitic, reniform, with tabular and pyramidal impressions; and crystallised in the following figures:

1. Rhomb,

1. Rhomb, in which the faces are sometimes cylindrically convex, sometimes cylindrically concave.
2. Lens, both common and saddle-shaped.
3. Flat double three-sided pyramid; is sometimes hollow.
4. Very acute single and double six-sided pyramid.

It also occurs in the following supposititious crystals:

1. Rhomb. 2. Double six-sided pyramid.

The true crystals are generally small and very small; the supposititious crystals large and middle-sized, and are either hollow, or lined with calcareous-spar.

The surface of the crystals is usually drusy, and is seldom shining, generally glistening or glimmering, and sometimes even dull.

Internally it alternates from shining to glistening, and the lustre is pearly.

The fracture is generally curved, seldom straight foliated, with a threefold oblique angular cleavage.

The fragments are rhomboidal.

The massive varieties occur in granular distinct concretions, of all the degrees of magnitude, but seldom fine granular: also in straight lamellar concretions, which are very much grown together.

It is more or less translucent, passing into translucent on the edges; and the crystals are sometimes perfectly translucent.

It is semi-hard; scratches calcareous-spar, but neither dolomite nor miemite.

It becomes greyish-white in the streak.

It is brittle.

It is easily frangible.

Specific gravity, 2.887, *Hauy.* 2.880, *Lichtenberg.*

Chemical

Chemical Characters.

It hardens, and becomes dark brownish-black before the blowpipe; and effervesces more or less briskly with acids, according to the quantity of manganese it contains.

Constituent Parts.

Lime, - -	43.0
Magnesia, - -	10.0
Oxide of Iron, -	8.0
Manganese, -	3.0
Water, and Carbonic Acid,	26.5

Berthier, Jour. des Mines, N° 103. p. 73.

Geognostic Situation.

It occurs principally in veins, when it forms either the predominating vein-stone, or is disseminated in the others. The most frequent accompanying vein-stone is calcareous-spar; besides which, it is often associated with heavy-spar, fluor-spar, quartz, sparry-ironstone, galena, iron-pyrites, native silver, and various ores of silver. Very often it rests on all the minerals of which the vein is composed: hence it is said to be the newest mineral in the vein; and we frequently observe thin crusts of it investing the surface of crystals, as of calcareous-spar, fluor-spar, heavy-spar, quartz, galena, &c. These crusts seldom invest the whole crystal, generally covering only a part of it; and it is observed, that it is the same side in all the crystals of the same cavity which are encrusted with the brown-spar; and also, that when the whole side is not covered, the crust has the same height, or is on the same level in all the crystals.

Geographic

Geographic Situation.

It occurs along with galena, and other ores of lead, in the lead-mines of Lead Hills and Wanlockhead in Lanarkshire; in the mines of Cumberland, Northumberland, and Derbyshire. On the Continent, it is found in Norway, Sweden, Saxony, Suabia, France, Hungary, and Transylvania.

Observations.

1. It is distinguished from *Calcareous-spar*, with which it has been confounded, by its colour-suite, inferior transparency, perfect pearly lustre, greater hardness, and higher specific gravity. It also in general effervesces less briskly with acids than calcareous-spar.

2. The straight lamellar variety has been mistaken for Heavy-spar, from which, however, it is distinguished, not only by its inferior weight, but also by its concretions being very closely aggregated, which is not the case with heavy-spar.

3. On exposure to the air, it changes, first to a light, then to a dark brown, bordering on black: hence the name *Brown-spar*, given to it by Werner.

Second Subspecies.

Fibrous Brown-Spar.

Fasriger Braunspath, *Werner*.

Fasriger Braunkalk, *Reuss*, b. ii. 2. s. 323.—Fasriger Braunspath, *Suck*. 1r th. s 629. *Id. Mohs*, b. ii. s. 121. *Id. Leonhard*, Tabel. s. 35. *Id. Karst.* Tabel. s. 50.—Fasriger Braunkalk, *Lenz*, b. ii. s. 722. *Id. Oken*, b. i. s. 394.

External Characters.

Its colours are reddish-white, flesh-red, pearl-grey, and yellowish-

yellowish-white. It is sometimes marked with ochre-yellow, and yellowish and blackish brown coloured spots.

It occurs massive, and in balls.

It is glimmering or glistening, and is pearly.

The fracture is straight, and either scopiform or diverging fibrous.

The fragments are splintery, or wedged-shaped.

It occurs in distinct concretions, which are either wedge shaped, prismatic, or coarse granular.

In other characters it agrees with the preceding subspecies.

Geognostic and Geographic Situations.

It has hitherto been found only in Lower Hungary, where it occurs in veins, as a member of different formations, along with foliated brown-spar, quartz, amethyst, iron-pyrites, and silver-glance or sulphuretted silver-ore.

Third Subspecies.

Columnar Brown-Spar.

Stänglicher Braunspath, *Klaproth.*

Stänglicher Braunspath, *Klaproth,* Beit. b. iv. s. 199. *Id. Karsten,* Tabel. s. 50. *Id. Lenz,* b. ii. s. 723.

External Characters.

Its colours are reddish-white, rose-red, and pearl-grey.

It is splendent, and appears pearly on the fracture-surface.

The fracture is foliated, but no distinct cleavage can be observed.

The

The fragments are wedge-shaped.

It occurs in wedge-shaped distinct concretions, which have glimmering and longitudinally streaked surfaces.

It is translucent.

It is brittle.

It is easily frangible.

Constituent Parts.

Carbonate of Lime, - -	51.50
Carbonate of Magnesia, -	32.00
Carbonate of Iron, - - -	7.50
Carbonate of Manganese, -	2.00
Water, - - - - -	5.00
	98.00

Klaproth, Beit. b. iv. s. 203.

Geographic Situation.

It is found at the mine named Segen Gottes at Gersdorf in Saxony; and that of Valenciana at Guanuaxuato in Mexico.

Observations.

It is distinguished from the other subspecies of Brownspar by fracture, fragments, distinct concretions, and transparency.

4. Gurhofite.

Gurhofian, *Karsten*.

Gurhofian, *Klaproth*, in Magazin der Gesellch. der Naturf. Freünde, b. i. s. 257.—Gurofian, *Karsten*, Tabel. s. 50. *Id. Klap.* Beit. b. v. s. 103. *Id. Lenz*, b. ii. s. 724.

External Characters.

Its colour is snow-white.

It

It occurs massive.
It is dull.
The fracture is flat conchoidal, passing into even.
The fragments are sharp-edged.
It is slightly translucent on the edges.
It is hard, bordering on semi-hard.
It is brittle.
It is rather difficultly frangible.
Specific gravity, 2.7600, *Karsten.*

Chemical Characters.

When pounded, and thrown into diluted and heated nitrous acid, it is completely dissolved with effervescence.

Constituent Parts.

Carbonate of Lime,	-	70.50
Carbonate of Magnesia,	-	29.50
		100

Klaproth, Gesellch. N. Fr. b. i. s. 258.

Geognostic and Geographic Situations.

It forms a vein in serpentine, in the rocks between Gurhof and Aggsbach, in Lower Austria.

Observations.

1. The name Gurhofite, is from the place near which it is found.

2. It was at one time considered to be a variety of semi-opal; but its greater weight distinguishes it from that mineral.

XXI LIMESTONE FAMILY.

THIS Family contains the following species: Tabular-Spar, Slate-Spar, Aphrite, Agaric Mineral, Chalk, Limestone, Lucullite, Marl, Bituminous Marl-Slate, and Arragonite.

1. Tabular-Spar.

Schaalstein, *Werner.*

Tafelspath, *Karsten.*

Tafelspath, *Reuss,* b. ii. s. 435. *Id. Lud.* b. ii. s. 144. *Id. Suck.* 1r th. s. 422.—Schaalstein, *Bert.* s. 166. *Id. Mohs,* b. ii. s. 1.–3.—Tafelspath, *Leonhard,* Tabel. s. 35. *Id. Karsten,* Tabel. s. 44.—Spath en tables, *Hauy,* Tabl. p. 66.—Schaalstein, *Lenz,* b. ii. s. 763.—Spathiger Conit, *Oken,* b. i. s. 392.

External Characters.

Its most common colour is greyish-white, which passes into greenish and yellowish white, and reddish-white.

It occurs massive, and crystallised in rectangular four-sided tables.

The lustre of the principal fracture is shining and pearly.

The principal fracture is foliated, with a single cleavage, and inclines to coarse fibrous and splintery.

It occurs in prismatic distinct concretions, which are long, thick, and broad, and promiscuously aggregated, and sometimes pass into granular.

It

It is translucent.
It is semi-hard.
It is brittle.
It is easly frangible.
Specific gravity, 2.855, *Stütz.* 2.86, *Hauy.*

Chemical Characters.

When put into nitrous acid, it effervesces for a moment, and then falls into grains. It is infusible before the blowpipe. It is phosphorescent when scratched with a knife.

Constituent Parts.

Silica, - -	50
Lime, - -	45
Water, - -	5
	100

Klaproth, Beit. b. iii. s. 291.

Geognostic and Geographic Situations.

It forms a bed in primitive limestone, where it is associated with brown garnets, blue-coloured calcareous-spar, tremolite, actynolite, and variegated copper-ore, at Dognatska in the Bannat of Temeswar.

2. Slate-Spar.

Schieferspath, *Werner*.

Shieferspath, *Wid.* s. 510.—Argentine, *Kirw.* vol. i. p. 105.—Schisto-spatho, *Nap.* p. 355.—Schifferspath, *Lam.* t. i. p. 385.—Le Spath schisteux, ou le Schieferspath, *Broch.* t. i. p. 558. Schieferspath, *Reuss,* b. ii. s. 50. *Id. Lud.* b. i. s. 152. *Id. Suck.* 1r th. s. 626. *Id. Bert.* s. 95. *Id. Mohs,* b. ii. s. 8. *Id. Hab.* s. 81. *Id. Leonhard,* Tabel. s. 34.—Chaux carbonatée nacré argentine, *Brong.* t. i. p. 232.—Verhærteter Aphrit, *Karsten,* Tabel. s. 50.—Chaux carbonatée nacré primitive, *Hauy,* Tabl. p. 6.—Schieferspath, *Lenz,* b. ii. s. 761.—Schieferige Schaumerde, *Oken,* b. i. s. 394.

External Characters.

Its colours are greenish-white, reddish-white, yellowish-white, greyish-white, and snow-white.

It occurs massive and disseminated.

The lustre is intermediate between shining and glistening, and is pearly.

The fracture is common curved, and undulating curved foliated, with a single cleavage; and in the large, it inclines to slaty.

The fragments are either indeterminate angular and blunt-edged, or are slaty.

It occurs in distinct concretions, which are large and coarse granular, and sometimes also thin and curved lamellar.

It is feebly translucent, or only translucent on the edges.

It is soft.

It is intermediate between sectile and brittle.

It

It is easily frangible.

It feels rather greasy.

Specific gravity, 2.647, *Kirwan*. 2.474, *Blumenbach*. 2.6300, *La Metherie*.

Chemical Characters.

It effervesces very violently with acids; but is infusible before the blowpipe.

Constituent Parts.

From Bremsgrün.		From Kongsberg.	
Lime, -	55.00	Lime, -	56.00
Carbonic Acid,	41.66	Carbonic Acid,	39.33
Oxide of Manganese,	3.00	Silica, -	1.66
	——	Oxide of Iron,	1.00
	Bucholz.	Water, -	2.00
			——
			Suersee.

Geognostic Situation.

It occurs in primitive limestone, along with calcareous-spar, brown-spar, fluor-spar, and galena; in metalliferous beds, associated with magnetic ironstone, galena, and blende; and in veins, along with tinstone.

Geographic Situation.

It occurs in Glen Tilt, Perthshire; and Assynt in Sutherland, in marble: in Cornwall; and near Granard in Ireland *. On the Continent, it is found along with tinstone, in the Saxon Erzgebirge; along with octahedrite, in a vein at St Christophe in Dauphiny; also in Norway, in metalliferous beds, and in limestone.

H 3 *Observations.*

* Greenough.

Observations.

It is characterised by its lustre, fracture, and distinct concretions. The lamellar concretions, according to Bournon, are but varieties of the primitive rhomboid of calcareous-spar; so that in this view, each concretion is a perfect crystal.

3. Aphrite.

Schaumerde, *Werner.*

This species is divided into three subspecies, viz. Scaly Aphrite, Slaty Aphrite, and Sparry Aphrite.

First Subspecies.

Scaly Aphrite.

Schaumerde, *Werner.*

Zerreiblicher Aphrit, *Karsten.*

Schaumerde, *Emm.* b. i. s. 484.—Silvery Chalk, *Kirw.* vol. i. p. 78.—L'Ecume de Terre, *Broch.* t. i. p. 557.—Ecume de Terre des Allemands, *Hauy,* t. iv. p. 360.—Schaumerde, *Reuss,* b. ii. 2. s. 317. *Id. Lud.* b. i. s. 152. *Id. Suck.* 1r th. s. 625. *Id. Bert.* s. 95. *Id. Mohs,* b. ii. s. 6. *Id. Hab.* s. 80. *Id, Leonhard,* Tabel. s. 34.—Chaux carbonatée nacré talquese, *Brong.* t. i. p. 252.—Zerreiblicher Aphrite, *Karsten,* Tabel. s. 50.. *Id. Haus.* s. 126.—Chaux carbonatée nacrée lamellaire, *Hauy,* Tabl. p. 6.—Schaumkalch, *Lenz,* b. ii. s. 757.—Erdige Schaumerde, *Oken,* b. i. s. 394.

External Characters.

Its colours are snow, yellowish, and reddish white, sometimes passing into silver-white.

It

It occurs massive, disseminated, in membranes, or extremely delicate crusts, and in small tuberose friable pieces.

It occurs sometimes solid, sometimes friable, and often of an intermediate degree of coherence.

Externally it is strongly glimmering; internally glistening, and pearly, inclining to semi-metallic.

The friable varieties consist of scaly parts; the more solid have a fracture which is curved slaty, passing into undulating foliated.

The more compact varieties very rarely appear in imperfect small granular concretions.

It is opaque.

It soils.

It is very soft, passing into friable.

It feels soft, and almost greasy.

It is very light; the scaly varieties supernatant.

Chemical Characters.

It falls into pieces, with a crackling noise, when put into water.

Owing to its loose texture, it effervesces most violently with acids. It is a nearly pure carbonate of lime.

Geognostic Situation.

It occurs in nests, disseminated, or in small veins, in the first flœtz limestone.

Geographic Situation.

It is found in Thuringia, and Hessia.

Observations.

1. This mineral was formerly confounded with Nacrite, from which it differs in external characters, geognostic situation, and chemical composition.

 2. According

2. According to Mohs, Tabular-Spar and Slate-Spar are connected, by means of this species, with Agaric Mineral and Chalk, and the transition from Chalk into Limestone is evident; so that thus the whole of these species form nearly a natural group.

Second Subspecies.

Slaty Aphrite.

Schaumschiefer, *Friesleben.*

Schaumschiefer, *Friesleben,* Geognostiche Beiträge, b. ii. s. 232

External Characters.

Its colours are snow-white, passing into yellowish, reddish, and silver white.

It occurs massive, seldom coarsely disseminated.

It is strongly glimmering, sometimes approaching to glistening, even to shining; and the lustre is pearly, which sometimes passes into semi-metallic.

It is slaty in the great, but undulating curved foliated in the small.

It splits very easily into extremely thin tabular fragments.

It is opaque; only very feebly translucent in the thinnest folia.

It soils pretty strongly, with scaly particles.

It is very soft.

It feels soft, and rather silky.

It is flexible in thin plates.

It is light.

Chemical

Chemical Characters.

It falls into pieces, with a crackling noise, when put into water. When touched with an acid, it effervesces with great violence, and is entirely dissolved in it.

Geognostic and Geographic Situations.

It occurs massive, imbedded, and in veins, in the first flœtz limestone, in Thuringia, and Hessia.

Observations.

1. The straight slaty variety passes into Slate-Spar, and into Scaly Aphrite.

2. Meinecke, and other old observers, described this mineral as a variety of Common Talc. It was first accurately examined and described by Friesleben.

Third Subspecies.

Sparry Aphrite.

Schaumspath, *Friesleben.*

Schaumspath, *Friesleben,* Geognostiche Beiträge, b. ii. s. 234.

External Characters.

Its colours are snow, yellowish, and greyish white.

It occurs seldom massive, generally disseminated; sometimes in flaky crusts, in veins, or imbedded in large crystals of selenite.

It is shining, sometimes inclining to splendent, sometimes to glistening; and the lustre is pearly, which inclines to vitreous in the splendent varieties.

The

The fracture is foliated, sometimes straight, sometimes curved, and the folia have a single distinct cleavage.

It is opaque; feebly translucent in thin pieces.

It occurs in large and small granular distinct concretions.

It soils slightly, with glimmering dusty particles.

It is soft.

It is sectile.

It is not particularly heavy.

Chemical Characters.

The same as the other subspecies.

Geognostic Situation.

It occurs in the first flœtz limestone, and first flœtz gypsum. According to Friesleben, it appears to be geognostically allied to selenite; and although it differs from that mineral in colour, transparency, lustre, sectility, feel, and effervescence with acids, yet it passes into it, and also into slaty aphrite, sometimes by simple gradations, sometimes by intermixture of the two minerals; and large lenticular crystals of selenite occur, which are pure at the edges, become gradually more opaque towards the centre, and in the centre are pure sparry aphrite.

Geographic Situation.

It occurs in Thuringia.

Observations.

It was first described and named by Friesleben, in his Geognostical Contributions.

4. Agaric

4. Agaric Mineral, or Rock-Milk.

Berg-Milch, *Werner*.

Agaricus mineralis, *Wall.* t. i. p. 30.—Bergmilch, *Wid.* s. 490. —Agaric Mineral, *Kirw.* vol. i. p. 76.—Bergmilch, *Estner*, ii. s. 914. *Id. Emm.* b. i. s. 430.—Agaric Mineral, *Nap.* p. 333. *Id. Lam.* p. 331.—Lait de Montagne, ou l'Agaric Mineral, *Broch.* t. i. p. 519.—Chaux carbonatée spongieuse, *Hauy*, t. ii. p. 167.—Bergmilch, *Reuss*, b. ii 2. s. 257. *Id. Lud.* b. i. s. 145. *Id. Suck.* 1r th. s. 582. *Id. Bert.* s. 87. *Id. Mohs*, b. ii. s. 8. *Id. Leonhard*, Tabel. s. 32.—Chaux carbonatée spongieuse, *Brong.* t. i. p. 210.—Montmilch, *Haus.* s. 127.—Bergmilch, *Karsten*, Tabel. s. 50.—Agaric Mineral, *Kid*, vol. i. p. 38.—Bergmilch, *Lenz*, b. ii. s. 727. *Id. Oken*, b. i. s. 411.

External Characters.

Its colours are snow-white, greyish-white, and yellowish-white.

It occurs frequently in crusts, also in loosely cohering tuberose pieces.

It is dull.

It is composed of fine dusty particles.

It soils strongly.

It feels meagre.

It adheres slightly to the tongue.

It is very light, almost supernatant.

Chemical Character.

It effervesces with acids, and is completely dissolved in them.

Constituent

Constituent Parts.

It is a pure Carbonate of Lime.

Geognostic and Geographic Situations.

It is found on the north side of Oxford, between the Isis and the Cherwell, and near Chipping-Norton, also in Oxfordshire *; and in the fissures of caves of limestone mountains in Switzerland, Austria, Salzburg, and other countries.

Uses.

In Switzerland, where it occurs abundantly, it is used for whitening houses.

Observations.

1. It is formed by water passing over and through limestone rocks, and afterwards depositing in holes, fissures, and on faces of rocks, the calcareous earth it had dissolved in its course.

2. It is named *Agaric Mineral*, from its sometimes adhering to rocks with the resemblance of a fungus or agaric: the name *Rock Milk* given to it by some mineralogists, is from its white appearance when oozing from the clefts of rocks; and the name *Lac Lunæ* is sometimes given to it, from the milky-like appearance it presents in a cave in Phrygia; this cave, according to the tradition of the neighbourhood, having been formerly frequented by Diana †.

5. Chalk.

* Kid's Mineralogy, vol. i. p. 39. † Ibid.

5. Chalk.

Kreide, *Werner*.

Creta alba, *Wall.* t. i. p. 27.—Kreide, *Wid.* s. 492.—Chalk, *Kirw.* vol. i. p. 77.—Kreide, *Estner,* b. ii. s. 917. *Id. Emm.* b. i. s. 433.—Creta commune, *Nap.* s. 331.—La Craie, *Broch.* t. i. p. 521.—Craie, *Hauy,* t. ii. p. 166.—Kreide, *Reuss,* b. ii. 2. s. 259. *Id. Lud.* b. i. s. 145. *Id. Suck.* 1r th. s. 583. *Id. Bert.* s. 87. *Id. Mohs,* b. ii. s. 9. *Id. Hab.* s. 70. *Id. Leonhard,* Tabel. s. 32.—Chaux carbonatée crayeuse, *Brong.* t. i. p. 208. *Id. Brard,* p. 29.—Kreide, *Karsten,* Tabel. s. 50. *Id. Haus.* s. 126.—Chalk, *Kid,* vol. i. p. 18.—Kreide, *Lenz,* b. ii. s. 728. *Id. Oken,* b. i. s. 410.

External Characters.

Its colour is yellowish-white, which sometimes passes to greyish-white, and snow-white. It is sometimes marked with yellowish-grey.

It occurs massive, disseminated, in crusts, and in extraneous external shapes.

It is dull.

The fracture is coarse and fine earthy.

The fragments are blunt-edged.

It is opaque.

It writes and soils.

It is rather sectile.

It is soft, and sometimes very soft.

It is very easily frangible.

It feels meagre and rough.

It adheres slightly to the tongue.

It is light.

Specific gravity, 2.252, *Mushenbroeck.* 2.315, *Kirwan.* 2.400, 2.675, *Gerhard.* 2.657, *Watson.*

Chemical

Chemical Characters.

It effervesces strongly with acids.

Constituent Parts.

		Chalk from Gallicia.			
Lime, -	56.5	Lime, -	47.00	Lime, -	53
Carbonic Acid,	43.0	Carbonic Acid,	33.00	Carbonic Acid,	42
Water, -	00.5	Silica, -	7.00	Alumina,	2
	——	Alumina,	2.00	Water, -	3
		Magnesia,	8.00		——
Bucholz, in Gehlen's Journal, b. iv. s. 416.		Iron, -	0.05	*Kirwan*, Min. vol. i. p. 77.	
			——		
		Hacquet.			

Geognostic Situation.

It constitutes one of the newer flœtz formations; is usually found in low situations, and frequently on sea coasts. It is stratified, and the strata in general are horizontal. It often contains flint, which is disposed either in interrupted beds in the chalk, or in globular, tuberose, or tabular masses imbedded in it. It abounds in organic remains, and these are principally of animals of the lower orders, such as echinites, belemnites, terebratulites, pinnites, &c. These petrifactions, are either in the state of carbonate, or are converted into flint, which latter is by far the most frequent. It cannot be considered as a metalliferous formation, as it contains nothing but small imbedded portions of iron-pyrites. Two principal kinds of chalk occur in chalk districts: the one is named *Hard*, the other *Soft Chalk*; the hard chalk always occurs undermost, is considerably harder than the other, and rarely contains petrifactions or flint; the soft chalk, on the contrary, rests upon the other, is softer, and abounds in flint and petrifactions.

Geographic

Geographic Situation.

It abounds in the south-eastern parts of England,—extends through several provinces in France,—occupies great tracts of country in Poland and Russia,—is met with on the shores of the Baltic,—and in the islands of Zeeland and Rugen.

Uses.

The uses of this mineral are various. The more compact kinds are employed as building-stones, when they are used either in a rough state, or are sawn into blocks of the requisite size and shape: it is burnt into quicklime, and used for mortar in different countries; thus, nearly all the houses in London are cemented with chalk-mortar *: it is also employed in great quantities in the polishing of glass and metals, and whitening the roofs of rooms, in the state of *whiting* †; in constructing moulds to cast metal in; by carpenters and others as a material to mark with. When perfectly purified, and mixed with vegetable colours, it forms a kind of pastil colour; thus, with litmus, turmeric, saffron, and sap-green, it forms durable colours, but vegetable colours that contain an acid, become blue when mixed with it. The *Vienna white* known to artists, is perfectly purified chalk. It is used by starch-makers and chemists to dry pecipitates on, for which

* According to Smeaton, it makes as good lime as the best limestone or marble.

† In the preparation of whiting, chalk is pounded, and diffused through water, and the finer part of the sediment is then dried; by this means, the siliceous particles are separated, which, by their hardness, would scratch the surface of metallic and other substances, in the polishing of which whiting is used.—*Aikin's Chem. Dictionary.*

which it is peculiarly qualified, on account of the remarkable facility with which it absorbs water. With isinglass or white of eggs, it forms a valuable lute or cement. In the gilding of wood, it is necessary, before laying on the gold, to cover it with a succession of coats of a mixture of whiting and size. The mineral is also used as a filtering-stone; and in a purified state, it is employed as a remedy to correct acidity in the stomach, and the morbid states which arise from this.

Observations.

It is conjectured that the name *Creta*, is derived from the island of Candia, (Creta of the ancients), where this mineral is said to occur. Ancient writers seem to use the word *creta* in different senses, as appears from the following observations: "The word *creta*, though applied by Wallerius and others to chalk, is generally used by the early naturalists to express clay: 'Proderit sabulosis locis *cretam* ingerere; cretosis ac nimium densis, *sabulum* *;' where, as *sabulum* certainly means sand, it is nearly evident, from the reciprocal use of the substances mentioned, compared with the opposite properties of sand and clay, that *creta* signifies the latter. 'Lateres non sunt e sabuloso, neque arenoso, multoque minus calculoso ducendi solo; sed e *cretoso* †.' Again, it may be observed, with respect to the following line,

'Hinc humilem Myconen, *cretosa*que rura *Cimoli* ‡,'

that the Cimolian earth is described in various passages of Pliny, &c. under characters peculiar to clay.

There

---

* Columella, p. 73.

† Plin. Nat. Hist. ed. Brot. vol. vi. p. 174.

‡ Ovid, Metam. lib. vii.

There are two passages in which *creta* seems to be applicable to chalk: one in Horace,

——— ' *creta* an carbone notandi *.'

The other in Pliny: ' Alia *creta* argentaria appellatur nitorem argento reddens †;' this being a common use of chalk at the present day."—*Kid's Mineralogy*, vol. i. p. 18, 19.

6. Limestone.

Kalkstein, *Werner*.

This species is divided into six subspecies, viz. Compact Limestone, Foliated Limestone, Fibrous Limestone, Lamellar Limestone, Tufaceous Limestone or Calc-tuff, and Pisiform Limestone or Pea-stone.

First Subspecies.

Compact Limestone.

Dichter Kalkstein, *Werner*.

This subspecies is divided into three kinds, viz. Common Compact Limestone, Blue Vesuvian Limestone, and Roestone.

* Horat. Sat. iii. lib. 2.

† Plin. Nat. Hist. ed. Brot. vol. vi. p. 184.

First Kind.

Common Compact Limestone.

Gemeiner Dichter Kalkstein, *Werner*.

Calcareus æquabilis, *Wall.* t. i. p. 122.—Dichter Kalkstein, *Wid.* s. 494.—Compact Limestone, *Kirw.* vol. i. p. 82.—Gemeiner Dichter Kalkstein, *Emm.* b. i. s. 437.—Pietra calcarea compacta, *Nap.* p. 33.—La pierre calcaire compacte commune, *Broch.* t. i. p. 523.—Chaux carbonatée compacte, *Hauy,* t. ii. p. 166.—Gemeiner Dichter Kalkstein, *Reuss,* b. ii. 2. s. 262. *Id. Lud.* b. i. s. 146. *Id. Suck.* 1r th. s. 585. *Id. Bert.* s. 88. *Id. Mohs,* b. ii. s. 14. *Id. Hab.* s. 71. *Id. Leonhard,* Tabel. s. 32.—Chaux carbonatée compacte, *Brong.* t. i. p. 199.—Gemeiner Kalkstein, *Haus.* s. 126.—Dichter Kalkstein, *Karsten,* Tabel. s. 50.—Chaux carbonatée compacte, *Hauy,* Tabl. p. 4.—Dichter gemeiner Kalkstein, *Lenz,* b. ii. s. 732.—Ungeformiter Kalk, *Oken,* b. i. s. 410.

External Characters.

Its most frequent colour is grey, of which the following varieties have been observed: yellowish, bluish, ash, pearl, greenish, and smoke grey; the ash-grey passes into greyish-black; the yellowish-grey into yellowish-brown, ochre-yellow, and into a colour bordering on cream-yellow. It also occurs blood-red, flesh-red, and peach-blossom-red, which latter colour is very rare.

It frequently exhibits veined, zoned, striped, clouded, and spotted coloured delineations; and sometimes also black and brown coloured arborisations.

It very rarely exhibits a beautiful play of colours, caused by intermixed portions of pearly shells.

It occurs massive, corroded, in large plates, rolled masses, and in various extraneous external shapes, of univalve, bivalve, and multivalve shells, of corals, and of fishes,

fishes, and more rarely of vegetables, as of ferns and reeds.

Internally it is dull, seldom glimmering, which is owing to intermixed calcareous-spar.

The fracture is small and fine splintery, which sometimes passes into large and flat conchoidal, sometimes into uneven, inclining to earthy, and it occasionally inclines to straight and thick slaty.

The fragments are indeterminate angular, more or less sharp-edged, but in the slaty variety they are tabular.

It is translucent on the edges.

It is semi-hard.

It is brittle.

It is easily frangible.

Specific gravity, Splintery, 2.600, 2.720, *Brisson.* Opalescent Shell Marble, 2.6732, *Leonhard.* Hartz Limestone, 2.489, *Lazuis.*

Chemical Characters.

It effervesces with acids, and the greater part is dissolved; and burns to quicklime, without falling to pieces.

Constituent Parts.

Rudersdorf.		Bluish-grey Limestone.		Limestone from Sweden.		Limestone from Ettersberg *	
Lime,	53.00	Lime,	49.50	Lime,	49.25	Lime,	33.41
Carbon. acid,	42.50	Carbon. acid,	40.00	Carbon. acid,	35.00	Carbon. acid,	42.00
Silica,	1.12	Silica,	5.25	Silica,	8.75	Silica,	10.25
Alumina,	1.00	Alumina,	2.75	Alumina,	2.50	Magnesia,	9.43
Iron,	0.75	Iron,	1.37	Iron,	2.75	Iron,	2.25
Water,	1.03	Water,	1.13	Loss,	1.75	Manganese,	1.25
						Loss,	1.41
	100		100		100		100
Simon, Gehlen's Jour. iv. s. 426.		*Simon*, Ib.		*Simon*, Ib.		*Bucholz*, Ib.	

* Some of the limestones in Fifeshire agree in composition with that of Ettersberg. Others, and these should have been mentioned under the article Magnesian Limestone, agree with that mineral in chemical composition.

Geognostic Situation.

This mineral occurs in vast abundance in nature, principally in flœtz formations, along with sandstone, gypsum, and coal. It is distinctly stratified, and the strata vary in thickness, from a few inches to many fathoms, and are from a few fathoms to many miles in extent. The strata generally incline to horizontal; sometimes, however, they are vertical, or variously convoluted, even arranged in concentric layers, thus presenting appearances illustrative of their chemical nature. Petrifactions, both of animals and vegetables, but principally of the former, abound in compact limestone: these are of corals, shells, fishes, and sometimes of amphibious animals. On a general view, it is to be considered as rich in ores of different kinds, particularly ores of lead and zinc: thus, nearly all the rich and valuable lead-mines in England are either situated in limestone, or the veins containing the ores are richest where they traverse the limestone.

Geographic Situation.

It abounds in the sandstone and coal formations, both in Scotland and England; and in Ireland, it is a very abundant mineral in all the districts where flœtz rocks occur. On the Continent of Europe, it is a very widely and abundantly distributed mineral; and forms a striking feature in many extensive tracts of country in Asia, Africa and America, as will be particularly described in the Geognostic part of this work.

Uses.

When compact limestone joins to pure and agreeable colours, so considerable a degree of hardness that it takes a good polish, it is by artists considered as a Marble; and if if it contains petrifactions mineralised, it is named *shell* or *lumachella*, and *coral* or *zoophytic marble*, according as the organic remains are testaceous or coralline *. In one particular variety of lumachella or shell marble, found at Bleiberg in Carinthia, the shells and fragments of shells, which belong to the nautilus tribe, are set in a brown-coloured basis, and reflect many beautiful and brilliant pearly inclining to metallic colours, principally the fire-red, green, and blue tints. It is named *opalescent* or *fire marble*. Another lumachella marble from Astracan, contains, in a reddish-brown basis, pearly shells of nautili, that reflect a very brilliant gold-yellow colour. In some compact marbles, the surface presents a beautiful arborescent appearance, and these are named *arborescent* or *dendritic marbles*. In different parts of Scotland, compact limestone is cut and polished as marble: this was the case in the parish of Cummertrees in Dumfriesshire,—in Cambuslang parish, in Lanarkshire,—in Fifeshire, &c. In England, many compact limestones are cut and polished as marbles; such are the limestones of Derbyshire, of Yorkshire, Devonshire, Somersetshire, and Dorsetshire. It is sometimes used as a building stone; and, in want of better materials, for paving streets, and making highways. When, by exposure to a high temperature, it is deprived of its carbonic acid, and converted into quicklime, it is used for mortar; also by the soap-maker, for rendering his alkalies caustic; by the tanner, for cleansing hides, or freeing them from hair, muscular substance, and fat; by the farmer, in the improvement of particular kinds of soil; and by the metallurgist, in the smelting of such

* The name *marmor*, is derived from the Greek μαρμαιρειν, *to shine*, or *glitter*, and was by the ancients applied, not only to limestone, but also to stones possessing agreeable colours, and receiving a good polish, such as gypsum, jasper, serpentine, and even granite and porphyry.

ores as are difficultly fusible owing to an intermixture of silica and alumina.

Second Kind.

Blue Vesuvian Limestone.

Blauer Vesuvischer Kalkstein, *Klaproth.*

Blauer Vesuvischer Kalkstein, *Klaproth,* Beit. b. v. s. 92. *Id. Lenz,* b. ii. s. 737.

External Characters.

Its colour is dark bluish-grey, partly veined with white.

Externally it appears as if it had been rolled; and the surface is uneven.

The fracture fine earthy, passing into splintery.

It is opaque.

It affords a white streak.

It is semi-hard in a low degree.

It is rather heavy.

Constituent Parts.

Lime, - -	58.00
Carbonic Acid, -	28.50
Water, which is somewhat ammoniacal, -	11.00
Magnesia, - -	0.50
Oxide of Iron, -	0.25
Carbon, - -	0.25
Silica, - -	1.25
	99.75

Klaproth, Beit. b. v. s. 96.

From

From this analysis, it appears, that the vesuvian limestone differs remarkably in composition from common compact limestone. In common compact limestone, 100 parts of lime are combined with at least 80 parts of carbonic acid; whereas in the vesuvian limestone, 100 parts of limestone are not combined with more than 50 parts of carbonic acid. Secondly, In common limestone, independent of the water which adheres to it accidentally, as far as we know, there is no water of composition; but in the vesuvian limestone, there are 11 parts of water of composition.

Geographic Situation.

This remarkable limestone is found in loose masses amongst unaltered ejected minerals in the neighbourhood of Vesuvius.

Observations.

It is known to some collectors under the name *Compact Blue Lava* of Vesuvius; and is employed by artists, in their mosaic work, to represent the sky.

Third Kind.

Roestone.

Roogenstein, *Werner.*

Stalactites oolithus, var. *b. c. d. Wall.* t. ii. p. 384.—Roogenstein, *Wid.* s. 511.—Oviform Limestone, *Kirw.* vol. i. p. 91. —Roogenstein, *Estner,* b. ii. s. 928. *Id. Emm.* b. i. s. 442. —Tufo oolitico, *Nap.* p. 353.—L'Oolite, *Broch.* t. i. p. 529. —Chaux carbonatée globuliforme, *Hauy,* t. ii. p. 171.—Roogenstein, *Reuss,* b. ii. 2. s. 270. *Id. Lud.* b. i. s. 148. *Id. Suck.* 1r th. s. 591. *Id. Bert.* s. 89. *Id. Mohs,* b. ii. s. 26. *Id. Hab.* p. 72. *Id. Leonhard,* Tabel. s. 32.—Chaux carbobonatée oolithe, *Brong.* t. i. p. 203.—Chaux carbonatée globuliforme, *Brard,* p. 31.—Erbsförmiger Kalkstein, *Karsten,* Tabel. s. 50.—Roestone, *Kid,* vol. i. p. 26.—Chaux carbonatée globuliforme, *Hauy,* Tabl. p. 4.—Roogenstein, *Lenz,* b. ii. s. 738. *Id. Oken,* b. i. s. 411.

External Characters.

Its colours are hair-brown, chesnut-brown, and reddish-brown, and sometimes yellowish-grey, and ash-grey.

It occurs massive.

Internally it is dull.

The fracture of the grains is fine splintery; but of the mass is round granular in the small, and slaty in the large.

The fragments in the large are blunt-edged.

It is composed of globular concretions, which have sometimes a rough or uneven, sometimes a smooth surface, and vary in size from that of a cherry-stone to that of a millet seed. These concretions have in general a central nucleus, which is either of compact limestone, or of iron-

ore

ore which has an exterior crust of black ironstone, and an interior mass of iron-ochre, and in some rarer instances, the centre is hollow, and is encircled with a crust of black ironstone. This nucleus is enveloped in concentric lamellar concretions: the globular concretions are in general imbedded in a base of marl, or are connected together without any intermediate substance.

It is opaque, or very faintly translucent on the edges.

It is semi-hard.

It is brittle.

It is very easily frangible.

Specific gravity, 2.6829, 2.6190, *Kopp.*

Chemical Character.

It dissolves with effervescence in acids.

Geognostic Situation.

It generally occurs in beds, from four to twelve inches thick, which alternate with calcareous sandstone, sandstone-slate, slaty variegated clayey loam, and rarely with red-coloured clay; and all of these belong to the second flœtz sandstone.

Geographic Situation.

It occurs in considerable quantity in the province of Thuringia; and it would appear, from Escher's observations, to be a production of the Jura Mountains *.

Uses.

The compact and fine granular varieties form a good building-stone: it is also used as a manure; but when burnt

* Escher's Brief, in Von Moll's Eph., der Berg und Hüttenkunde. St. 3. s. 433.

burnt into quicklime, the marly varieties afford rather an indifferent mortar; but those mixed with sand a better mortar.

Observations.

1. It passes into Sandstone, Limestone, and Marl.

2. Some naturalists, as Daubenton, Saussure, Spallanzani, and Gillet Lamont, conjecture, that Roestone is carbonate of lime, which has been granulated in the manner of gunpowder, by the action of water: the most plausible opinion is that which attributes the formation of this mineral to crystallization from a state of solution.

3. It is named *Roestone*, from its resemblance in form to the roe of a fish.

Second Subspecies.

Foliated Limestone.

Blättriger Kalkstein, *Werner*.

This subspecies is divided into two kinds, viz. Foliated Granular Limestone, and Calcareous-Spar.

First

First Kind.

Foliated Granular Limestone.

Blättriger Körniger Kalkstein, *Werner.*

Calcareus micans, *Wall.* t. i. p. 126.; Calcareus inæquabilis, *Id.* p. 128.; Marmor unicolor album, *Id.* p. 133.—Körniger Kalkstein, *Wid.* s. 496.—Foliated and Granular Limestone, *Kirw.* vol. i. p. 84.—Körniger Kalkstein, *Estner,* b. ii. s. 931. *Id. Emm.* b. i. s. 445.—Pierre calcaire grenue, *Broch.* t. i. p. 531.—Chaux carbonatée saccaroide, *Hauy,* t. ii. p. 164. —Körniger Kalkstein, *Reuss,* b. ii. 2. s. 273. *Id. Lud.* b. i. s. 148. *Id. Suck.* 1r th. s. 593.—Kleinblättricher Kalkstein, *Bert.* s. 89.—Körnigblattricher Kalkstein, *Mohs,* b. ii. s. 28. *Id. Hab.* s. 74.—Gemeiner körniger Kalkstein, *Leonhard,* Tabel. s. 32.—Chaux carbonatée saccaroide, *Brong.* t. i. p. 192. *Id. Brard,* p. 28.—Marmor, *Haus.* s. 126.—Körniger Kalkstein, *Karst.* Tabel. s. 50.—Marble, *Kid,* vol. i. p. 4.—Chaux carbonatée lamellaire, et Chaux carbonatée saccaroide, et Chaux carbonatée sub-granulaire, *Hauy,* Tabl. p. 5.—Körniger Kalkstein, *Lenz,* b. ii. s. 739.—Halbgeförmiter Kalk, *Oken,* b. i. s. 409.

External Characters.

Its most common colour is white, of which it presents the following varieties: snow-white, yellowish-white, greyish-white, and greenish-white, seldom reddish-white: from greyish-white it passes into bluish-grey, greenish-grey, ash-grey, smoke-grey; and from this latter into greyish-black? From reddish-white it passes into pearl-grey, and flesh-red; from yellowish-white into cream-yellow; and from greenish-white into siskin-green, and olive-green.

It

It has generally but one colour; sometimes it is spotted, dotted, clouded, striped, veined, and arborescent.

It occurs massive.

Internally it alternates from shining to glistening and glimmering, and the lustre is intermediate between pearly and vitreous.

The fracture is foliated, but sometimes inclines to splintery.

The fragments are blunt-edged.

It occurs in distinct concretions, which are coarse, small, and fine granular: the small granular sometimes pass into compact, and then the concretions are only distinguishable by their glimmering lustre.

It is more or less translucent.

It is semi-hard.

It is brittle.

It is easily frangible.

Specific gravity, Parian Marble, 2.8376, *Brisson.* Carrara Marble, 2.717. Scottish, 2.716, *Kirwan.* 2.658, 2.711, *Karsten.*

Chemical Characters.

It generally phosphoresces when pounded, or when thrown on glowing coals. It is infusible before the blowpipe. It dissolves with effervescence in acids.

Constituent Parts.

Lime,	- -	56.50
Carbonic acid,	-	43.00
Water,	- -	0.50

Bucholz, in Neuen Journal der Chem. iv. s. 419.

Geognostic

Geognostic Situation.

This mineral occurs in beds, in granite, gneiss, mica-slate, clay-slate, syenite, greenstone, grey-wacke, and rarely in some of the flœtz rocks. It is observed, that the varieties which occur in the older rocks are in general more highly crystalline than those which are found in the newer. It frequently contains imbedded minerals of different kinds, such as quartz, mica, hornblende, tremolite, sahlite, asbestus, steatite, serpentine, galena, blende, iron-pyrites, and magnetic ironstone: of these the quartz and mica are the most frequent.

Geographic Situation.

This mineral occurs in all the great ranges of primitive and transition rocks that occur in Europe, and in such as have been examined in Asia, Africa, and America. It is observed, that in general the white and grey varieties abound in primitive countries; the variegated in those composed of transition rocks. All of them receive a good polish, and hence are known to artists as marbles.

Uses.

All the varieties of this subspecies may be burnt into quicklime; but it is found, that in many of them, the concretions exfoliate and separate during the volatilization of their carbonic acid, so that by the time when they are rendered perfectly caustic, their cohesion is destroyed, and they fall into a kind of sand,—a circumstance which will always render it improper to use such varieties in a common kiln. But the most important use of this mineral is as marble. The marbles we are now to mention,

tion, have in general purer colours, more translucency, and receive a higher polish than those of compact limestone. They have been known from a very early period; and ancient statuaries have immortalised their names, by the master-pieces of art which they have executed in them. To give a full description of all the ancient and modern marbles enumerated by mineralogists, would much exceed the limits of this article; and besides, it would encroach on the more complete economical history of them, intended to be given in another work. We shall here notice only some of the more remarkable ancient marbles, and a few of the modern marbles found in this country, and on the Continent of Europe.

Ancient, or Antique Marbles.

Under this head, we include those marbles which were made use of by the ancients, and the quarries of many of which are no longer known.

1. *Parian Marble.*—Its colour is snow-white, inclining to yellowish-white, and it is fine granular, and when polished, has somewhat of a waxy appearance. It hardens by exposure to the air, which enables it to resist decomposition for ages. Varro and Pliny inform us, that it was named *Lychnites* by the ancients, from its being hewn in the quarry by the light of lamps, from λυχνις, *a lamp*; but Hill is of opinion, that the appellation is from the verb λυχνειν, *to be very bright*, or *shining*, from the shining lustre of this marble: the etymological derivation of Varro and Pliny is that which is generally adopted. Dipœnus, Scyllis, Malas, and Micciades, employed this marble, and were imitated by their successors. This preference was justified by the excellent qualities of this marble; for it receives with accuracy the most delicate touches of the chisel, and it retains for ages, with all the

the softness of wax, the mild lustre even of the original polish. The finest Grecian sculpture which has been preserved to the present time, is generally of Parian marble.—The Medicean Venus, the Diana venatrix, the colossal Minerva (called Pallas of Velletri), Ariadne (called Cleopatra), Juno (called Capitolina), &c. It is also Parian marble on which the celebrated tables at Oxford are inscribed.

2. *Pentelic Marble*, from Mount Pentelicus, near Athens. This marble very closely resembles the preceding, but is more compact, and finer granular, sometimes combined with splintery. At a very early period, when the arts had attained their full splendour, in the age of Pericles, the preference was given by the Greeks, not to the marble of Paros, but to that of Mount Pentelicus, because it was whiter, and also, perhaps, because it was found in the vicinity of Athens. The Parthenon was built entirely of Pentelic marble. Many of the Athenian statues, and the works carried on near to Athens during the administration of Pericles, (as, for example, the temples of Ceres or Eleusis), were executed in the marble of Pentelicus *. Among the statues of this marble in the Royal Museum in Paris, are the Torso; a Bacchus in repose; a Paris; the Discobolus reposing; the bas-relief known by the name of the Sacrifice; the throne of Saturn; and the Tripod of Apollo. It is remarked by Dr Clarke, that while the works executed in Parian marble remain perfect, those which were finished in Pentelican marble have been decomposed, and sometimes exhibit a surface as earthy and as rude as common limestone. This is principally owing to veins of extraneous substances which intersect the Pentelican quarries, and which appear

* Clarke's Travels, vol. iii.

appear more or less in all the works executed in this kind of stone.

3. *Greek White Marble,—Marmo Greco* of Italian artists. Its colour is snow-white; is fine granular; and is rather harder than the other white marbles; hence it takes a higher polish. This is one of those varieties, which, being found near the river Coralus in Phrygia, was called *Corallitic* or *Corallic Marble* by the ancients. The Greek marble was obtained from several islands of the Archipelago, such as Scio, Samos, &c.

4. *White Marble of Luni*, on the coast of Tuscany. It is of a snow-white colour, small granular, and very compact; it takes a fine polish, and may be employed for the most delicate work: hence it is said to have been preferred by the Grecian sculptors, both to the Parian and Pentelic marbles. It is the general opinion of mineralogists, that the Belvidere Apollo is of Luni marble; but the Roman sculptors look upon it as Greek marble *. The Antinous of the Capitol, preserved in the Royal Museum in Paris, is also of this marble.

5. *White Marble of Carrara*.—It is of a beautiful white colour, but is often traversed by grey veins, so that it is difficult to procure middle-sized pieces without them. It is not so subject to turn yellow as the Parian. This marble, which is almost the only one in use by modern sculptors, was also quarried and wrought by the ancients. Its quarries are said to have been opened in the time of Julius Cæsar. In the centre of the blocks of marble, beautiful rock-crystals are found, which are called *Carrara Diamonds*.

6. *White*

* Dr Clarke says it is of Parian marble. Vid. Travels, vol. iii.

6. *White Marble of Mount Hymettus* in Greece.—This marble has a greater intermixture of grey than any of the varieties already mentioned. The statue of Meleager in the Royal Museum in Paris, is of this marble.

7. *Translucent White Marble,—Marmo statuario* of the Italians.—This marble much resembles that of Paros, but differs from it in being more translucent. There are at Venice, and in several other towns in Lombardy, columns and altars of this marble, the quarries of which are unknown.

8. *Flexible White Marble.*—This is a fine granular, greyish-white coloured marble, which is flexible in a considerable degree. It was dug up in the feod of Mandragone. In the Borghese Palace in Rome, there are five or six tables of it.

These are the chief white marbles which the ancients used for the purposes of architecture and sculpture.

9. *Red antique Marble,—Rosso antico* of the Italians,—*Ægyptum* of the ancients.—This marble, according to antiquaries, is of a deep blood-red colour, here and there traversed by white veins, and, if closely inspected, appears to be sprinkled over with minute white dots, as if it were strewed with sand. The Egyptian Antinous in the Royal Museum in Paris, is of this marble. But the most highly prized variety of antique red marble, is that of a very deep red, without veins, such as is seen in the Indian Bacchus in the same Museum. The white points, which are never wanting in the true red antique marble, distinguish it from others of the same colour. It is not known from whence the ancients obtained this marble: the conjecture is, that it was brought from Egypt.

10. *Green antique Marble,—Verde antico* of the Italians.—This beautiful marble is an indeterminate mixture of white marble and green serpentine. It was known

to the ancients under the name *Marmor Spartum* or *Lacedæmonium*.

11. *Yellow antique Marble,—Giallo antico* of the Italians.—This marble is of a yellowish-brown, sometimes inclining to a cream-yellow colour, and is either of an uniform colour, or marked with black or deep yellow-coloured rings. It is found only in small detached pieces, and in antique inlaid-work. The Sienna marble is a good substitute for it.

12. *Antique Cipolin Marble.*—Cipolin is a name given to all such white marbles as are marked with green-coloured zones, caused by talc or chlorite. It was much used by the ancients. It takes a fine polish, but its green coloured stripes always remain dull, and are that part of the marble which first decomposes, when exposed to the open air. There are modern Cipolins as fine as that used by the ancients.

13. *African breccia Marble,—Antique African Breccia.*—It has a black ground, in which are imbedded fragments or portions of a greyish-white, of a deep red, or of a purple wine colour. This is said to be one of the most beautiful marbles hitherto found, and has a superb effect when accompanied with gilt ornaments. Its native place is not known with certainty: it is conjectured to be Africa. The pedestal of Venus leaving the Bath, and a large column, both in the Royal Museum in Paris, are of this marble.

Scottish Marbles.

The Marbles of this part of Great Britain have hitherto been but little attended to, although it is highly probable that many valuable varieties occur in the different primitive and transition districts. At present, we shall rest satisfied

satisfied with enumerating a few of the best known varieties.

1. *Tiree Marble.*—Of this marble, there are two varieties, viz. the Red and White.

a. Red Tiree Marble.—This is one of the most highly prized of the Scottish marbles. Its colours are red, of various tints, such as rose-red, and flesh-red; also reddish-white: its lustre is glimmering; and the fracture is minute foliated, accompanied with splintery. It is very faintly translucent, or only highly translucent on the edges. It is always intermixed with different other earthy minerals, that add to its beauty, and give it a peculiar appearance. The most frequent of the imbedded minerals is common hornblende; the others are pale green sahlite, blackish-brown mica, and green chlorite. In some varieties, the hornblende is so abundant, that at first sight they might be confounded with syenite: in others, where nearly the whole mass is of hornblende, it would be considered as a variety of hornblende rock.

b. White Tiree Marble.—Its colours are greyish-white and bluish-white: it contains scales of mica, and crystals or grains of common hornblende; which latter, when minutely diffused, give the marble a green or yellowish-green colour, and when very intimately combined with the mass, form beautiful yellowish-green spots.

2. *Iona Marble.*—Its colours are greyish-white and snow-white. Its lustre is glimmering, and fracture minute foliated, combined with splintery. It is harder than most of the other marbles. It is an intimate mixture of limestone and tremolite; for if we immerse it in an acid, the carbonate of lime will be dissolved, and the fibres of tremolite remain unaltered. It is sometimes intermixed with steatite, which gives it a green or yellow colour, in spots. These yellow or green coloured portions receive

a considerable polish, and have been erroneously described as nephritic stone, and are known also under the name of *Iona* or *Icolmkill Pebbles*. The marble itself does not receive a high polish: this, with its great hardness, have brought it into disrepute with artists.

3. *Skye Marble.*—In the Island of Skye, in the property of Lord Macdonald, there are several varieties of marble, deserving of notice. One variety is of a greyish inclining to snow-white colour: another greyish-white, veined with ash-grey; and a third is ash-grey, or pale bluish-grey, veined with lemon-yellow or siskin-green *.

4. *Assynt.*—The following varieties of marble found in Sutherland, have been introduced into commerce by Mr Joplin of Gateshead.

a. White marble, which acquires a smooth surface on the polisher, but remains of a dead hue, like the marble of Iona: hence its uses as an ornamental marble are much circumscribed.

b. White mottled with grey, and capable of receiving a high polish, and is not deficient in beauty.

c. Grey coloured, and highly translucent and crystalline, and capable of being applied to the purposes of ornament in sepulchral sculpture.

d. Dove-coloured, compact, translucent, and receiving a good polish.

e. Pure white, and translucent, and capable of being used in plain ornaments, but too translucent for sculpture.

f. White, with irregular yellow marks, from being intermixed with serpentine. It is very compact.

g. White variety, with layers of slate-spar.

5. *Glen*

* Mineralogy of Scottish Isles, vol. ii.

5. *Glen Tilt Marble.*—I am informed that the limestone of Glen Tilt, first mentioned by Dr Macknight, in his description of that valley *, has of late attracted the notice of the Duke of Athole, through the suggestion of Dr Macculloch. The marbles are white and grey, and veined or spotted with yellow or green: they vary in the size of the grain or concretion, and also in the degree and kind of polish they receive. As soon as the superficial and decayed portions of the rock are removed, it will be seen how far the marble or limestone will answer for the arts.

6. *Marble of Ballichulish.*—This marble is of a grey or white colour, and is very compact. It may be raised in blocks of considerable size.

7. *Boyne Marble.*—Its colours are grey or white, and it receives a pretty good polish.

8. *Blairgowrie Marble.*—Mr Williams, in his Natural History of the Mineral Kingdom, mentions a beautiful saline marble, of a pure white colour, which occurs near Blairgowrie in Perthshire, not far from the road side. According to him, it may be raised in blocks and slabs, perfectly free of blemishes, and in every respect fit to be employed in statuary and ornamental architecture.

9. *Glenavon Marble,*—is of a white colour, and the concretions are large granular. It is mentioned by Williams as a valuable marble; but he adds, that its situation is remote, and difficult of access.

English Marbles.

Hitherto but few marbles of granular foliated limestone have been quarried in England; the greater number

 of

* Wernerian Memoirs, vol. i. p. 862.

of varieties belonging to the flœtz limestone. One of the most remarkable of the English marbles of the present class, is that of Anglesea, named *Mona Marble*, which is not unlike the *Verde Antico*. Its colours are greenish-black, leek-green, and sometimes purple, irregularly blended with white; but they are not always seen together in the same piece. The white part is limestone: the green shades are said to be owing to serpentine and asbestus. The Black Marbles found in England, are varieties of Lucullite.

Irish Marbles.

The Black Marbles of Ireland, now so generally used by architects, are Lucullites. In the county of Waterford, different kinds of marble are known; as at Toreen, a fine variegated sort, of various colours, viz. chesnut-brown, white, yellow, and blue, and which takes a good polish: a grey marble, beautifully clouded with white, susceptible of a good polish, has been found near Kilcrump, in the parish of Whitechurch, in the same county. At Loughlougher, in the county of Tipperary, a fine purple marble is found, which, when polished, is said to be beautiful. Smith describes several variegated marbles in the county of Cork; but whether these, and others now enumerated as Irish marbles, are granular limestone, I cannot discover, as I have neither met with good descriptions of them, nor seen any specimens. Thus, he mentions one with a purplish ground, and white veins and spots, found at Churchtown: a bluish and white marble from the same place; and several fine ash-coloured varieties, as that of Castle Hyde, &c. The county of Kerry affords several variegated marbles, such as that found near Tralee. Marble of various colours is found

in

in the same county, in the islands near Dunkerron, in the river Kenmare: some are purple and white, intermixed with yellow spots; and some beautiful specimens have been seen, of a purple colour, veined with dark green.

French Marbles *.

A great many different kinds of marble are quarried in the different Departments of the kingdom of France, and of these we shall mention the following.

1. *Griotte Marble.*—Its colour is deep brown, with blood-red oval spots, produced by shells. This marble has obtained its name from its brownish colour, being similar to that of a variety of cherries, likewise called *griotte*; but it also sometimes contains large white veins, which traverse the other spots, and which, as destroying the harmony of the other tints, are considered as a defect. Some of the ornaments of the Triumphal Arch of the Carousel, are made of griotte; which is now much employed in the decoration of public monuments, and of splendid furniture. It is sold at about 200 francs the cubic foot. It is found in the Department of Herault.

2. *Marble of Languedoc, or of St Beaume.*—It is of a bright red colour, and is marked with white and grey zones, formed by madrepores. The eight columns which adorn the triumphal arch in the Carousel at Paris, are of this marble. The quarries are at St Beaume, in the Department of Aude.

* As I have not seen all the varieties of foreign marble now to be described, I cannot pretend to say with certainty that the whole of them belong to the Granular Foliated Limestone. The descriptions are from Brard's Treatise on Precious Stones.

3. *Campan Marble.*—This is a mixture of granular foliated limestone and a green talcky mineral, which forms veins on its surface. There are three varieties of Campan, which, however, are often united in the same piece: the first, called *Green Campan*, is of a pale sea-green colour, and exhibits on its surface lines of a much deeper green, and forming a kind of net-work; the second, called *Isabel Campan*, is of a delicate rose-colour, and, like the first, is furnished with undulating veins of green talc: the third variety, the *Red Campan*, is of a deep red colour, with veins of a still deeper red, and in some degree resembles parts of the griotte. In order to form a correct idea of the Campan marble, properly speaking, we must imagine that these three varieties are united, so as to form large stripes, from a few inches, to two, three, or even six feet broad, which produce a very grand and pleasing effect, when viewed in large masses. When, therefore, the Campan marble can be employed in the large way, it may be looked upon as one of great beauty and splendour. It should not, however, be exposed to the weather, since, by so doing, the talcose substance exfoliates, and leaves hollow spaces, which render its surface uneven and rough; but it answers extremely well in the interior of buildings, for chimney-pieces, slabs for tables, &c There are immense quarries of this valuable marble at Campan, near Bagnere, in the High Pyrenees.

4. *Sarencolin Marble*—It exhibits on its surface large straight zones, and angular spots, of a yellow or blood-red colour, so that at first view it bears some resemblance to the marble called Sicilian. The finer varieties have become very scarce. It is found at Sarencolin, in the High Pyrenees.

5. *Breccia*

5. *Breccia Marble of the Pyrenees.*—One variety contains, in a brownish-red basis, black, grey, and red, middle-sized spots. It admits a good polish. Another variety has an orange-yellow coloured basis, containing small fragments of a snow-white colour. Both varieties are found in the High Pyrenees.

Italian Marbles.

1. *Sienna Marble, or Brocatello di Siena.*—It has a yellowish colour, and disposed in large irregular spots, surrounded with veins of bluish-red, passing sometimes into purple. It is by no means uncommon in Sienna. At Montarenti, two leagues from Sienna, another yellow marble is found, which is traversed by black and purplish-black veins. This is frequently employed throughout Italy.

2. *Mandelato Marble.*—It is a light red marble, with yellowish-white spots, found at Lugezzana in the Veronese. Another variety, bearing the same name, occurs at Preosa. They are both employed for columns, and various other works.

3. *Green Marble of Florence.*—It is of a green colour, which it owes to an intermixture of steatite.

4. *Verdi di Prado Marble.*—It is a green marble, marked with dark green spots, having greater intensity than the base or ground. It is found near the little town of Prado in Tuscany.

5. *Rovigo Marble.*—It is of a white colour, but is inferior in quality to those of Carrara and Genoa. It is found at Padua.

6. *Luni Marble.*—It is of a white colour, with red-coloured spots and dots. It is found at Luni, on the coast of Tuscany.

7. *Venetian*

7. *Venetian Marble.*—It is white, with red and yellow spots and veins. It is found in the Venetian territory.

8. *Lago Maggiore Marble.*—It is white, with black spots and dots, and is of great beauty. It has been employed for decorating the interior of many churches in the Milanese.

9. *Breche d'Italie.*—It has a reddish-brown ground, veined with white. It is a beautiful marble, but requires much care in keeping, since it becomes soon spotted, by coming into contact with greasy substances.

10. *Bretonico Marble.*—This beautiful marble is composed of yellow, grey, and rose-coloured portions or fragments. It is found near the village of Bretonico, in the Veronese.

11. *Bergamo Marble.*—It is composed of grey and black fragments, in a green basis.

Sicilian Marbles.

The island of Sicily abounds in marbles. Baron Borch describes upwards of a hundred varieties. Of these, the best known in this country is that named *Sicile Antique*, or by English artists *Sicilian Jasper*. It is red, with large stripes like ribbons, white, red, and sometimes green. Among the Sicilian breccia marbles, are those of Gallo, the one of a light grey colour, presenting elegant rose-coloured spots, of different shades; and the other also grey, veined yellow, and exhibiting on its surface white translucent spots. The breccia marble of Monte Alcano is light grey, with round rose-coloured spots. That of Taormina has a deep red ground, and presents on its surface yellow and greyish-white spots.

Spanish

Spanish Marbles.

Spain abounds in beautiful marbles. The vicinity of Valencia, Cadiz, Burgos, Grenada, Molina, and Carthagena, offer a great number of them; and the Tagus, in its course, winds through hills of marble. Hence it is that the monuments in Spain, those of the middle ages, and of modern times, are profusely decorated with indigenous marbles. The vault of the beautiful theatre of Toledo, is supported by 350 marble columns. The Mosque of Cordova, erected by Caliph Abdoulrahman III. is ornamented with 1200 columns, most of which are of Spanish marble. Among the ruins of ancient Merida which was built twenty-eight years before the commencement of the Christian era, fragments of the most valuable marble are still discovered; and the Church of the Escurial, and the Palace itself, are decorated with the most beautiful marbles; and the same may be said of the principal churches in Madrid. The following are some of the principal marbles found in Spain.

1. *White Marble.*—Near Cordova, there is a white fine granular marble, which takes a good polish, and is very fit for sculpture. Near Filabres, three leagues from Almeria in Grenada, there is a hill of about a league in circumference, and 2000 feet in height, which is said to be entirely composed of the purest white marble, capable of the finest polish; and the rocks which surround the town of Molina in New Castile, are composed of a white marble, which has been employed in the Palace of the Alhambra at Grenada.

2. *Red Marble.*—There is a beautiful red variety, with shining red and white spots and veins, called *Red Seville Marble.* There is also a flesh-coloured variety, veined with white, from Santiago. A dull red marble, with minute

nute black veins, is found in Meguera in Valencia, and much used in Spain for tables. The mountains of Guipuscoa afford a red marble, veined with grey, and closely resembling that of Sarencolin.

3. *Tortosa Marble.*—Its basis or ground is violet, and it is spotted with bright yellow.

4. *Grenada Marble.*—It is of a green colour, and very much resembles the Verde Antico. It is found at Grenada.

5. *Spanish Brocatello Marble.*—This is a well known and very beautiful variety of marble.

6. *Breccia Marble.*—Several beautiful varieties of this marble occur in Spain. At Riela in Arragon, there is a beautiful breccia marble, composed of angular portions or fragments of a black marble, imbedded in a reddish-yellow base. The breccia marble of Old Castile is of a bright red, dotted with yellow and black, and incloses middle-sized fragments of a pale yellow, brick-red, deep brown, and blackish-grey.

Portuguese Marbles.

Few marbles have hitherto been discovered in Portugal, and none of them equal in beauty the finer varieties found in Spain.

Swiss Marbles.

Granular foliated limestone occurs abundantly in Switzerland, but it has not hitherto been much used as a marble.

German Marbles.

Germany abounds in marbles, and affords many varieties, remarkable either for their beauty or singularity. They are quarried in great quantity, and carried to different

ferent parts of that vast country, or are exported into the neighbouring states. The varieties are so numerous, that we cannot, in the very brief view we are now taking, pretend to notice even the more remarkable of them, but must refer, for the particular descriptions, to the economical department of this work.

Norwegian Marbles.

Norway is poor in marbles, almost the only quarry of this stone being that of Gillebeck, in the district of Christiania.

Swedish Marbles.

Sweden does not afford many kinds of marble, and none of them are eminently distinguished for the beauty of their appearance. The principal marble is that of Fagernech, which is white, with veins of green talc.

Russian and Siberian Marbles.

The vast Empire of Russia affords a great many different kinds of marble. Georgi, in his Description of the Russian Empire, enumerates white, grey, green, blue, yellow, and red varieties; and Patrin has given the following account of the Siberian marbles. "The Uralian Mountains furnish the finest and most variegated marbles. The greater part is taken from the neighbourhood of Catharinenburg, where they are wrought, and from thence transported into Russia, particularly to Petersburgh. The late Empress caused an immense palace to be built in her capital for Orloff, her favourite, which is entirely coated with these fine marbles, both inside and outside. The Empress built the church of Isaac with the same marbles, on a vast space, near the statue of Peter

Peter the Great." Patrin found no white statuary marble in the Uralian Mountains; but in that part of the Altain Mountains which is traversed by the river Irtish, he in two places saw enormous rocks of marble, perfectly white and pure, from which blocks might be hewn.

Asiatic Marbles.

At present we are very imperfectly acquainted with the marbles of Asia.

Shaw mentions a red marble from Mount Sinai: Russell, in his Natural History of Aleppo, gives an imperfect account of the marbles of Syria; and some Persian marbles are noticed by Chardin. Mr Morier, in his Journey through Persia, mentions a very beautiful marble, under the name *Marble of Tabriz*, and informs us, that the tomb of the celebrated Persian poet Hafitz is constructed with it, and that the wainscotting of the principal room of the Hafl-ten, near Schiraz, is likewise of this marble. Its colours are described as light green, with veins, sometimes of red, sometimes of blue, and it has great translucency. It is cut in large slabs; for Mr Morier saw some that measured nine feet in length, and five feet in breadth. He says, that it is not procured near the city of Tabriz, or taken from a quarry, but is said to be rather a petrifaction, found in large quantities, and in immense blocks, on the borders of the Lake Shahee, near the town of Meraugheh. If it is a mere calcareous deposition, formed in the way of calcareous-alabaster or calc-sinter, it must be considered, not as marble, but a variety of that mineral.

The marbles of Hindostan, Siam, and China, are almost unknown to us. Authors speak of a quarry of white marble in the neighbourhood of Pekin; and of a similar marble in the vicinity of the capital of Siam.

African

African Marbles.

No account has hitherto been published of the marbles of this quarter of the globe.

American Marbles.

A good many different marbles have been discovered in the United States. The principal quarries are at Stockbridge and Lanesborough, Massachussets: in Vermont and Pennsylvania: in New-York; and in Virginia. According to Professor Hall, as mentioned by Mr Kœnig, marble has been found in many places on the west side of the Green Mountains in Vermont. A few years since, a valuable quarry was opened in Middleburg, a town situated on Otter Creek, eleven miles above Vergennes. The marble is of different colours in different parts of the bed. The principal colour, however, is bluish-grey. It takes a good polish, and is in general free of admixture of any substance that might affect its polish.

We know even less of the marbles of South America than of North America. Almost the only accounts we possess, are those of Molina, in his History of Chili, from which it appears, that a very considerable number of different varieties, both of simple and variegated marbles, occurs in that country, and amongst these is the white statuary marble.

Second

Second Kind.

Calcareous-Spar, or Calc-Spar.

Kalkspath, *Werner.*

Spathum, *Wall.* t. i. p. 140.—Körniger Kalkstein, var. *Wid.* s. 427.—Common Spar, *Kirw.* vol. i. p. 86.—Kalkspath, *Estner,* b. ii. s. 941. *Id. Emm.* b. i. s. 456.—Spatho calcareo, *Nap.* p. 341.—Calcaire cristallisé, *Lam.* t. i. p. 29.—Chaux carbonatée cristallisé, *Hauy,* t. ii. p. 127.—Le Spath calcaire, *Broch.* t. i. p. 536.—Spathiger Kalkstein, *Reuss,* b. ii. 2. s. 284.—Kalkspath, *Lud.* b. i. s. 149. *Id. Suck.* 1r th. s. 600. Grossblättricher Kalkstein, *Bert.* s. 90.—Kalkspath, *Mohs,* b. ii. s. 31. *Id. Hab.* s. 76.—Chaux carbonatée, *Lucas,* p. 3. —Gemeiner spathiger Kalkstein, *Leonhard,* Tabel. s. 33.—Chaux carbonatée pure spathique, *Brong.* t. i. p. 189.—Chaux carbonatée, *Brard,* p. 26.—Kalkspath, *Haus.* s. 125. —Spathiger Kalkstein, *Karsten,* Tabel. s. 50.—Crystallised Carbonate of Lime, *Kid,* vol. i. p. 50.—Chaux carbonatée, *Hauy,* Tabl. p. 2.—Kalkspath, *Lenz,* b. ii. s. 742.—Geformter Kalk, *Oken,* b. i. s. 408.—Calcareous-Spar, *Aikin,* p. 139.

External Characters.

Its most frequent colour is white, of which the following varieties occur, viz. reddish, snow, greyish, greenish, and yellowish white. From reddish-white, it passes on the one side into pearl-grey, brick-red, flesh-red, rose-red, and brownish-red ; and on the other side into pale violet-blue : from greyish-white, it passes into smoke-grey, ash-grey, yellowish-grey, and greenish-grey : from greenish-grey, it passes into apple, asparagus, olive, and leek green : from yellowish-grey, it passes into a colour intermediate between wax and ochre yellow, and into honey-yellow ; and from honey-yellow, into yellowish-brown, and greyish-black.

The

The white-coloured transparent varieties are often iridescent.

It occurs massive, disseminated, and very frequently crystallised. Its suite of crystals forms a circle, in which the figures can be brought under three divisions, and the last figure of the last division, joins with the first figure of the first division, and these all pass into each other. The divisions are the following:

1. *Acute six sided Pyramid.*
2. *Six-sided Prism.*
3. *Three-sided Pyramid.*

I. *Six-sided Pyramid* *.

When perfect, it is always acute, and two and two lateral planes meet under obtuser angles than the others.

It is generally obliquely streaked, but the streaks run from the acute towards the obtuse edges.

It occurs,

A. Single.

B. Double. The lateral planes of the one, set obliquely on the lateral planes of the other, so that the edge of the common base forms a zig-zag line.

These pyramids occur, either perfect, or in the following varieties:

1. The apex acuminated with three planes, which are set on the obtuse lateral edges. These are parallel with the cleavage
2. The apex flatly acuminated with three convex faces, which are set on the acute lateral edges. The convexity is in the direction of the axis of the double pyramid.

 3. Double

* The primitive form is an oblique rhomboidal prism, the alternate lateral edges of which are 105° 5′ and 74° 55′.

3. The apex flatly acuminated with six planes, which are set on the lateral planes.
4. The angles on the common base of the double pyramid truncated, thus forming a transition into the six-sided prism.
5. The acute lateral edges of the double pyramid sometimes truncated.
6. Twin-crystal.

The double six-sided pyramids apparently pushed into each other, in the direction of their length, in which they are either

(1) *Unchanged* in position, when the acute edges rest on the obtuse edges; or they are

(2.) *Turned around* one-sixth of their periphery, so that obtuse edges are set on obtuse edges, and acute edges on acute edges; and the alternate angles on the common base have broken re-entering angles.

7. Aggregated in single rows, so that there is formed either an acute double, or a single four-sided pyramid.

II. *Six sided Prism.*

It is equiangular, but generally with alternate broad and narrow lateral planes. It originates from the pyramid N° 4.; and hence it presents the following varieties:

1. The six-sided prism, acutely acuminated with six planes, of which two and two meet under obtuse angles, and each is set obliquely on the lateral edges.
2. The preceding figure, in which the six-planed acumination is again flatly acuminated with three planes,

planes, which are set on the obtuse lateral edges of the six-planed acumination.

3. When the planes of the three-planed acumination increase so much that those of the six-planed acumination disappear, a six-sided prism is formed, flatly acuminated with three planes, which are set on the alternate lateral planes in an unconformable position.
4. In some rare instances, the three-planed acumination is acute. In general, the angle of the acumination varies much in magnitude. When the prism becomes very low, there is formed a double three-sided pyramid, which must vary in its summit angle, according as the acumination is flat or acute. These prisms are often pyramidally aggregated.
5. The six-sided prism, acuminated with six planes, is often truncated on the apices: when these truncating planes increase very much in magnitude, a six-sided prism is formed, in which the angles are more or less deeply truncated.
6. When the prism becomes very low, it may be view-as an equiangular six-sided table, which is sometimes aggregated in a rose-like form.
7. Sometimes the six-sided prism is truncated on the lateral edges.

III. *Three-sided Pyramid.*

It originates from the prism N° 4. and has frequently remains of the prism on the common basis. It is generally double, and the lateral planes of the one are set on the lateral edges of the other: Sometimes also single; and we meet with nearly all the varieties that occur in the

 acuminations

acuminations of the prism, because it originates from it. The varieties are the following.

1. Flat double three-sided pyramid, often aggregated in rows, or in the pyramidal form. When the pyramids are aggregated in a straight direction, there is formed a six-sided prism, flatly acuminated with three planes, which are set in an unconformable position on the lateral edges.
2. The same pyramid, deeply truncated on the apices. From this originates the six-sided table, in which the terminal planes are set alternately oblique on the lateral planes.
3. The double three-sided pyramid sometimes becomes rather acute, and approaches to the cube; generally the two angles on the extremities of the axis of the double pyramid are truncated.
4. Very acute double three-sided pyramid: when the two apices of the double pyramid are so deeply truncated that the truncating planes meet the higher angles of the common basis, an octahedron is formed.
5. The acute single and double three-sided pyramid, with convex faces. These sometimes exhibit divided planes, and give rise to the six-sided pyramid of the first division.

It would extend this description too much, were we to attempt to give an account of every variety of form exhibited by these crystals; and besides, we have already enumerated the principal ones.

The crystals occur of various magnitudes, as large, small, and very small.

Externally it alternates from splendent to glimmering: internally it alternates from smooth and specular-splendent to shining.

The

The lustre is generally vitreous, which, however, in some varieties, inclines to resinous, in others to pearly.

The fracture is perfect foliated, with a threefold equiangular cleavage; the folia meet under oblique angles, and are generally straight, but occasionally curved. A concealed foliated fracture is to be observed, in which there is a threefold cleavage, running towards the acute angles of the rhomboidal fragments, and thus dividing them into obtuse angular three-sided pyramidal fragments. In the Iceland spar, a perfect conchoidal fracture is sometimes to be observed.

The massive varieties occur in distinct concretions, which are often large and coarse, but seldom small granular; also in thick and thin wedge-shaped prismatic concretions.

The fragments are rhomboidal.

It occurs transparent, semi-transparent, and occasionally only translucent. It refracts double *.

It is semi-hard; it scratches gypsum, but is scratched by fluor-spar.

It is brittle, and very easily frangible.

Specific gravity, 2.715.

Chemical Characters.

It is infusible before the blowpipe, but becomes caustic, losing by complete calcination about 43 *per cent.*; effervesces violently with acids.

* The double refracting power of calcareous-spar was first observed by Erasmus Bartholin.

Constituent Parts.

	Iceland Spar.	Iceland Spar.	Iceland Spar.	From Andreasberg.
Lime, -	56.15	55.50	56.50	55.9802
Carbonic Acid,	43.70	44.00	43.00	43.5635
Water, - -		0.50	0.50	0.1000
Oxide of Manganese, with trace of Iron,	0.15			0.3562
	100.00	100.00	100.00	100.0000
	Stromyer, Gilbert's Annalen for 1813, p. 217.	*Philips*, Phil. Mag. xiv. 290.	*Bucholz*, Gehl. Journ. iv. 412.	*Stromyer*, Gilbert's An. for 1813, p. 217.

Geognostic Situation.

It never occurs in mountain-masses, but venigenous in almost every rock, from granite to the newest flœtz formation. The oldest formation of this mineral is that in veins, where it is accompanied with felspar, rock-crystal, probably also with epidote, sphene, and chlorite. It occurs also in beds, along with augite, hornblende, garnet, and magnetic ironstone; and frequently in veins in different metalliferous formations. Thus, it is associated with nearly all the metallic minerals contained in gneiss, mica-slate, clay-slate, syenite, porphyry; seldomer in granite, more frequently, again, in grey-wacke, and along with cobalt and copper ores in the oldest flœtz limestone. Veins, almost entirely composed of calcareous-spar, abound in the newest limestone formations; and it is a common mineral, either in veins, or in cotemporaneous masses, in the various rocks of the flœtz-trap series.

Geographic Situation.

This mineral is so common in every country, as to render any account of its geographic distribution unnecessary.

cessary. It may, however, be remarked, that it occurs very abundantly in Fifeshire, where it occasionally appears in amygdaloidal masses, several feet square. It is probable, that the beautiful variety of calcareous-spar named *Iceland spar*, from the country where it is found, occurs in amygdaloidal rocks, because it is there associated with zeolite *.

Third Subspecies.

Fibrous Limestone, or Satin-Spar.

Gemeiner Fasriger Kalkstein, *Werner.*

Pierre calcaire fibreuse, *Broch.* t. i. p. 549.—Gemeiner Fasriger Kalkstein, *Reuss,* b. ii. 2. s. 304. *Id. Mohs,* b. ii. s. 85. *Id. Leonhard,* Tabel. s. 34.—Chaux carbonatée fibreuse, *Brong.* t. i. p. 218.—Fasriger Kalkstein, *Karsten,* Tabel. s. 50.—Satin-spar, *Kid,* vol. i. p. 49.—Chaux carbonatée fibreuse-conjointe, *Hauy,* Tabl. p. 3.—Gemeiner Fasriger Kalkstein, *Lenz,* b. ii. s. 750.—Fibrous Carbonate of Lime, *Aikin,* p. 140.

External Characters.

Its colours are greyish, reddish, and yellowish-white.

It occurs massive.

Its lustre is glistening or shining, and pearly.

The fracture is parallel, sometimes coarse, sometimes delicate, and occasionally undulated fibrous.

The fragments are splintery.

It is feebly translucent.

 It

* The purest and most beautiful Iceland-spar, is found in Iceland, on the east side of the island, near the harbour of Rödefiord, where it is said to form a mass fourteen feet thick.

It is semi-hard; harder than calc-sinter.
It is brittle.
It is easily frangible.
Specific gravity, 2.70, *Pepys.*

Constituent Parts.

Lime,	- -	50.8
Carbonic acid,	-	47.6
		98.4

Pepys, in Kid's Min. vol. i. p. 49.

Stromyer says that fibrous limestone contains some *per cents*. of gypsum.

Geognostic and Geographic Situations.

It occurs in thin layers in clay-slate at Aldstone Moor in Cumberland: in layers and veins in the middle district of Scotland, as in Fifeshire.

Uses.

It is sometimes cut into necklaces, crosses, and other ornamental articles.

Fourth Subspecies.

Calc-Sinter *.

This subspecies is divided into two kinds, viz. Fibrous Calc-Sinter, and Lamellar Calc-Sinter.

First

* This is the Alabaster of the ancients, and is by the moderns named *Calcareous Alabaster*, to distinguish it from another mineral, gypsum, which they name *Gypseous Alabaster*.

First Kind.

Fibrous Calc-Sinter.

Kalk-sinter, *Werner*.

Sintricher fasriger Kalkstein, *Reuss*, b. ii. 2. s. 306.—Kalksinter, *Lud.* b. i. s. 150. *Id. Suck.* 1r th. s. 618. *Id. Bert.* s. 93. *Id. Mohs*, b. ii. s. 86. *Id. Hab.* s. 78.—Fasriger Kalksinter, *Leonhard*, Tabel. s. 34.—Sintriger Kalkstein, *Karsten*, Tabel. s. 50.—Chaux carbonatée concretionnée, *Hauy*, Tabl. p. 4.—Sintricher Kalkstein, *Lenz*, b. ii. s. 751.

External Characters.

Its most common colour is white, of which it presents the following varieties: snow, greyish, greenish, and yellowish white. The yellowish-white passes into wax, honey, and wine yellow, and yellowish-brown. It occurs also siskin, pistachio, asparagus, mountain, and verdigris green; which latter passes into sky-blue. Sometimes it is flesh-red, peach-blossom-red, and reddish-brown. It is occasionally spotted, and striped.

It occurs massive, and in many different particular external shapes, as in crusts, reniform, botryoidal, tuberose, stalactitic, and tubiform *.

The surface is generally dull and rough, often drusy, seldom smooth, and glimmering, passing into glistening.

Internally it is glimmering, or glistening and pearly.

The fracture is scopiform, or stellular fibrous, and sometimes uneven, passing into splintery.

The

* Very rarely, the longish external shapes are terminated with a three-sided pyramid.

The fragments are indeterminate angular, splintery, or wedge-shaped.

It very often occurs in lamellar concretions, which are either straight, or fortification-wise bent, or are concentric: seldom large and coarse granular.

It is more or less translucent.

It is semi-hard, inclining to soft.

It is brittle.

It is very easily frangible.

It is not particularly heavy; inclining to light.

Constituent Parts.

Lime, - - -	56.0
Carbonic Acid, -	43.0
Water, - -	1.0
	100.0

Bucholz, in Gehlen's Journal, b. iv. s. 425.

Geognostic and Geographic Situations.

It is found encrusting the roofs, walls, and floors of caves, particularly those situated in limestone rocks.

Second Kind.

Lamellar Calc-Sinter.

Schaaliger Kalkstein, *Karsten.*

Schaaliger Kalkstein, *Reuss*, b. ii. 2. s. 309. *Id. Karsten*, Tabel. s. 50. *Id. Lenz*, b. ii. s. 754.

External Characters.

Its colours are snow-white, yellowish-white, greenish-white, and cream-yellow.

It

It occurs in thick stalactitic, tubular, and claviform masses, which have a rough surface.

Internally it is splendent or shining, and pearly.

The fracture is foliated, frequently inclining to broad radiated.

The fragments are indeterminate angular.

It occurs in lamellar concretions.

It is translucent.

It is soft inclining to semi-hard.

Constituent Parts.

It is a nearly pure Carbonate of Lime.

Geognostic and Geographic Situations.

This mineral is generally associated with the fibrous calc-sinter. Both minerals are formed from water holding carbonate of lime in solution. Nothing is more common than the presence of carbonic acid in water; and when a superabundance of this acid is present, the acid is capable of holding in solution a portion of carbonate of lime; but the carbonic acid makes its escape, and thus deprives the water of its solvent power, when the solution comes to be agitated, or exposed to the atmosphere, or to a change of temperature. Water thus impregnated with carbonate of lime, oozes slowly through rocks of any kind, until it reaches the walls and roofs of caves: there some time elapses before a drop of sufficient size to fall by its own weight is formed, and in this interval some of the calcareous particles are separated from the water, owing to the escape of the carbonic acid, and adhere to the roof. In this manner, successive particles are separated, and attached to each other, until a *stalactite* is formed. If the percolation of the water containing

ing calcareous particles is too rapid to allow time for the formation of a stalactite, the earthy matter is deposited from it after it has fallen from the roof upon the floor of the cave; and in this case, the deposition is called a *stalagmite*. In some cases, the separation of the calcareous matter takes place both at the roof and on the floor of the cave; and in the course of time, the substance of each deposition increasing, they both meet, and form pillars, often of great magnitude, and that appear destined to support the roof of the cave. Water charged with calcareous earth also oozes through the walls of these caves, and deposites in them a crust of calc-sinter, of various forms; so that in this manner the whole comes to be encrusted with calcareous matter; and if the infiltration continues, the cave in the process of time is entirely filled up.

Caves of this kind occur in almost every country. Maccallister's Cave, in the island of Skye, and those in the limestone hills of Derbyshire, are the most striking appearances of this kind hitherto observed in Scotland and England. But the most celebrated stalactitic cave, is that of Antiparos in the Archipelago, and which has been particularly described by Tournefort. Similar caves occur in Germany, France, Switzerland, and Spain, in the United States of America, and other countries.

Italy, which is so rich in fine marble, is not less so in beautiful calc-sinter or calcareous alabaster: the territory of Volterra in Tuscany, alone, furnishes no fewer than twenty different varieties. Sicily is also abundant in calc-sinter; and of these, the rose-coloured variety of Trapani is much admired.

Spain is next to Italy the most productive country of calcareous alabaster. The environs of Granada and Ma-

laga

laga are particularly remarkable for the beautiful varieties of this mineral which they afford.

Persia also abounds in highly prized varieties of calcareous alabaster.

Uses.

Calc-sinter or Calcareous Alabaster, is used for the same purposes as marble, and is cut into tables, columns, vases, drapery for marble figures, and sometimes also into statues. It was also used by the ancients in the manufacture of their unguentary vases. A vessel of this kind is mentioned in the 26th chapter of Matthew's Gospel, where it is said, "There came unto him a woman, having an *alabaster* box of precious ointment." The most beautiful calcareous alabasters, those used by the ancients, are conjectured to have been brought from the mountains of the Thebaid, situated between the Nile and the Red Sea, near the city of Alabastron. In the National Museum in Paris, there is a colossal figure of an Egyptian deity, cut in this rare kind of alabaster. Many different varieties of this mineral are described by authors: the following are enumerated by Brard.

I. *Alabaster of One Colour.*

1. *Antique white Calcareous Alabaster.*—This variety is very rare: it is now only found amongst the ruins of ancient monuments, and particularly at Ortée, not far from Rome; but we are ignorant of the place from whence the ancients procured it.

2. *Yellowish-white, inclining to rose, or Oriental Alabaster.*—The Egyptian statue already mentioned, is made of this beautiful variety of alabaster. It is supposed that the Egyptians procured it from Upper Egypt; but the same

same variety is found at present in the vicinity of Alicant and Valencia in Spain, and of Trapani in Sicily.

3. *Alabaster of Sienna.*—Its colour is honey-yellow, and it is nearly transparent. A similar variety is found in the island of Malta, of which statues of considerable size are made.

II. *Striped Alabaster,—Onyx Marble of the Ancients.*

The ancients procured these alabasters from the mountains of Arabia, and also from several districts in Germany.

1. *Striped Alabaster from Malaga.*—Two leagues from Malaga, there is a cave filled with wax-yellow alabaster, which, when cut perpendicularly, appears agreeably striped with two different yellow tints; but when cut in another direction, it only presents large irregular spots. The Palace of Madrid is ornamented with this alabaster.

2. *Alabaster from Montreal in Sicily.*—This variety is marked with bright red and yellow stripes.

3. *Alabaster from Caputo in Sicily.*—It is marked with yellow and white stripes.

4. *Alabaster from Mount Pellegrino.*—The stripes are narrow, and of two colours; the one yellow, the other extremely deep black.

5. *Maltese Alabaster.*—Several varieties of alabaster are quarried in Malta: amongst others, one exhibits wax-yellow and white stripes, and another black, brown, and white stripes.

III. *Spotted Alabaster.*

These spots are often produced by the manner in which the stone is cut. There are two very rich columns of this variety, in what used to be called the Hall of the Emperor,

Emperor, in the National Museum in Paris. They were discovered in the year 1780, amidst the ruins of the ancient city of Gabii, four leagues from Rome.

Observations.

1. Some varieties of calc-sinter are so porous, as to allow water to percolate through them, and are on that account used as filtering-stones.

2. At the springs of St Philippi in Tuscany, moulds of different kinds are suspended on the walls of the basins into which the calcareous water of these springs falls: after a certain period, these moulds are removed, covered with a very solid incrustation of white and very fine calc-sinter, which is easily separated from the moulds, and is found to present excellent impressions of the moulds, whatever they are. It is said that vases, and even statues, are formed in the above manner from calcareous springs, near Guanca-Velica in Peru.

3. Some of the great caves in limestone countries, are formed by masses of limestone irregularly heaped on each other, and connected together by means of calc-sinter.

4. Authors differ as to the derivation of the word *Alabaster*. It does not appear to have originated from *albus*, as some pretend, because the white varieties were rare, and it was the yellow kinds that were most highly prized by the ancients: others are of opinion, and it is the most plausible one, that it is derived from the Greek word αλαβαστρον, which is by some derived from α *neg*. and λαμβανειν or λαβειν, *to hold*, because the vessels made of this mineral were without handles, and very smooth, and were therefore difficult to lay hold of. Vessels used for holding ointment or perfume, were made of this stone, and were named *Alabastron*. Afterwards, the name alabastron was applied to ointment vessels made of other substances. Thus,

Thus, in Theocritus, Idyll. xv. lin. 114. we have χρυσεα αλαβαστρα, *golden alabastra* Raphelius remarks, on Mathew xxvi. 7., that Herodotus, among the presents sent by Cambyses to the King of Æthiopia, mentions μυρου αλαβαστρον; and Cicero, Academ. lib. ii speaks of "alabastrum unguenti plenum." Mat. xxvi. 7.; Mark xiv. 3.; Luke vii. 37.

Fifth Subspecies.

Tufaceous Limestone, or Calc-Tuff *.

Kalk-Tuff, *Werner*.

Tuff Kalkstein, *Reuss*, b. ii. 2. s. 314.—Kalktuff, *Lud.* b. i. s. 157. *Id. Suck.* 1r th. s. 623. *Id. Mohs*, b. ii. s. 105. *Id. Leonhard*, Tabel. s. 34.—Tuffartiger Kalkstein, *Karsten*, Tabel. s. 50.—Calcareous Tufa, *Kid*, vol. i. p. 24.—Chaux carbonatée concretionnée incrustante, *Hauy*, Tabl. p. 4.—Kalchtuff, *Lenz*, b. ii. s. 755.—Tuffa, *Aikin*, p. 141.

External Characters.

Its colours are yellowish-grey, which sometimes approaches to smoke-grey; also ash-grey, and ochre-yellow; and sometimes spotted brown and yellow.

It occurs massive, perforated, ramose, spongy, tubular, claviform, botryoidal, globular, cellular, and in crusts; inclosing vegetable stems and leaves; also bones of animals, as of elephants and rhinoceroses, and land shells; and

* The term *tufa*, appears to be derived from the verb τύφω, which, in its original signification, is appropriate to volcanic productions, especially to such as are of a spongy or porous texture.—*Kid.*

and also with frequent impressions of leaves, mosses, and roots.

Internally it is dull, or very faintly glimmering.

The fracture is fine grained uneven, inclining to earthy; and sometimes splintery.

The fragments are indeterminate angular.

The globular variety is composed of lamellar distinct concretions.

It is opaque, or translucent on the edges.

It is sometimes semi-hard, sometimes soft, and is frequently soft, inclining to friable.

It feels rough.

It is brittle.

It is easily frangible.

It is light.

Constituent Parts.

It is a nearly pure Carbonate of Lime.

Geognostic Situation.

It occurs in beds, generally in the neighbourhood of lakes and rivers; also encrusting rocks, and enveloping animal and vegetable remains in the vicinity of calcareous springs.

Geographic Situation.

It is a frequent mineral in the neighbourhod of all the calcareous springs in this country, as in those at Starly Burn in Fifeshire, and other places; and on the Continent of Europe, it is also a frequent mineral.

Uses.

The hardest kinds are used for building-stones, and are also burnt into quicklime. It is sometimes also used as a filtering-stone.

Observations.

The substance called *Osteocolla* or *Beinbruch* by the older mineralogists, from its resemblance to a mass of agglutinated bones, is nothing more than a deposition of calc-tuff, which has taken place around small branches and twigs of trees.

Sixth Subspecies.

Pisiform Limestone, or Pea-stone.

Erbsenstein, *Werner.*

La pierre de Pois, ou la Pisolite, *Broch.* t. i. p. 555.—Erbsenstein, *Lud.* b. i. s. 151. *Id. Suck.* 1r th. s. 621. *Id. Bert.* s. 93. *Id. Mohs,* b. ii. s. 93. *Id. Hab.* s. 79.—Dichter Kalksinter, *Leonhard,* Tabel. s. 34.—Chaux carbonatée concretionnée, Pisolithe, *Brong.* t. i. p. 213.—Erbsförmiger Kalkstein, *Karsten,* Tabel. s. 50.—Pisolithus, or Pea-stone, *Kid,* vol. i. p. 27.—Chaux carbonatée concretionnée globuliforme-testacée, *Hauy,* Tabl. p 4.—Erbsenstein, *Lenz,* b. ii. s. 754. Peastone, *Aikin,* p. 140.

External Characters.

Its most common colour is yellowish-white, which sometimes approaches to snow-white; from yellowish-white it passes into reddish-brown, and even inclines to flesh-red.

It occurs massive.

Internally it is dull, or very feebly glimmering.

The fracture is even.

The fragments are blunt-edged.

It is composed of round granular concretions, that generally lie in a basis of calc-sinter: these concretions are composed

composed of very thin curved lamellar concretions, which generally include a nucleus of quartz, or of fragments of felspar, or of granite, and very rarely double six-sided pyramids of rock-crystal.

It is opaque, or feebly translucent on the edges.

It is soft.

It is brittle, and very easily frangible.

It is not particularly heavy.

Constituent Parts.

It is Carbonate of Lime, slightly coloured with iron.

Geognostic and Geographic Situations.

It is found in great masses in the vicinity of the Hot Springs at Carlsbad, in Bohemia. According to Werner, it is formed in the following manner: Particles of sand are raised in the water, by means of air-bubbles, and become covered with calcareous earth, which is deposited around them in lamellar concretions; at length, the globular concretions thus formed, acquire so much weight that they fall down, and being agglutinated, by means of calc-sinter, form peastone. What renders this explanation very probable, is the almost constant occurrence of particles of quartz-sand in the centre of these globular concretions. In some rare instances, the centre of the concretions is empty. A mineral resembling peastone, occurs at the Baths of St Philippi in Tuscany; also at Perscheesberg in Silesia; and in Hungary.

Use.

It is sometimes cut into plates, for ornamental purposes.

 7. Lucullite.

7. Lucullite.

This species is divided into three subspecies, viz. Compact Lucullite, Prismatic Lucullite, and Foliated Lucullite.

First Subspecies.

Compact Lucullite.

Dichter Lucullan, *John.*

Lapis suillus, *Wall.* t. i. p. 148.—Swinestone, *Kirw.* vol. i. p. 89. —Stinkstein, *Wid.* s. 521. *Id. Estner,* b. ii. s. 1023. *Id. Emm.* b. i. s. 487.—Pierre calcaire puante, ou pierre puante, *Lam.* t. ii. p. 58.—Chaux carbonatée fetide, *Hauy,* t. ii. p. 188. —La pierre puante, *Broch.* t. i. p. 567.—Gemeiner Stinkstein, *Reuss,* b. ii. 2. s. 335. *Id. Lud.* b. i. s. 155. *Id. Suck.* 1r th. s. 638. *Id. Bert.* s. 111. *Id. Mohs,* b. ii. s. 126.—Gemeiner Stinkstein, *Leonhard,* Tabel. s. 36.—Chaux carbonatée fetide, *Brong* t. i. p. 236.—Gemeiner Stinkstein, *Haus* s. 128. *Id. Karst.* Tabel. s. 50.—Swinestone, *Kid,* vol. i. p. 29.—Chaux carbonatée fetide, *Hauy,* Tabl. p. 6.; et Chaux carbonatée bituminifere, *Id.* p. 6.—Gemeiner Stinkstein, *Lenz,* b. ii. s. 767.—Dichter Stinkstein, *Oken,* b. i. s. 407.—Swinestone, *Aikin,* p. 14.

This subspecies is divided into two kinds, viz Common Compact Lucullite or Black Marble; and Stinkstone.

First

First Kind.

Common Compact Lucullite, or Black Marble.

Dichter Lucullan; Schwarzer Marmor, *John*, Chemisches Laboratorium, t. ii. s. 227. *Id. Lenz*, b. ii. s. 765.

External Characters.

Its colour is greyish-black.

It occurs massive.

Internally it is strongly glimmering, inclining to glistening

The fracture is fine-grained uneven, and large conchoidal.

The fragments are indeterminate angular, and rather sharp-edged.

It is opaque.

It is semi-hard.

It yields a dark ash-grey coloured streak.

It is brittle, and easily frangible.

Specific gravity, 3.000, *John*.

When two pieces are rubbed against each other, a smell resembling that of sulphureted hydrogen is exhaled, the intensity of which is increased when we at the same time breathe on them.

Chemical Characters.

It is infusible without addition. When exposed to a high temperature in an open crucible, it burns white. With sulphuric acid, it forms a black-coloured mass: it dissolves in nitrous and muriatic acids, but leaves an insoluble black-coloured substance. During the solution and escape of the carbonic acid, a smell resembling that of sulphureted hydrogen is evolved.

Constituent Parts.

Lime, - - -	53,38
Carbonic Acid, - -	41.50
Black Oxide of Carbon, -	0.75
Magnesia, and Oxide of Manganese,	0.12
Oxide of Iron, - -	0.25
Silica, - - -	1.13
Sulphur, - - -	0.25
Potash, combinations of Muriatic and Sulphuric Acids, and Water,	2.62
	100.00

John, Chem. Laborat. b. ii. s. 240.

Geognostic Situation.

The geognostic relations of this mineral are still but little known: it is said to occur in beds in transition and older flœtz rocks.

Geographic Situation.

Hills of this mineral occur in the district of Assynt in Sutherland. It is the *Assynt* or *Sutherland Marble* of artists *. Varieties of it are met with at Ashford, Matlock, and Monsaldale, in Derbyshire: at Kilkenny; at Crayleath, in the county of Down; at Kilcrump, in the county of Waterford; at Churchtown, in the county of Cork; and in the county of Galway, in Ireland. The black marbles of Dinan and Namur, in the Netherlands, are of the same nature. Faujas St Fond is said to have discovered the old quarries of this mineral worked by the ancients, two leagues from Spa, not far from Aix-la-Chapelle.

Uses.

* Geological Transactions, vol. ii. p. 408, 409, 410.

Uses.

The finer varieties of this mineral have been highly prized, and used as marble from a very remote period. It was so much admired and esteemed by the Consul Lucullus, that he gave it his name. Pliny observes: " Post hunc Lepidum ferme quadriennio L. Lucullus Consul fuit, qui nomen (ut apparet ex re) *Luculleo Marmori* dedit, admodum delectatus illo, primusque Romam invexit, atrum alioqui, cum cætera maculis aut coloribus commendentur. Nascitur autem in Nili insula, solumque horum marmorum ab amatore nomen accepit *." It is said that Marcus Scaurus ornamented his palace with columns thirty-eight feet high of lucullite; and Ferber describes busts and pedestals of it in the Capitol, and at Albani. The mausoleum of Frederick-William, father of Frederick the Great, at Potsdam, is of black marble. The Chinese cut it into bars, and use it along with other minerals in the construction of their musical instrument named *king*. The *Paragone* mentioned by Ferber as a variety of black marble, is said to be basalt. Under the title *Nero antico*, the Italians include all the fine antique lucullites, which are now very rare, and are only to be met with in polished slabs or pieces.

The finest varieties of lucullite met with in trade in this island, are the black marbles of Sutherlandshire, Kilkenny, and Galway.

Observations.

1. It is distinguished from other *Marbles* and *Limestones* by its deep black colour, the strong sulphureous smell it emits when rubbed, and higher specific gravity.

 2. It

* Plin. Hist. Nat.

2. It has been confounded by Boetius de Boot, and Agricola, with several other minerals, as Obsidian, Basalt, and Lydian-stone.

3. It was first described as as a particular mineral by Dr John, in the Memoirs of the Society of the Friends of Natural History in Berlin, and afterwards in his work intituled Chemical Laboratory.

Second Kind.

Stinkstone, or Swinestone.

Stinkstein, *Werner.*

External Characters.

Its colours are yellowish and greyish white, smoke-grey, ash-grey, bluish-grey, and brownish-grey, pitch-black, and cream-yellow, which passes into wood-brown, hair-brown, yellowish-brown, liver-brown, and blackish-brown.

It is sometimes dendritic on the surface, or clouded with greyish-black.

It occurs massive, disseminated through gypsum, in plates, and in grains, which are either perfect globular, lenticular, or amygdaloidal. The grains are imbedded in a basis, which varies from nearly earthy to sparry.

Internally it is dull or glimmering.

The fracture is sometimes small splintery, sometimes imperfect conchoidal, and fine grained uneven, which passes into earthy, or straight slaty.

The fragments are indeterminate angular, or slaty.

It sometimes occurs in very small granular distinct concretions, and the globular varieties in thin concentric lamellar concretions.

It

It is opaque, and only the cream-yellow varieties translucent on the edges.

It is semi-hard.

It affords a greyish-white coloured streak; and when rubbed, emits the smell of sulphureted hydrogen gas.

It is brittle.

It is easily frangible.

Specific variety, 2.750, slaty variety from Bottendorf.

Chemical Characters.

Nearly the same as in the preceding kind.

Constituent Parts.

From Bottendorf.	
Carbonate of Lime, -	148—149.00
Silica, - - -	7.00
Alumina, - - -	5.25
Oxide of Iron, - -	2.50
Oxide of Manganese, - -	1 00
Oxide of Carbon, and a little Bitumen,	0.50
Lime *, - - - -	1.00
Sulphur, Alkali, Salt, Water, -	3.75
	170.00

John, Chem. Laborat. t. ii. s. 242.

Geognostic Situation.

This mineral occurs in beds, in the first flœtz limestone, and occasionally alternates with the first flœtz gypsum, and with beds of clay. In some places, the strata are quite straight, in others have a zig-zag direction

* I have copied the above analysis from Dr John's work; yet I do not see how it is possible that 1 part of Lime could be discovered along with 149 of Carbonate of Lime.

tion, or are more or less deeply waved, and they are occasionally disposed in a concentric manner, like the concentric lamellar concretions of greenstone. Some strata contain angular pieces of stinkstone, which at first sight might be taken for fragments; and even whole beds occur, which are composed throughout of angular portions, either connected together by means of clay, or immediately joined without any basis. These various appearances do not seem to have been occasioned by any mechanical force acting upon the strata after their formation, but are rather to be viewed as original varieties of structure, which have taken place during the formation of the strata.

It has been also met with in beds in shell limestone, and in the newer coal formation.

Geographic Situation.

It occurs in the vicinity of North Berwick in East Lothian, resting on old red sandstone; and in the parish of Kirbean in Galloway. On the Continent, it is a frequent rock in Thuringia and Mansfeld.

Uses.

In ancient times, it was used as a medicine in veterinary practice: at present, it is principally employed as a limestone, and when burnt, affords an excellent lime, both for mortar and manure. In some districts, as in Thuringia, it is used as a paving-stone, and also cut into troughs, steps for stairs, door-posts, and other similar purposes.

Observations.

The names *Stinkstone* and *Swinestone* given to this mineral, are from the disagreeable odour it emits when pounded or rubbed.

Second

Second Subspecies.

Prismatic Lucullite.

Stänglichter Lucullan, *John.*

Madreporit, *Klaproth,* Beit. b. iii. s. 272. *Id. Hauy,* t. iv. p. 378. *Id. Lucas,* p. 200.—Madreporstein, *Leonhard,* Tabel. s. 36.—Chaux carbonatée Madreporite, *Brong.* t. i. p. 229.—Madreporite, *Brard,* p. 413.—Stänglicher Anthraconit, *Haus.* s. 128.—Madreporstein, *Karst.* Tabel. s. 50.—Chaux carbonatée bacillaire-fasciculée gris-noiratre, *Hauy,* Tabl. p. 3. —Stänglichter Lucullan, *John,* Chem. Laborat. b. ii. s. 243. *Id. Lenz,* b. ii. s. 770.—Stängliger Stinkstein, *Oken,* b. i. s. 407.

External Characters.

Its colours are greyish-black, pitch-black, and smoke-grey.

It occurs massive.

The external surface is sometimes delicately longitudinally streaked.

Externally it is sometimes dull, sometimes glistening: internally it is shining and splendent, and the lustre is intermediate between vitreous and resinous.

The fracture is minute and curved foliated; sometimes it approaches to conchoidal.

The fragments are indeterminate angular, sometimes inclining to rhomboidal.

It occurs in slightly diverging prismatic distinct concretions.

It is translucent on the edges, or opaque.

It is semi-hard.

It affords a grey-coloured streak.

It is brittle.

It is easily frangible.

When

When rubbed, it emits a strong smell of sulphureted hydrogen gas.

Specific gravity, 2.653, 2.688, 2.703, *John.*

Chemical Characters.

When pounded and boiled in water, it gives out a hepatic odour, which continues but for a short time. The filtrated water possesses weak alkaline properties, and contains a small quantity of a muriatic and sulphuric salt. It does not appear to be affected by pure alkalies. It dissolves with effervescence in nitrous and muriatic acids, and leaves behind a coal-black or brownish-coloured residuum.

Constituent Parts.

From Stavern in Norway.		From Greenland.		From Garphytta, in Nericke in Sweden.	
Carbonic Acid,	41.50	Carbonic Acid,	41.53	Carbonic Acid,	41.75
Lime,	53.37	Lime,	53.00	Lime,	54.00
Oxide of Manganese,	0.75	Oxide of Manganese,	1.00	Oxide of Manganese,	0.50
Oxide of Iron,	1.25	Oxide of Iron,	0.75	Oxide of Iron,	0.75
Oxide of Carbon,	1.25	Oxide of Carbon,	1.00	Brown Oxide of Carbon,	0.75
Sulphur,	0.25	Sulphur,	0.50	Sulphur, Alkali, Alkaline Muriate and Sulphate, Water,	2.25
Alumina,	1.25	Alumina,	0.75		
Silica,	1.25	Silica, Alkali, Alkaline Muriate, Water,	1.47		
Alkali, Alkaline Muriate, Water, Magnesia, Zirconia,	2.13				
	100.00		100·00		100,00
John, Chem. Laborat. b. ii. s. 246.		*John*, ib. s. 248.		*John*, ib. s. 250.	

Geognostic and Geographic Situations.

At Stavern in Norway, it appears to occur in transition rocks: in alum-slate at Garphytta in Nericke: in Greenland; and in the Russbachthal in Salzburg.

Observations.

Observations.

This mineral, which was first discovered by Von Moll, in the Russbachthal in Salzburg, was named by him *Madreporite*, on account of the resemblance of its prismatic concretions to certain lithophytes. It was first described by Schroll, and analysed by Heim *. According to Heim, it contains, Lime, 63.250, Alumina, 10.125, Silica, 12.500, Oxide of iron, 10.988. The same result is said to have been obtained in the School of Mines in Paris †; but both differ so much from the analysis of Klaproth, that we do not hesitate in considering them as erroneous. The publication of Klaproth's chemical examination, induced Von Moll to name it *Anthraconite*, on account of the carbon which it contains ‡; and Dr John, from its intimate connection, both mineralogical and chemical, with Common Lucullite and Stinkstone, arranges it in the system along with these minerals.

Third Subspecies.

Foliated or Sparry Lucullite.

Späthiger Lucullan, *John.*

Späthiger Lucullan, *John,* Chem. Laborat. b. ii. s. 250. *Id.* *Lenz,* b. ii. s. 772.

External Characters.

Its colours are yellowish, greyish, and greenish-white; also bluish-grey, and greyish and velvet black.

It

* Von Moll's Jahrbücher der Berg und Hüttenkunde, 1ster Band, s. 291.—304.

† Hauy, Traité de Mineralogie, t. iv. p. 378, 379.

‡ Ephem. der Berg. und Hütt. 2. b. ii. liæf s. 305.

It occurs massive, disseminated, and crystallised in acute six-sided pyramids.

Internally it alternates from glimmering to shining.

The fracture is small and minute foliated, and sometimes curved, sometimes scaly foliated.

The fragments are generally rhomboidal.

It occurs in small and fine granular concretions.

It is translucent, or translucent on the edges.

It is semi-hard, approaching to soft.

It is brittle, and easily frangible.

When rubbed, it emits an urinous smell.

Specific gravity, 2.650, *John*.

Chemical Characters.

They agree with those of the preceding subspecies: in its solution in acids, there remains a minute black-coloured residuum.

Constituent Parts.

From Moscau.	From Garphytta. Translucent variety.	From Garphytta. Black variety.
Carbonate of Lime, 96.50	Carbonate of Lime, 99.1	Carbonate of Lime, 95.0
Carbonate of Manganese, Magnesia and Iron, 1.50	Carbonate of Manganese, Magnesia and Iron, - 0.9	Carbonate of Manganese, Magnesia and Iron, 1.5
Oxide of Iron, 1.00	A trace of Carbon, and of an odorous substance.	Mixture of Alum-slate, and of Iron-pyrites, as constituent parts, 3.5
Lime, Alumina, Carbon, Silica, and Water, 1.00		
100.00	100.0	100.00
John, Chem. Laborat. b. iii. s. 90.	*Hisinger* and *Berzelius*, in Afhandlingar i Fysik Kemi och Mineralogi.	*Hisinger* and *Berzelius*, Ib.

Geognostic

Geognostic and Geographic Situations.

It occurs in veins, and also in small cotemporaneous masses, in a bed of limestone in clay-slate, at Andreasberg in the Hartz : in veins of silver-ore in hornblende-slate at Kongsberg in Norway : also in transition alum-slate in larger and smaller elliptical masses, the centre of which is of iron-pyrites, and the periphery sparry lucullite, at Andrarum in Schonen, Garphytta in Nericke, and Christiania in Norway.

8. Marl.

Mergel, *Werner*.

This species is divided into two subspecies, viz. Earthy Marl, and Compact Marl.

First Subspecies.

Earthy Marl.

Mergel Erde, *Werner*.

Mergel Erde, *Wid.* s. 523.—Earthy Marle, *Kirw.* vol. i. p. 94.—Mergelerde, *Estner*, b. ii. s. 1027. *Id. Emm.* b. i. s. 491.—Marna terrosa, *Nap.* p. 360 —La Marne terreuse, *Broch.* t. i. p. 569.—Erdiger Mergel, *Reuss*, b. ii. 2. s. 339 —Mergelerde, *Lud.* b. i. s. 156. *Id. Suck.* 1r th. s. 643. *Id. Bert.* s. 114. *Id. Mohs*, b ii. s. 129.—Erdiger Mergel, *Hab.* s. 73. *Id. Leonhard*, Tabel. s. 36. *Id. Karst.* Tabel. s. 50.—Mergel Erde, *Haus.* s. 127. *Id. Lenz*, b ii. s. 777.—Erd Mergel, *Oken*, b. i. s. 406.

External Characters.

Its colours are ash-grey, pale smoke-grey, which passes into

into yellowish-white, and greyish-white. These are the colours it exhibits when dry: when moist, and in its original repository, its colours are pale blackish brown or brownish-black, also dark yellowish, hair, and wood brown; seldomer dark ash-grey, smoke-grey, or greenish-grey; more rarely of a colour intermediate between hair or liver brown and yellowish-grey, which passes into muddy cream-yellow.

It consists of fine sandy or dusty particles, which are loosely cohering.

It is dull or feebly glimmering.

The dusty particles feel very fine, and soft.

It soils strongly.

When it is rather compact, it becomes shining in the streak.

It emits a strong urinous smell.

Chemical Characters.

It effervesces strongly with acids.

Constituent Parts.

It is said to be composed of Lime, Alumina, Silica, and Bitumen.—*Friesleben.*

Geognostic and Geographic Situations.

It occurs in beds in the first and second flœtz limestone and gypsum formations, along with stinkstone, in Thuringia and Mansfeld.

Observations.

1. It passes into Limestone, Stinkstone, and Black Clay.

2. It is sometimes mixed with mica and calcareous-spar, also with iron-ochre, and seldomer with pure clay, quartz, sand, gypsum, and aphrite.

3. Masses.

3. Masses of stinkstone and limestone, of various sizes and shapes, occur in the beds of marl-earth, which at first sight might be confounded with fragments, although, when closely examined, they prove to be of cotemporaneous formation with it.

Second Subspecies.

Compact or Indurated Marl.

Verhärteter Mergel, *Werner.*

Verhärteter Mergel, *Wid.* s. 524.—Indurated Marl, *Kirw.* vol. i. p. 95.—Verhärteter Mergel, *Emm.* b. i. s. 493.—Marna indurita, *Nap.* p. 361.—La Marne endurcie, *Broch.* t. i. p. 571. —Argile calcifere, ou Marne, *Hauy,* t. iv. p. 445.—Verhärteter Mergel, *Reuss,* b. ii. 2. s. 341. *Id. Lud.* b. i. s. 156. *Id. Suck.* 1r th. s. 642. *Id. Bert.* s. 115. *Id. Mohs,* b. ii. s. 130. *Id. Hab.* s. 73. *Id. Leonhard,* Tabel. s. 36. *Id. Karst.* Tabel. s. 50. *Id. Haus.* s. 127. *Id. Lenz,* b. ii. s. 779. —Stein Mergel, *Oken,* b. i. s. 407.—Marl, *Aikin,* p. 141.

External Characters.

Its colours are yellowish and greyish white; smoke-grey, bluish-grey, and greenish-grey; greyish-black; mountain-green, verdigris-green, and asparagus-green; and cream yellow. It is sometimes spotted reddish and brownish in the rents, and marked with dendritic delineations.

It occurs massive, in blunt angular pieces, vesicular, in flattened balls; and frequently contains petrifactions of fishes and crabs, also gryphites, belemnites, chamites, pectinites, ammonites, terebratulites, ostracites, musculites, and mytulites.

 It

It is dull, both externally and internally, and only glimmering when intermixed with foreign parts.

The fracture is generally earthy, which approaches sometimes to splintery, sometimes to conchoidal; in the great inclines to slaty.

The fragments are angular and blunt-edged, and sometimes tabular.

It is soft.

It is opaque.

It is brittle.

It is easily frangible.

It is not particularly heavy.

Chemical Characters.

Before the blowpipe, it intumesces, and melts into a greenish-black slag. It effervesces briskly with acids.

Constituent Parts.

Carbonate of Lime, -	50
Silica, - - -	12
Alumina, - -	32
Iron and Oxide of Manganese,	2

Kirwan.

Geognostic Situation.

It occurs in beds in the flœtz limestone and coal formations; also in the new flœtz formations that rest upon chalk.

Geographic Situation.

It frequently occurs in the coal formations in Scotland and England, and in the new flœtz formation which rests upon chalk in the south of England. On the Continent

tinent of Europe, it abounds in the flœtz limestone and coal formations; and also in the new formation that rests upon the chalk in different parts of France, as in the vicinity of Paris.

Uses.

Several different kinds of compact marl occur in nature: these are calcareous marl, in which the calcareous earth predominates; clay marl, in which the aluminous earth is in considerable quantity; and ferruginous marl, in which the mass contains a considerable intermixture of oxide of iron. This latter kind occurs in spheroidal concretions, called *septaria* or *ludi Helmontii*, that vary from a few inches to a foot and a half in diameter. When broken in a longitudinal direction, we observe the interior of the mass intersected by a number of fissures, by which it is divided into more or less regular prisms, of from three to six or more sides, the fissures being sometimes empty, but oftener filled up with another substance, which is generally calcareous-spar. From these septaria are manufactured that excellent material for building under water, known by the name of *Parker's Cement* * The calcareous and aluminous marls are used for improving particular kinds of land; also for mortar; in some kinds of pottery; and in the smelting of particular ores of iron.

 Observations.

* These marly septaria abound in the Isle of Shepey, in the Medway, and often contain in their interior globular portions of heavy-spar, having a diverging fibrous fracture. Similar septaria occur in Derbyshire, and in the county of Durham, in which latter district, the internal fissures are filled with quartz.

Observations.

1. All compact marls fall into powder when exposed to the air, but some more readily than others.

2. The *Florence Marble*, or *Ruin Marble*, as it is sometimes called, appears to be a very compact marl, inclining to compact limestone. It presents angular figures of a yellowish-brown, on a base of a lighter tint, and which passes to greyish-white. Seen at a distance, slabs of this stone resemble drawings done in bister. "One is amused (says Brard) to observe in it kinds of ruins: there it is a Gothic castle half destroyed, here it presents ruined walls; in another place old bastions; and, what still adds to the illusion, is, that, in these sorts of natural paintings, there exists a kind of aërial perspective, which is very sensibly perceived. The lower part, or what forms the first plane, has a warm and bold tone; the second follows it, and weakens it as it increases the distance; the third becomes still fainter; while the upper part, agreeing with the first, presents in the distance a whitish zone, which terminates the horizon, then blends itself more and more as it rises, and at length reaches the top, where it forms sometimes as it were clouds. But approach close to it, all vanishes immediately, and these pretended figures, which, at a distance, seemed so well drawn, are converted into irregular marks, which present nothing to the eye." To the same compact marl, may be referred the variety called *Cottam Marble*, from being found at Cottam, n ar Bristol. It resembles in many respects the Landscape Marble.

3. Sartorius describes a mineral under the name *Leutrite*, which he considers to be nearly allied to Marl, and of which the following are the characters:—Its colours are dark greyish-white; also yellowish-white, and ochre-

ochre-yellow. It occurs massive. It is dull. The fracture is uneven, and earthy. The fragments are blunt-edged. It is soft, passing into friable. It feels meagre and rough. It is not particularly heavy. When rubbed with any body, it shines in the dark: it even becomes phosphorescent by rubbing with paper. Its phosphorescence is increased by heating. It occurs in a bed on the Leutra, near Jena.

9. Bituminous Marl-Slate.

Bituminöser Mergelschiefer, *Werner*.

Bituminöser Mergelschiefer, *Wid.* s. 526.—Bituminous Marlite, *Kirw.* vol. i. p. 103.—Bituminöser Mergelschiefer, *Estner*, b. ii. s. 1035 *Id. Emm.* b. i. s. 498.—Schisto marno bituminoso, *Nap.* p. 363.—Le Schiste marneuse bitumineux, *Broch* t. i. p. 574.—Bituminöser Mergelschiefer, *Reuss*, b. ii. 2. s. 376. *Id. Lud.* b. i. s. 157. *Id. Suck.* 1r th. s. 646. *Id. Bert.* s. 116. *Id. Mohs*, b. ii. s. 132. *Id. Hab.* s. 74. *Id. Leonhard*, Tabel. s. 36. *Id. Karst.* Tabel. s. 50. *Id. Haus.* s. 127. *Id. Lenz*, b. ii. s. 786. *Id. Oken*, b. i. s. 405.

External Characters.

Its colour is intermediate between greyish-black and brownish-black.

It occurs massive, in whole beds.

Its lustre is glimmering, glistening, or shining.

Its fracture is straight or curved slaty.

The fragments are slaty in the large, but indeterminate and rather sharp angular in the small.

It is opaque.

It is shining and resinous in the streak.

 It

It is soft.

It feels meagre.

It is rather sectile.

It is easily frangible.

It is not particularly heavy.

Constituent Parts.

It is said to be a Carbonate of Lime united with Alumina, Iron, and Bitumen.

Geognostic Situation.

It occurs in the first flœtz limestone, and generally forms the lowest bed of that formation. It frequently contains ores of copper, particularly of copper-pyrites, copper-glance, variegated copper-ore, and, more rarely, native copper, copper green, and azure-copper-ore. It contains abundance of petrified fishes, and these are said to be most numerous in those situations where the strata are basin-shaped. Many attempts have been made to ascertain the genera and species of these animals, but hitherto with but little success. It would appear, that the greater number resemble fresh-water species, and a few marine species. It also contains fossil remains of lizards, shells, corals, and of cryptogamous fresh-water plants.

Geographic Situation.

Europe.—It abounds in the Hartzgebirge; also in Magdeburg, and Thuringia. It is a frequent mineral in Upper and Lower Saxony: it occurs also in Franconia, Bohemia, Bavaria, Silesia, Suabia, Hesse-Cassel, and Switzerland.

America.—It is said to occur in the Cordilleras of South America.

Use.

Use.

When it is much impregnated with copper-ore, it is named *copper-slate*, and is smelted as an ore of copper.

10. Arragonite.

Arragon, *Werner.*

This species is divided into three subspecies, viz. Common Arragonite, Columnar Arragonite, and Acicular Arragonite.

First Subspecies.

Common Arragonite.

Gemeiner Arragon, *Werner.*

Arragon-Spar, *Kirw.* vol. i. p. 87.—Arragon, *Estner,* b. ii. s. 1039. *Id. Emm.* b. v. s. 357.—L'Arragonite, *Broch.* t. i. p. 576. *Id. Hauy,* t. iv. p. 337.—Excentrischer Kalkstein, *Reuss,* b. ii. 2. s. 300. *Id. Lud.* b. i. s. 158.—Excentrischer Kalkstein, *Suck.* 1r th. s. 615.—Arragon, *Bert.* s. 97. *Id. Mohs,* b. ii. s. 98.—Arragonite, *Lucas,* p. 192.—Exzentrischer Kalkstein, *Leonhard,* Tabel. s. 34.—Chaux carbonatée Arragonite, *Brong.* t. i. p. 221.—Arragonite, *Brard,* p. 403.—Gemeiner Arragonit, *Haus.* s. 127.—Arragon, *Karsten,* Tabel. s. 50.—Arragon-Spar, *Kid,* vol. i. p. 55. —Arragonite, *Hauy,* Tabl. p. 6.—Gemeiner Arragon, *Lenz,* b. ii. s. 787.—Arragonite, *Oken,* b. i. s. 404.—Arragonite, *Aikin,* p. 142.

External Characters.

Its colours are greyish-white, greenish-white, and yel-

 lowish-

lowish-white; also pearl-grey; pale mountain-green, and pale violet-blue.

It occurs crystallised in the following figures:

1. Perfect equiangular six-sided prism, either equilateral, or with two opposite sides broad, and four smaller lateral planes; or two opposite sides narrow, and four larger lateral planes. Sometimes the prism is so much compressed that it appears like a table *.
2. Six-sided prism, bevelled on the extremities, the bevelling planes set sometimes on the broader lateral planes, more rarely on the edges formed by the meeting of the smaller lateral planes, and the lateral edges and bevelling edges are sometimes truncated.
3. Oblique four-sided prism, bevelled on the extremities, the bevelling planes set on the acute lateral edges, and the bevelling edge more or less deeply truncated.
4. When the bevelling planes of the preceding figure increase very much, an acute octahedron is formed.

The crystals are middle-sized and small; they are generally attached by their terminal planes, seldomer by their lateral planes; sometimes imbedded, and are to be observed intersecting each other.

The lateral planes of the crystals are sometimes smooth, more frequently more or less deeply streaked or grooved. The terminal planes are seldom smooth, generally uneven and rough, and sometimes also deeply notched.

The

* According to Bournon, the primitive form of this mineral is a rhomboidal four-sided prism, in which the lateral planes meet under angles of 117° 2' and 62° 58.

The external lustre varies from dull to shining, and is vitreous: internally it is shining and glistening, and vitreous, inclining to resinous.

The fracture is foliated, with a fourfold cleavage, in which three of the cleavages are parallel with the lateral planes, and the fourth with the terminal planes.

The fragments are indeterminate angular, and rather sharp-edged.

It is translucent, passing into semi-transparent, and refracts double.

It is semi-hard: it scratches fluor-spar, and even glass, but with difficulty.

It is brittle.

It is easily frangible.

Specific gravity, 2.9465, *Hauy*. 2.883, 2.928, *Karsten*, 2.9267, *Biot*. 2.891, *Wiedeman*. 2.912, *Bournon*.

Chemical Characters.

If we expose a small fragment to the flame of a candle, it almost immediately splits into white particles, which are dispersed around the flame. This change takes place principally with fragments of transparent crystals, fragments of the other varieties becoming merely white and friable. Fragments of calcareous-spar, when placed in a similar situation, undergo no alteration.

It is completely soluble, with effervescence, in the nitric and muriatic acids.

Constituent

Constituent Parts.

Lime, 58.5	Lime, 56.327	Lime, 54.5	Lime, 55.5
Carbon. acid, 41.5	Carbon. acid, 43.045	Carbon. acid, 41.5	Carbon. acid, 43.7
100.0	Water, 0.628	Water, 3.5	Water, 0.8
		0.5	
		100.0	100.0
Fourcroy & Vauquelin, Annal. du Mus. t. iv. p. 405.	*Biot & Thenard*, N. Bull. des Sciences, de la Soc. Ph. t. i. n. 32.	*Bucholz*, Gehlen's Jour. b. iii. s. 80.	*Holme*, Annals of Phil. vol. I p. 384.

From Molina in Arragon.		From Bastanes.	
Carbonate of Lime,	94.5757	Carbonate of Lime,	94.8249
Carbonate of Strontian,	3.9662	Carbonate of Strontian,	4.0836
Hydrate of Iron, -	0.7060	Protoxide of Manganese, with a trace of Iron,	0.0939
Water of crystallization,	0.3000	Water of crystallization,	0.9831
	99.5489		99.9855

Stromeyer, in Gilbert's Annalen der Physik, xiv. 217. October 1813.

Geognostic and Geographic Situations.

It occurs in Spain, imbedded in granular and fibrous gypsum, along with reddish-brown coloured quartz crystals; in trap rocks in Auvergne; and in similar rocks at Caupenne in the vicinity of Dax; and at Bastenes in Bearn, also in France.

Second

Second Subspecies.

Columnar Arragonite.

Stänglicher Arragon, *Werner.*

Stänglicher Arragon, *Haus.* s. 127.—Iglit, *Leonhard,* Tabel. s. 33.—Stänglicher Arragon, *Lenz,* b. ii. s. 790.

External Characters.

Its colours are yellowish-white, yellowish-grey, and pearl-grey.

It occurs crystallised, in rather long equiangular six-sided prisms, which are sometimes rounded.

Internally it is shining.

The fracture is small and fine-grained uneven.

The fragments are indeterminate angular, and rather sharp-edged.

It occurs in distinct concretions, which are thin and thick columnar; sometimes wedge-shaped, and scopiformly aggregated.

It is strongly translucent, passing into transparent.

It is semi-hard; harder than calcareous-spar.

It is rather brittle.

It is easily frangible.

It is not particularly heavy.

Specific gravity, 2.8500.

Constituent

Constituent Parts.

	From Vertaison in Auvergne.
Carbonate of Lime, -	97.7227
Carbonate of Strontian, -	2.0552
Hydrate of Iron, -	0.0098
Water of crystallization, -	0.2104
	99.9774

Stromeyer, Gilbert's Annalen der Physik, xlv. 217. October 1813.

Geognostic and Geographic Situations.

It occurs in trap rocks in Scotland: in the fissures of basalt at Vertaison, in the Department of Puy de Dome in France: in Iceland; and also, according to some naturalists, in the lava of Vesuvius.

America.—In Peru.

Third Subspecies.

Acicular, or Needle Arragonite.

Spiessiger oder Nadelformiges Arragon, *Werner.*

Iglit, *Leonhard,* Tabel. s. 33.—Spiessiger Arragon, *Lenz,* b. ii. s. 791.

External Characters.

Its colours are greyish and yellowish white, which pass into yellowish-grey, and yellowish-brown; also greenish-grey, which passes asparagus-green, celandine-green, and verdigris-green.

It occurs globular, coralloidal; and crystallised in the following figures:

1. Very

1. Very acute double six-sided pyramid, in which the lateral planes of the one are set on the lateral planes of the other. Sometimes two opposite planes are broader than the others, and the pyramid ends in a line.
2. Acute double six-sided pyramid truncated on the extremities.
3. Acute double six-sided pyramid, rather flatly acuminated with three planes, which are set on the alternate lateral edges; or flatly acuminated with four planes, which are set on the opposite lateral edges. It sometimes happens, that two opposite acuminating planes become so large in comparison of the others, that the pyramid appears to be bevelled at the extremities.

These pyramids sometimes approach to the prismatic form, sometimes to the acicular, which latter passes into the capillary.

The pyramidal crystals are sometimes scopiformly aggregated; and the acicular crystals are scopiformly and stellularly aggregated.

The lateral planes of the crystals are transversely streaked.

Externally the lustre is shining or glistening, and vitreous, slightly inclining to pearly: internally the lustre is splendent, and is vitreous, inclining to resinous.

The fracture is stellular or scopiform radiated, and fibrous: the cross fracture small and fine-grained uneven, which sometimes inclines to splintery, sometimes to conchoidal.

The fragments are indeterminate angular, and rather sharp-edged.

It is translucent and transparent.

It is semi-hard.

It

It is brittle.

It is easily frangible.

Specific gravity, 2.858, Iglo, *Esmark*. 2.8500, Yellowish-white from Neumarkt, *Kopp*. 2.9870, Yellowish-white from Auvergne, *Kopp*. 2.7500, 3.000, Greyish-white, and Verdigris-green from Schwaz, *Kopp*.

Geognostic and Geographic Situations.

It occurs in trap rocks in Scotland. On the Continent of Europe, it occurs at Schwaz in the Tyrol, along with copper-green, grey copper-ore, ochry brown ironstone, iron-pyrites, quartz, and calcareous-spar; near Iglo in Hungary, associated with, calcareous-spar, small botryoidal calcedony, ochry-brown ironstone, and copper-green; near Schemnitz, accompanied with brown-spar, brittle silver-ore, lead-glance, &c.; at Saalfeldt, with compact brown ironstone; also at Joachimsthal in Bohemia; in a vein in basaltic amygdaloid, at the ruins of the castle of Limburg, near the Kaiserstuhle in the Breisgaw; in the Leogang in Salzburg, in acicular crystals, superimposed on brown ironstone, along with azure copper-ore, and pyramidal calcareous-spar; in common compact limestone on the Wolfstein, near Neumarkt in the Upper Palatinate; in a limestone quarry, along with brown-spar, at Marienberg, in the Saxon Erzgebirge, &c. The beautiful coralloidal variety known under the name *Flos ferri*, and which is by some considered as a calc-sinter, occurs in the iron mines of Stiria and Carinthia; at St Marie aux Mines; and in Dufton Fell in Cumberland.

America.—In Peru.

Asia.—It occurs in the trap rocks of Van Diemen's Land, and the neighbouring islands.

Africa.

Africa.—In the lavas or trap rocks of the Isle de Bourbon.

Observations.

1. Arragonite is distinguished from *Calcareous-spar*, by its crystallizations, fracture, lustre, superior hardness, and specific gravity.

2. This mineral has received several names at different periods: the common and prismatic varieties have been named *Arragonian Apatite*, *Arragonian Calc-spar*, and *Hard Calcareous-spar*; and the pyramidal varieties have been described under the names *Iglit* or *Igloit*, and *Flos ferri*.

3. The discovery of Strontian Earth in arragonite, reconciles the differences that formerly existed amongst chemists and mineralogists, in regard to the claim of that mineral to the rank of a distinct species.

XXI. APA-

XXII. APATITE FAMILY.

This Family contains the following species: Apatite and Phosphorite.

1. Apatite.

Apatit, *Werner*.

This species is divided into two subspecies, viz. Common Apatite, and Conchoidal Apatite.

First Subspecies.

Common Apatite.

Gemeiner Apatit, *Werner*.

Phosphorite, *Kirw.* vol. i. p. 128.—Gemeiner Apatit, *Wid.* s. 528.—Phosphorit, *Estner,* b. ii. s. 1049. *Id. Emm.* b. i. s. 502.—Fosforite lamellare, *Nap.* p. 367.—Apatit, *Lameth.* t. ii. p. 85.—L'Apatite, *Broch.* t. i. p. 580.—Chaux phosphatée, *Hauy,* t. ii. p. 234.—Gemeiner & Blättricher Apatit, *Reuss,* b. ii. 2. s. 355. & 362.—Apatit, *Lud.* b. i. s. 159. *Id. Suck.* 1r th. s. 655. *Id. Bert.* s. 99. *Id. Mohs,* b. ii. s. 139. Chaux phosphatée, *Lucas,* p. 11.—Apatit, *Leonhard,* Tabel. s. 38.—Chaux phosphatée Apatite, *Brong.* t. i. p. 240.—Chaux phosphatée, *Brard,* p. 44.—Blättriger Apatit, *Haus.* s. 123.—Apatit, *Karsten,* Tabel. s. 52.—Crystallised Phosphate of Lime, *Kid,* vol. i. p. 80.—Chaux phosphatée, *Hauy,* Tabl. p. 7.—Gemeiner Apatit, *Lenz,* b. ii. s. 804.—Blättriger Apatit, *Oken,* b. i. s. 397.—Apatit, *Aikin,* p. 136.

External Characters.

Its most frequent colours are snow-white, yellowish-white,

white, reddish-white, and greenish-white: from greenish-white it passes into mountain-green, celandine-green, leek-green, emerald-green? asparagus-green, pistachio-green, and olive-green. It occurs also hyacinth-red, flesh-red, rose-red, and pearl-grey, from which it passes into violet-blue, and lavender-blue, and seldom into indigo-blue. Sometimes it is pale wine-yellow, and yellowish-brown. Frequently several of these colours occur in the same piece.

It seldom occurs massive and disseminated; generally crystallised in the following figures:

I. Prism.

Six-sided prism *.

a. Perfect, fig. 125.

b. With truncated lateral edges †, fig. 126.

c. With truncated lateral and terminal edges ‡, fig. 127.

d. With truncated terminal edges ||, fig. 128.

e. With truncated terminal edges and angles §, fig. 129.

f. With bevelled terminal edges and truncated angles.

g. With rounded edges, so that the prism appears cylindrical.

h. Acuminated on one extremity with six planes, which are set on the lateral planes. In this figure, the apex of the acumination, all the angles,

* Chaux phosphatée primitive, Hauy.

† Chaux phosphatée peridodecaedre, Hauy.

‡ Chaux phosphatée emarginée, Hauy.

|| Chaux phosphatée annulaire, Hauy.

§ Chaux phosphatée unibinains, Hauy.

angles, and the alternate lateral edges, are slightly truncated.

i. Acuminated on both extremities with six planes, the apices, lateral edges, and angles occasionally truncated.

II. Table.

a. Six-sided table, in which the edges and angles are sometimes truncated.

b. Eight-sided table, in which four of the terminal edges are truncated.

The crystals are small, very small, and middle-sized; and occur sometimes single, sometimes many irregularly superimposed on each other.

The lateral planes are generally longitudinally streaked; the truncating planes smooth.

Externally it is splendent; internally glistening, and the lustre is resinous, inclining to vitreous.

The longitudinal fracture is imperfect foliated, with a fourfold cleavage, in which three of the cleavages are parallel with the lateral planes, and one with the terminal planes of the prism. The cross fracture is intermediate between imperfect conchoidal and even.

The fragments are indeterminate angular, and not particularly sharp-edged.

The massive varieties occur in coarse and small granular distinct concretions; and the concretions seldom incline to lamellar.

It is generally translucent; seldom nearly transparent, when it refracts double.

It is semi-hard; harder than calcareous-spar, but not so hard as fluor-spar.

It is brittle.

It is easily frangible.

Specific gravity, 3.179, *Lowry*.

Physical

Physical Characters.

It becomes electric by heating, and also by being rubbed with woollen cloth.

Chemical Characters.

When thrown on glowing coals, it emits a pale grass-green phosphoric light. It dissolves very slowly in the nitric acid, and without effervescence. It gradually loses its colour, when heated before the blowpipe, but its lustre and transparency are heightened. It is infusible without addition.

Geognostic Situation.

It occurs in tinstone veins, and also imbedded in talc.

Geographic Situation.

It occurs in yellow foliated talc, and, along with fluor-spar, in the mine called Stony Gwynn, in Cornwall: at Schlackenwald in Bohemia, in tinstone veins, along with tungsten, wolfram, topaz, and fluor-spar; at Ehrenfriedersdorf in Saxony, along with tinstone, copper, and arsenical pyrites, fluor-spar, steatite, lithomarge, talc, and quartz; imbedded in lepidolite near Rosena in Moravia; in a mixture of quartz and felspar at Nantes, Four-au-Diable, in department of the Lower Loire; at Arendal in Norway, along with magnetic ironstone, garnet, hornblende, and limestone; in veins, on St Gothard in Switzerland; and in Estremadura in Spain, in small tables, along with phosphorite.

Second Subspecies.

Conchoidal Apatite or Asparagus-Stone.

Muschlicher Apatit, *Hausmann.*

Spargelstein, *Werner.*

Chrysolith ordinaire, ou proprement dite, *De Lisle,* t. ii. p. 271. —Chrysolithe, *De Born.* t. p. 68. 2. E. *a.* 3.—Spargelstein, *Emm.* b. iii. s. 359.—La Pierre d'Asperge, *Broch.* t. i. p. 586.—Muschlicher Apatit, *Reuss,* b. ii. 2. s. 358.—Chaux phosphatee crysolithe, *Brong.* t. i. p. 240.—Muschlicher Apatit, *Haus.* s. 123. *Id. Lenz,* b. ii. s. 808. *Id. Oken,* b. i. s. 397.

External Characters

Its colours are mountain-green, leek-green, grass-green, pistachio-green, asparagus-green, and siskin-green, which passes into sulphur-yellow, and wine-yellow. It also occurs sky-blue, greenish and yellow grey, and clove-brown.

It occurs sometimes massive and disseminated; but most frequently crystallised, and in the following figures:

1. Equilateral, longish, six-sided prism, rather acutely acuminated with six planes, which are set on the lateral planes *, fig. 130.
2. Sometimes the acumination ends in a line †.
3. The same figure, truncated on the lateral edges of the prism, fig. 131.

The crystals are middle-sized, small, and very small; sometimes longitudinally streaked, and sometimes traversed by cross rents.

Externally

---

* Chaux phosphatée pyramidée, Hauy.

† Chaux phosphatée cuneiforme, Hauy.

† Chaux phosphatée didodecaedre, Hauy.

Externally the crystals are shining and splendent, and vitreous: internally shining or glimmering, and resinous or vitreous.

The longitudinal fracture is imperfect foliated; the cross fracture flat and perfect conchoidal.

The fragments are blunt-edged.

It occurs in large and small granular distinct concretions.

It alternates from transparent to translucent.

It is semi-hard.

It affords a greyish-white streak.

It is brittle, and easily frangible.

Specific gravity, 3.200, from Uto, *Klaproth*. 3.190, from Zillerthal, *Klaproth*.

Chemical Characters.

Some varieties of this subspecies do not phosphoresce when exposed to heat.

Constituent Parts.

Apatite from Utö.		From Zillerthal.	
Phosphate of Lime,	92.00	Lime, -	53.75
Carbonate of Lime,	6.00	Phosphoric Acid,	46.25
Silica, -	1.00		——
Loss in heating,	0.50		100
Manganese a trace.	——	*Klaproth*, Beit. b. iv. s. 197.	
	99.50		
Klaproth, Beit. b. v. s. 181.			

Geognostic and Geographic Situations.

Europe.—It is found in a porous iron-shot limestone near Cape de Gate, in Murcia in Spain; in granite, near Nantes, and in basalt at Mont Ferrier, in France; imbedded in green talc, in the Zillerthal in Salzburg; in

beds

beds of magnetic ironstone, along with sphene or rutilite, calcareous-spar, hornblende, quartz, and augite, at Arendal in Norway.

America.—Imbedded in granite at Baltimore *; and in mica-slate in West Greenland †.

Observations.

1. This mineral was at one time described as a kind of Schor; afterwards as a variety of Beryl, on account of colours and figure; and some authors have arranged it with Fluor-spar, and others along with Chrysolite. Werner ascertained that it was a distinct species from any of those just enumerated, and named it *Apatite*, from the Greek word *απαταω, to deceive,* on account of its having been confounded with so many other minerals. It was Klaproth who first analysed it.

2. The conchoidal subspecies has been considered by some authors as a mere variety of the common apatite; whilst others have raised it to the rank of a species. thus, many of the French mineralogists arrange it with the varieties of common apatite; while some German mineralogists describe the asparagus-green varieties under the name *Asparagus-stone*, and certain green and blue varieties under the name *Moroxite*. The name moroxite given to this mineral by Karsten, is borrowed from the Morochites of Pliny, concerning which, that author says, "Est gemma, per se porracea viridisque, trita autem candicans."—*Histor. Natur.* l. xxxvii.

3. Apatite is distinguished from Beryl, Schorl, and Chrysolite, by its inferior hardness: its greater hardness and non-effervescence with acids, distinguish it from Calcareous-spar.

2. Phosphorite.

---

* Gilmor, in Bruce's Mineralogical Journal, p. 228.

† Mr Giesecké.

2. Phosphorite.

Phosphorit, *Werner*.

This species is divided into two subspecies, viz. Common Phosphorite, and Earthy Phosphorite.

First Subspecies.

Common Phosphorite.

Gemeiner Phosphorit, *Karsten*.

Gemeiner Phosphorit, *Haus.* s. 123. *Id. Karst.* Tabel. s. 52. —Chaux phosphatée terreuse, *Hauy*, Tabl. p. 8.—Gemeiner Phosphorit, *Lenz*, b. ii. s. 801.—Massive Apatite, *Aikin*, p. 137.

External Characters.

Its colours are yellowish, reddish, and greenish white, and pearl-grey, inclining to flesh-red; sometimes spotted with brown, and ochre-yellow, and rarely tarnished on the fissures with brown and black.

It occurs massive, stalactitic, reniform, in crusts, with impressions, and crystallised in six-sided tables.

The surface is uneven and drusy.

It is dull or glistening.

The fracture is imperfect floriform foliated.

It occurs in curved lamellar distinct concretions.

The fragments are blunt-edged.

It is opaque, or feebly translucent on the edges.

It is soft, and very soft.

It is brittle.

It is easily frangible.

Specific gravity, 2.81, *Pelletier.*

Chemical Characters.

It becomes white before the blowpipe, and, according to Proust, melts with difficulty into a white-coloured glass. When rubbed in an iron mortar, it emits a green-coloured phosphoric light; and the same effect is produced when it is pounded and thrown on glowing coals.

Constituent Parts.

Lime, - - -	59.0
Phosphoric Acid, - -	34.0
Silica, - - -	2.0
Fluoric Acid, - -	2.5
Muriatic Acid, - -	0.5
Carbonic Acid, - -	1.0
Oxide of Iron, - -	1.0
	100.0

Pelletier, Journal des Mines, N. 166.

Geognostic and Geographic Situations.

It occurs in crusts, and crystallised, along with apatite and quartz, at Schlackenwald in Bohemia, but most abundantly near Lagrofan, in the province of Estremadura in Spain, where it is sometimes associated with apatite, and frequently with amethyst, and forms whole beds, that alternate with limestone and quartz.

Second

Second Subspecies.

Earthy Phosphorite.

Erdiger Phosphorit, *Karsten.*

Erdiger Phosphorit, *Haus.* s. 123. *Id. Karsten,* Tabel. s. 52.—Chaux phosphatée pulverulente, *Hauy,* Tabl. p. 8.—Erdiger Phosphorit, *Lenz,* b. ii. s. 802.

External Characters.

Its colours are greyish-white, greenish-white, and pale greenish-grey.

It consists of dull dusty particles, which are partly loose, partly cohering, and which soil slightly, and feel meagre and rough.

Chemical Characters.

It phosphoresces when laid on glowing coals.

Constituent Parts.

Earthy Phosphorite from Marmarosch.

Lime, - - -	47.00
Phosphoric Acid, -	32.25
Fluoric Acid, - -	2.50
Silica, - - -	0.50
Oxide of Iron, - -	0.75
Water, - - -	1.00
Mixture of Quartz and Loam,	11.50
	95.50

Klaproth, Beit. b iv. s. 373.

Geognostic

Geognostic and Geographic Situations.

It occurs in a vein, in the district of Marmarosch in Hungary.

Observations.

1. It was for some time described in systems of mineralogy as a variety of Fluor-spar.

2. The celebrated Prussian chemist John, has published the description and analysis of a mineral under the name *Ratofkite*, from Ratofka, near Werea in Russia, which appears to be nearly allied to this subspecies, and may possibly prove to be a new species, intermediate between fluor and apatite. The following is the description and analysis of it, as given by John, in the second continuation of his work intituled, " Chemische Untersuchungen," &c.

" Its colour is lavender-blue; and it is composed of loose dusty dull particles, that soil slightly, and are not particularly heavy. It is contained in aphrite. *Constituent Parts*: Fluat of Lime, 49.50. Phosphate of Lime, 20.00. Muriatic Acid, 2.00. Phosphate of Iron, 3.75. Water, 10.00.

XXIII. FLUOR

XXIII. FLUOR FAMILY.

This Family contains but one species, viz. Fluor.

Fluor.

Fluss, *Werner*.

It is divided into three subspecies, Compact Fluor, Fluor-Spar or Sparry Fluor, and Earthy Fluor.

First Subspecies.

Compact Fluor.

Dichter Fluss, *Werner*.

Fluor solidus, *Wall.* t. i. p. 542.?—Dichter Fluss, *Wid.* s. 542. —Compact Fluor, *Kirw.* vol. i. p. 127.—Dichter Fluss, *Estner*, b. ii. s. 1067. *Id. Emm.* b. i. s. 516.—Fluorite compatta, *Nap.* p. 374.—Le fluor compacte, *Broch.* t. i. p. 594. —Dichter Fluss, *Reuss*, b. ii. 2. s. 379. *Id. Lud.* b. i. s. 161. *Id. Suck.* 1r th. s. 663. *Id. Bert.* s. 103. *Id. Mohs*, b. ii. s. 150. *Id. Hab.* s. 89.—Chaux fluatée massive compacte, *Lucas*, p. 247.—Dichter Fluss, *Leonhard*, Tabel. s. 38.— Chaux fluatée compacte, *Brong.* t. i. p. 245.—Dichter Fluss, *Haus.* s. 124. *Id. Karsten*, Tabel. s. 52.—Chaux fluatée compacte, *Hauy*, Tabl. p. 9.—Dichter Fluss, *Lenz*, b. ii. s. 823. *Id. Oken*, b. i. s. 899.—Compact Fluor, *Aikin*, p. 136.

External Characters.

Its colours are greyish-white and greenish-grey; which latter sometimes inclines to blue, sometimes to greenish-white

white. It is sometimes spotted or striped with verdigris-green, lavender and dark violet blue, and also brownish-red.

It occurs massive and corroded.

Externally it is dull, or feebly glimmering: internally faintly glimmering, or dull, and vitreous, approaching to resinous.

The fracture is even, which passes on the one side into small splintery, on the other into flat conchoidal.

The fragments are rather sharp-edged.

It is translucent in small pieces.

It is semi-hard; harder than calcareous-spar.

It affords a snow-white coloured streak.

It is brittle.

It is easily frangible.

It is not particularly heavy.

Chemical Characters.

When exposed to heat, it becomes feebly phosphorescent.

Constituent Parts.

It is said to be Fluate of Lime, intermixed with a small portion of Phosphate of Lime.

Geognostic and Geographic Situations.

It is found in veins, associated with fluor-spar, at Stolberg in the Hartz: the veins traverse rocks of grey-wacke, and besides fluor-spar, contain heavy-spar and copper-pyrites. It has also been met with at Kongsberg in Norway; in Salzburg; and at Adon-tschalong in Dauria.

Second

Second Subspecies.

Fluor-Spar or Sparry Fluor.

Fluss-Spath, *Werner.*

Fluor spathosus; Fluor granularis, et Fluor cristallisatus, *Wall.* t. i. p. 180, 183.—Spath fusible ou vitreux, *Romé de Lisle,* t. ii. p. 1.—Chaux fluorée, *De Born.* t. i. p. 355.—Fluss-spath, *Wid.* s. 558.—Foliated or Sparry Fluor, *Kirw.* vol. i. p. 127. —Fluss-spath, *Estner,* b. ii. s. 1070. *Id. Emm.* b. i. s. 519. —Fluorite lamellare, *Nap.* p. 375.—Fluor, *Lam.* t. i. p. 78. —Chaux fluatée cristallisée, *Hauy,* t. ii. p. 247.—Le Spath Fluor, *Broch.* t. i. p. 595.—Spathiger Fluss, *Reuss,* b. ii. 2. s. 381. *Id. Lud.* b. i. s. 162. *Id. Suck.* 1r th. s. 664. *Id. Bert.* s. 103. *Id. Mohs,* b. ii. s. 151. *Id. Hab.* s. 83.—Chaux fluatée, *Lucas,* p. 12.—Spathiger Fluss, *Leonhard,* Tabel. s. 38.—Chaux fluatée spathique, *Brong.* t. i. p. 243.—Chaux fluatée, *Brard,* p. 47.—Gemeiner, Stänglicher, Schaaliger, & Körniger Fluss-spath, *Haus,* s. 123, 124.—Spathiger Fluss, *Karsten,* Tabel. s. 52.—Fluat of Lime, or Fluor-spar, *Kid,* vol. i. p. 73.—Chaux fluatée, *Hauy,* Tabl. p. 9.—Fluss-spath, *Lenz,* b. ii. s. 824. *Id. Oken,* b. i. s. 398.—Crystallised Fluor, *Aikin,* p. 135.

External Characters.

It occurs yellowish, greyish, reddish, and greenish white; yellowish, greenish, and pearl grey; bluish black; azure, violet, plum, lavender, smalt, and sky blue; verdigris, celandine, mountain, leek, emerald, grass, asparagus, pistachio, and olive green; wine, honey, and wax yellow; carmine, crimson, flesh, and rose red; and yellowish, and clove brown.

The

The colours are of all degrees of intensity, and sometimes pieces occur spotted or striped.

It occurs massive, disseminated; and crystallised in the following figures:

1. Cube, which is the most frequent crystallization, and is also the fundamental figure of the species, fig. 132. *.
2. Cube, truncated on all the edges †, fig. 133. When these truncating planes increase so much as to cause the faces of the cube to disappear, there is formed
3. The rhomboidal or garnet dodecahedron ‡, fig. 134.
4. Cube, with truncated angles ||, fig. 135. When these truncating planes increase, so as to cause the faces of the cube to disappear, there is formed an
5. Octahedron, or double four-sided pyramid §, fig. 136. This figure is sometimes truncated on the edges, as fig. 137. or on the edges and angles at the same time; and varieties of it occur, in which the planes or faces are cylindrically convex ¶.
6. Cube, with bevelled edges **, fig. 138. When the bevelling planes enlarge so much, as to cause the original faces of the cube to disappear, a tessular crystal, with 24 triangular planes, is formed, fig. 139.
7. Cube, in which all the angles are acuminated with three planes, which are set on the lateral planes.
8. Cube, in which all the angles are acuminated with six planes, which are set on the lateral planes.

9. Tetra-

* Chaux fluatée cubique, Hauy.

† Chaux fluatée cubo-dodecaedre, Hauy.

‡ Chaux fluatée dodecaedre, Hauy.

|| Chaux fluatée cubo-octaedre, Hauy.

§ Chaux fluatée primitive, Hauy.

¶ Chaux fluatée spheroidale, Hauy.

Chaux fluatée bordée, Hauy.

9. Tetrahedron, truncated on the edges.

The crystals are large, middle-sized, small, and very small, and are placed on one another, or side by side.

The surface is smooth and splendent, or drusy and rough.

Internally the lustre is specular-splendent, or shining, and is vitreous, inclining to pearly, and sometimes to adamantine.

The fracture is either perfect or imperfect foliated, with a fourfold equiangular cleavage, which is parallel with the planes of an octahedron or tetrahedron.

The fragments are octahedral or tetrahedral.

It occurs in distinct concretions, which are sometimes large, coarse, and small granular, sometimes thick and curved lamellar, and occasionally columnar, which latter sometimes exhibits fortification-wise curved violet stripes.

It alternates from translucent to transparent, and refracts single.

It is semi-hard; harder than calcareous-spar.

It is brittle.

It is very easily frangible.

Specific gravity, 3.0943,—3.1911, *Hauy*. 3.148, *Gellert*. 3 092, *Brisson*. 3.156,—3.184, *Muschenbroeck*. 3.138, 3.228, *Karsten*.

Chemical Characters.

Before the blowpipe, it generally decrepitates, gradually loses its colour and transparency, and melts, without addition, into a greyish-white glass. When two fragments are rubbed against each other, they become luminous in the dark. When gently heated, or laid on glowing coal, it phosphoresces, (particularly the sky-blue, violet-blue, and green varieties,) partly with a blue, partly with a green light. When brought to a red-heat, it is is deprived of its phosphorescent property. The violet-blue variety from Nertschinskoi, named *Chlorophane*, when placed on glowing coals, does not decrepitate, but soon throws out a beautiful verdigris-green and apple-green light, which gradually disappears as the mineral cools, but may be again excited, if it is heated; and this may be repeated a dozen of times, provided the heat is not too high. When the chlorophane is exposed to a red-heat, its phosphorescent property is entirely destroyed. Pallas mentions a pale violet-blue variety spotted with green, from Catharinenburg, which is so highly phosphorescent, that when held in the hand for some time, it throws out a pale whitish light; when placed in boiling water, a green light; and exposed to a higher temperature, a bright blue light. When sulphuric acid is added to heated fluor-spar, in the state of powder, a white penetrating vapour (the fluoric acid) is evolved, which has the property of corroding glass.

Constituent Parts.

	From Northumberland.	From Gersdorf.
Lime, - -	67.34	67.75
Fluoric Acid, -	32.66	32 25
	100.00	100.00
	Thomson, in Wern. Mem. vol. i. p. 11.	*Klaproth*, Beit. b. iv. s. 365.

Geognostic Situation.

It occurs principally in veins, that traverse, primitive, transition, and sometimes flœtz rocks; also in beds, associated with other minerals; in kidneys in flœtz limestone; and in drusy cavities in trap-rocks. It sometimes forms the petrifying mineral in fossil organic remains: thus

thus, it has been found penetrating and encrusting bivalve shells *; and entrochi have been described, in which the one-half was calcareous carbonate, having the natural texture of the entrochus, the other half pure violet-coloured fluor-spar †.

Geographic Situation.

Europe.—Fluor-spar is a rare mineral in Scotland, having been hitherto found only in two places, viz. near Monaltree in Aberdeenshire, where it is contained in a small vein of galena or leadglance which traverses granite; and in the island of Papa Stour, one of the Zetlands, in small quantity, in a trap rock ‡. It occurs much more abundantly in England, being found in all the galena veins that traverse the coal formation in Cumberland and Durham; in great quantities, and often associated with galena, in veins or kidney, in flœtz limestone in Derbyshire; and it is the most common vein-stone in the copper, tin, and lead veins, that traverse granite, clay-slate, &c. in Cornwall and Devonshire. It is also a frequent mineral on the Continent of Europe, being generally associated with ores of different kinds, either in beds or veins, but principally in the latter: thus, in the Saxon Erzgebirge, it occurs in beds in primitive rocks, along with tinstone: also in veins, along with silver and lead ores, and straight lamellar heavy-spar; and in veins of another formation, along with silver-ores, cobalt, and nickel. In the lower district of the Hartz, there is a remarkable formation, in great veins, which can be traced for ten or twelve miles

* Kid, vol. i. p. 74.

† Bournon describes a specimen of entrochites from Derbyshire, having the characters above described.—Vid. *Cat. Min.* p. 11.

‡ Mineralogical Travels, vol. ii. p. 207.

miles in length: in these, the fluor-spar occurs in great quantity, along with galena, iron-pyrites, copper-pyrites, much sparry iron-ore, calcareous-spar, and quartz: at Kongsberg in Norway, in veins that traverse mica-slate and hornblende-slate, along with ores of silver, lead, zinc, copper, iron, and arsenic, and the crystals are most commonly octahedral or polyhedral: in mica-slate at Jönkoping in Oeland, in Sweden: in the Bohemian and Saxon Erzgebirge, in veins, along with tinstone, arsenic-pyrites, iron-pyrites, and copper-pyrites, quartz, and apatite: in Switzerland, in very small veins, along with felspar, rock-crystal, and other minerals that characterise what are considered as veins of the oldest formation: near Regensburg in Lower Bavaria, along with quartz, inclining to calcedony, in veins traversing granite: at Freiburg in the Breisgau, in veins traversing gneiss: in newer granite, at Baveno in Italy; and it has been lately discovered imbedded in the coarse limestone which rests over chalk in the neighbourhood of Paris; and in unaltered ejected masses at Somma, near Naples. The beautiful carmine-red octahedral variety is found in the neighbourhood of Mont Blanc; the remarkably phosphorescent varieties known under the name Chlorophane, are found in Cornwall, in the mine named Pednandrae. Fluor-spar also occurs in Franconia, Austria, France, Denmark, Hessia, Silesia, Russia, Hungary, and Transylvania.

Asia.—The chlorophane variety is found at Catharinenburg and Nertschinskoi: other varieties are found in granite, in the neighbourhood of the lake Gussino-Osero, on the Mongol frontier; also at Schlangenberg, in the silver mine Zimeof, in the Altain range, &c. It is also enumerated as a production of the island of Ceylon.

America.

America.—West Greenland; California; Mexico; and in New Jersey, Connecticut, New Hampshire, and Virginia, in the United States *.

Uses.

On account of the variety and beauty of its colours, its transparency, the ease with which it can be worked, and the high polish it receives, it is cut into vases, pyramids, and other ornamental articles. The largest, and most beautiful varieties for use, are found in Derbyshire, and it is in that country that all the ornamental articles of fluor-spar are manufactured. It is also used by the metallurgist as a flux for ores, particularly those of iron and copper; and hence the name *fluor* given to it. The acid it contains, has been employed in the way of experiment for engraving upon glass.

Observations.

1. It is distinguished from *Calcareous-spar*, by its greater hardness and weight, and its not effervescing with acids: from *Gypsum*, by its superior hardness, and its decrepitating in the fire, whilst gypsum exfoliates, and becomes white; and from *Heavy-spar*, by its inferior specific gravity.

2. The red varieties have been named *False Ruby*, the yellow *False Topaz*, the green *False Emerald*, and the blue *False Sapphire* and *Amethyst*.

3. The name *Chlorophane*, given to the varieties that easily become phosphorescent, is from the green light they exhibit.

* Bruce and Barton, in American Mineralogical Journal, N. i. p. 32, 33.

Third Subspecies.

Earthy Fluor.

Erdiger Fluss, *Karsten.*

Le Fluor terreuse, *Broch.* t. i. p. 593.—Erdiger Fluss, *Reuss,* b. ii. 2. s. 378. *Id. Lud.* b. i. s. 161. *Id. Suck.* 1r th. s. 662. *Id. Leonhard,* Tabel. s. 38.—Chaux fluatée terreuse, *Brong.* t. i. p. 245. *Id. Brard,* p. 48.—Erdiger Fluor, *Haus.* s. 124. *Id. Karsten,* Tabel. s. 52.—Earthy Fluor, *Kid,* vol. i. p. 78. —Chaux fluatée terreuse, *Hauy,* Tabl. p. 9.—Fluserde, *Lenz,* b. ii. s. 829.

External Characters.

Its colours are greyish-white, and violet-blue, which is sometimes so deep as almost to appear black.

It occurs generally in crusts, investing some other mineral.

It is dull.

It is earthy.

It is friable, passing into very soft.

Constituent Parts.

It is said to be a compound of Lime and Fluoric Acid.

Geognostic and Geographic Situations.

It occurs in vein , along with fluor-spar, at Beeralston in Devonshire ; in limestone, along with fluor-spar and arragonite, in Cumberland ; at Freyberg in Saxony; and Kongsberg in Norway. At Beeralston, the white variety is regularly interposed between the layers of the octahedral

octahedral fluor, without in the least disturbing its crystallization.

Observations.

The erroneous analysis of the Earthy Phosphorite of Marmarosch by Pelletier, by which it appeared to contain 28 *per cent.* of fluoric acid, has led several authors to confound it with the earthy fluor.

XXIV. GYPSUM FAMILY.

This Family contains the following species: Gypsum, Anhydrite, Vulpinite, and Glauberite.

1. Gypsum *.

Gyps, *Werner*.

This species is divided into five subspecies, viz. Earthy Gypsum, Compact Gypsum, Fibrous Gypsum, Foliated Gypsum, and Sparry Gypsum or Selenite.

First

---

* Gypsum is from the Greek word Γυψος. The following explanation of the term γυψος, shews that it was applied by the ancients to an earthy substance that had been exposed to the action of fire: Γύψος οἱονεὶ γήεψος τὶς οὖσα· ἡ ἑψηθεῖσα γῆ (*a*): in which it corresponds with the gypsum of the moderns. The ancient naturalists sometimes seem to apply the term to sulphate of lime, the gypsum of the present day, and sometimes to a calcined carbonate of lime, or quicklime, which they called *calx*. In the following passage, it is applied to a sulphate of lime: "Cognata calci res gypsum est. Qui coquitur lapis non dissimilis alabastritæ esse debet: omnia autem optimum fieri compertum est e *lapide speculari*, squamamve talem habente (*b*):" the term *lapis specularis* applying very closely to our selenite, which is a sulphate of lime. "Gypsoma dicto statim utendum est, quoniam *celerrime* coit (*c*):" the word *celerrime* being more applicable to the comparatively rapid consolidation of calcined gypsum than to that of common mortar. There is a passage in Theophrastus, in which a ship is said to have been set on fire, in consequence of the moistening of its cargo, which consisted of gypsum and wearing-apparel: in this case, there can be little doubt, that the substance called gypsum could not have been of the same nature with the gypsum of the present day, which in no instance, perhaps, contains such a proportion of carbonate of lime, as when calcined, would be sufficient to produce this effect.—*Kid's Min.* vol. i. p. 69, 70, 71.

(*a*) Vid. Etymolog. Magn.

(*b*) Plin. Hist. Nat. lib. xxxvi.

(*c*) Plin. Hist. Nat. lib. xxxvi.

First Subspecies.

Earthy Gypsum.

Gyps-erde, *Werner.*

Gypsum terrestre farinaceum; Farina fossilis, *Wall.* t. i. p. 36. —Gypserde, *Wid.* s. 543.—Farinaceous Gypsum, *Kirw.* vol. i. p. 120.—Gypserde, *Estner,* b. ii. s. 1095 *Id. Emm.* b. i. s. 527.—Gesso terroso, *Nap.* p. 379.—Le Gypse terreux, *Broch.* t. i. p. 601.—Chaux sulphatée terreuse, *Hauy,* t. ii. p. 278.—Erdiger Gyps, *Reuss,* b. ii. 2. s. 391. *Id. Lud.* b. i. s. 163. *Id. Suck.* 1r th. s. 669. *Id. Bert.* s. 105. *Id. Mohs,* b, ii. s. 178.—Chaux sulphatée terreuse, *Lucas,* p. 13.—Erdiger Gyps, *Leonhard,* Tabel. s. 37.—Chaux sulphatée Gypse terreux, *Brong.* t. i. p. 174.—Chaux sulphatée terreuse, *Brard,* p. 52.—Erdiger Selenit, *Haus.* s. 125.—Erdiger Gyps, *Karsten,* Tabel. s. 52.—Farinaceous Gypsum, *Kid,* vol. i. p. 65. —Chaux sulphatée terreuse, *Hauy,* Tabl. p. 10.—Gypserde, *Lenz,* b. ii. s. 833.—Earthy Gypsum, *Aikin,* p. 138.

External Characters.

Its colour is yellowish-white, which passes into yellowish-grey, and sometimes inclines to snow-white.

It is composed of loose dusty particles.

It is dull.

It feels meagre, and rather rough.

It soils slightly.

It is light.

Geognostic Situation.

It is found immediately under the soil, in beds several feet thick, resting on gypsum, both of the first and se-

cond

cond flœtz formations: it also occurs in nests or cotemporaneous masses in these formations. It is conjectured to have been formed in some instances by the decay of previously existing gypsum beds; in others, it appears to be an original deposite, of cotemporaneous formation with the solid kinds of gypsum.

Geographic Situation.

It is found in Saxony, Salzburg, and Norway.

Use.

In some districts, it is used as a manure.

Observations.

It is distinguished from *Agaric Mineral* by its colour, rough feel, and its soiling feebly; and from *Earthy Heavy-spar*, by its inferior specific gravity.

Second

Second Subspecies.

Compact Gypsum.

Dichter Gyps, *Werner*.

Gypsum alabastrum, et Gypsum æquabile, *Wall.* t. i. p. 161, 162.—Dichter Gyps, *Wid.* s. 544.—Compact Gypsum, *Kirw.* vol. i. p. 121.—Dichter Gyps, *Estner*, b. ii. s. 1098. *Id. Emm.* b. i. s. 529.—Gesso compatto alabastro, *Nap.* p. 384. —Alabastrite, *Lam.* t. ii. p. 76.—Chaux sulphatée compacte, *Hauy*, t. ii. p. 266.—Le Gypse compacte, *Broch.* t. i. p. 602. —Dichter Gyps, *Reuss*, b. ii. 2. s. 393. *Id. Lud.* b. i. s. 163. *Id. Suck.* 1r th. s. 670. *Id. Bert.* s. 105. *Id. Mohs*, b. ii. s. 179. *Id. Hab.* s. 86.—Chaux sulphatée compacte, *Lucas*, p. 13.—Dichter Gyps, *Leonhard*, Tabel. s. 37.—Chaux sulphatée Gypse compacte, *Brong.* t. i. p. 174.—Chaux sulphatée compacte, *Brard*, p. 52.—Dichter Selenit, *Haus.* s. 125.— Dichter Gyps, *Karsten*, Tabel. s. 52.—Chaux sulphatée compacte, *Hauy*, Tabl. p. 10.—Dichter Gyps, *Lenz*, b. ii. s. 834. —Massive Gyspum, *Aikin*, p. 138.

External Characters.

Its colours are snow, yellowish, reddish, and greyish white; smoke, yellowish, ash, bluish and greenish grey; pale sky-blue, and violet-blue; a colour intermediate between brownish and brick red, seldom flesh-red; sometimes honey-yellow. Frequently several colours occur in the same piece, and these are either spotted, flamed, striped, or veined.

It occurs massive.

It is generally dull, seldom feebly glimmering.

The fracture is even, passing into fine splintery.

The

The fragments are indeterminate angular, and blunt-edged.

It is translucent on the edges.

It is very soft.

It is sectile.

It is easily frangible.

Specific gravity, 2.1679, before absorption of water; 2.2052, after absorption of water, *Brisson.* 1.875, *Muschenbroeck.* 1.872,—2.288, *Kirwan.*

Chemical Characters.

All the different varieties of gypsum, when exposed to heat, are deprived of their water of crystallization, become opaque, fall into a powder, which, when mixed with water, speedily hardens on exposure to the air. They are difficultly fusible before the blowpipe, without addition, and melt into a white enamel: when heated with charcoal, they are converted into sulphuret of lime.

Constituent Parts.

Lime,	34	
Sulphuric Acid,	48	
Water,	18	
	100	*Gerhard.*

Geognostic Situation.

It occurs in beds, along with granular gypsum, selenite and stinkstone, in the first flœtz formation.

Geographic Situation.

It occurs in Derbyshire; Ferrybridge in Yorkshire; and Nottinghamshire: on the Continent, in Mansfeldt, Thuringia, Bavaria, Switzerland, and France.

Third

Third Subspecies.

Fibrous Gypsum.

Fasriger Gyps, *Werner.*

Gypsum, striatum, *Wall.* t. i. p. 167.—Fasriger Gyps, *Wid.* s. 546.—Fibrous Gypsum, *Kirw.* vol. i. p. 122.—Fasriger Gyps, *Estner*, b. i. s. 1105. *Id. Emm.* b. i. s. 536.—Gesso fibroso, *Nap.* p. 386.—Chaux sulphatée fibreux, *Hauy*, t. ii. p. 266.—Le Gypse fibreuse, *Broch.* t. i. p. 604.—Fasriger Gyps, *Reuss*, b. ii. 2. s. 396. *Id. Suck.* 1r th. s. 678. *Id. Bert.* s. 106. *Id. Mohs*, b. ii. s. 182.—Chaux sulphatée fibreuse, *Lucas*, p. 13.—Fasriger Gyps, *Leonhard*, Tabel. s. 37.—Chaux sulphatée Gypse fibreux, *Brong.* t. i. p. 174.—Chaux sulphatée fibreuse, *Brard*, p. 52.—Fasriger Gyps, *Karst.* Tabel. s. 52. *Id. Haus.* s. 125.—Fibrous Gypsum, *Kid*, vol. i. p. 65.—Chaux sulphatée fibreuse-conjointe, *Hauy*, Tabl. p. 10.—Fasriger Gyps, *Lenz*, b. ii. s. 844.—Fibrous Gypsum, *Aikin*, p. 138.

External Characters.

Its principal colours are white, grey, and red: of white, it possesess the following varieties, viz. yellowish, greyish, snow, and reddish white; from reddish-white, it passes into brick-red, flesh-red, and brownish-red; the yellowish-white passes into yellowish-grey, wine-yellow, and honey-yellow.

Sometimes several colours occur together in the same specimen.

It occurs massive, and dentiform.

Its lustre passes from glistening, through shining to splendent, and is pearly.

The fracture is parallel, and sometimes curved fibrous, and passes from delicate to coarse fibrous, bordering on radiated;

radiated: sometims it is foliated in one direction, and fibrous in others.

The fragments are long splintery.

It is translucent.

It is very soft.

It is sectile.

It is very easily frangible.

Constituent Parts.

Lime, - -	27.0
Sulphuric Acid, -	41.5
Carbonate of Lime, -	9.0
Water, - -	22.0
	99.5 *John.*

This analysis is mentioned by Lenz, as having been made by Dr John; but I do not find it in any of the volumes of the Chemisches Laboratorium of that chemist.

Geognostic Situation.

It occurs both in the first and second flœtz gypsum formations of Werner, but most abundantly in the second formation: it is also a member of the newest or bony gypsum formation, that which rests on chalk, and contains bones of quadrupeds.

Geographic Situation.

It occurs in red sandstone, near Moffat; in red clay, on the banks of the Whitadder in Berwickshire; also in Cumberland, Yorkshire, Cheshire, Worcestershire, Derbyshire, Somersetshire, and Devonshire *. On the Continent of Europe, it is met with in Thuringia, Mansfeldt, Bavaria, Salzburg, Switzerland, France, &c.

Uses.

* Greenough.

Uses.

When cut *en cabachon*, and polished, it reflects a light not unlike that of the cat's-eye, and is sometimes sold as that stone. It is also cut into necklaces, ear-pendants, and crosses, and in this form, it is often sold for a harder mineral, the Fibrous Limestone, or even imposed on the ignorant for that variety of felspar named Moonstone.

Observations.

It might be confounded with Fibrous Limestone, and Asbestus, but is readily distinguished from these minerals by its inferior hardness, and the alteration it undergoes at a low red heat.

Fourth Subspecies.

Foliated Granular Gypsum.

Blættriger Gyps, *Werner.*

Gypsum lamellare, *Wall.* t. i. p. 165.—Blættriger Gyps, *Wid.* s. 548.—Granularly Foliated Gypsum, *Kirw.* vol. i. p. 123.—Blættriger Gyps, *Estner,* b. ii. s. 1109. *Id. Emm.* b. i. s. 532. —Gesso lamellare, *Nap.* p. 381.—Le Gyps lamelleux, *Broch.* t. i. p. 606.—Körniger Gyps, *Reuss,* b. ii. 2. s. 400.—Blættriger Gyps, *Lud.* b. i. s. 163. *Id. Suck.* 1r th. s. 673. *Id. Bert.* s. 107. *Id. Mohs,* b. ii. s. 180.—Körnigblättriger Gyps, *Hab.* s. 85.—Körniger Gyps, *Leonhard,* Tabel. s. 37. *Id. Karst.* Tabel. s. 52.—Schupiger Gyps, *Haus.* s. 125.—Körniger & Klein blättriger Gyps, *Lenz,* b. ii. s. 833.—Massive Gypsum, with a granularly lamellar structure, *Aikin,* p. 138.

External Characters.

Its most common colours are white, grey, and red; seldomer

seldomer yellow, brown, and black. The white colours are, snow, greyish, yellowish, and reddish white; from reddish-white, it passes into flesh-red, blood-red, and brick-red; the greyish-white passes into ash-grey, and smoke-grey, and greyish-black; and the yellowish-grey passes into wax-yellow. It seldom occurs of a hair-brown colour, and this only when it is intermixed with stink-stone. The colours sometimes occur in spotted, striped, and veined delineations.

It occurs massive; and crystallised in the following figures:

1. Common lens, in which the surface is rough.
2. Six-sided prism, flatly bevelled on the terminal planes; frequently two prisms are joined together in a determinate manner, thus forming a species of twin-crystal.

The lustre passes from shining through glistening to glimmering, and is pearly.

The fracture is perfect, and slightly curved foliated, with a single cleavage; sometimes it is narrow and short, and generally stellular radiated *.

The fragments are very blunt-edged.

It occurs in coarse, small, and fine granular distinct concretions; the radiated varieties in prismatic concretions.

It is translucent.

It is very soft.

It is sectile.

It is very easily frangible.

Specific gravity, 2.2741,—2.3108, *Brisson*.

Constituent

* The radiated varieties have been frequently confounded with the fibrous varieties, which generally occurs in a newer formation.

Constituent Parts.

Lime,	- -	32
Sulphuric Acid,	-	30
Water,	- -	38
		98 *Kirwan.*

A reddish-white variety, found near Lüneburg, according to Hausmann, afforded, besides Sulphate of Lime, 4 parts of Muriate of Lime.—*Haus.* Nord. Deutsch. Beit. st. ii. s. 98.

Geognostic Situation.

It occurs in beds in primitive rocks, as gneiss, and mica-slate: in a similar respository in transition clay-slate; but most abundantly in beds in the rocks of the flœtz class. The first or oldest flœtz gypsum formation, which is principally composed of foliated granular gypsum, and compact gypsum, also contains selenite, rounded cotemporaneous portions of radiated gypsum, stinkstone, saline or muriatiferous clay, rock-salt, and some fibrous gypsum. It is contained in the first flœtz limestone formation, or simply rests on it, and is covered either with the second sandstone or second limestone formations. In the second flœtz gypsum formation, the principal rocks are foliated granular, fibrous, and radiated gypsum, with small portions of selenite, and earthy gypsum; and it either alternates with the second sandstone formation, and its clay, or is superimposed on it. The third, and, as far as we know at present, the newest flœtz gypsum formation, contains a fine granular foliated variety, with smaller portions of fibrous gypsum, in which, in some places, bones of quadrupeds are contained; in others, crystals of boracite, small masses of rock-salt, flint, quartz, and sulphur.

Geographic

Geographic Situation.

Europe.—It occurs in Cheshire and Derbyshire; at the Segeberg, near Kiel; and at Lüneburg, where it contains crystals of boracite, and sometimes of quartz. It is associated with flœtz rocks in Thuringia, Mansfeldt, Silesia, Suabia, Bavaria, Austria, Switzerland, the Tyrol, Poland, Spain, and France. At Airolo, in the St Gothard group, it occurs in beds in gneiss; at St Meul in the Valais, it alternates with hornblende-slate; and with mica-slate on Mount Cenis: in Salzburg, it is associated with transition limestone, and clay-slate, and there it sometimes contains sparry gypsum or selenite, and also grey copper-ore, copper-pyrites, iron-pyrites, galena, and cinnabar.

Asia.—It is found in Persia, Caramania, and in different parts of Siberia, both in flœtz and primitive mountains. Pallas mentions his having met with granular gypsum, along with mica, serpentine, and felspar, in Siberia.

America.—It is found near Athapuscau Lake, where rock-salt occurs; also in the United States of America; in Nova Scotia; and at the foot of the Andes, in South America.

Uses.

The foliated and compact subspecies of gypsum, when pure, and capable of receiving a good polish, are by artists named simply *Alabaster*, or, to distinguish them from calc-sinter, or what is called calcareous alabaster, *Gypseous Alabaster*. The finest white varieties of granular gypsum, are selected by artists for statues and busts: the variegated kinds are cut into pillars, and various ornaments for the interior of halls and houses; and the most beautiful variegated sorts are cut into vases, columns, plates,

plates, and other kinds of table furniture. Those varieties that contain imbedded portions of selenite, when cut across, exhibit a beautiful iridescent appearance, and are named *Gypseous Opal.* In Derbyshire, and also in Italy, the very fine granular varieties, are cut into large vases, columns, watch-cases, plates, and other similar articles. If a lamp is placed in a vase of snow-white translucent gypsum, a soft and pleasing light is diffused from it through the apartment. It is said the ancients being acquainted with this property, used gypsum in place of glass, in order that the light in their temples might be pale and mysterious, and in harmony with the place. The *phengites* of the ancients would appear to have been foliated gypsum. According to Pliny, it was employed instead of glass in windows, on account of its translucency; and in the Temple of Fortune, and the gilded palace of Nero, the chambers were lined with the finest and most highly prized kinds of phengites. Domitian, who towards the close of his life became suspicious and distrustful of all around him, had the portico in which he used to walk lined with phengites, in order that he might see what was passing behind him. Both subspecies are used in agriculture. Much difference of opinion has prevailed among agriculturists with respect to the uses of gypsum. It is said to have been very advantageously employed in America; and also in the county of Kent; but it has failed in most of the other counties of England, though tried in various ways, and for different crops. When peat-ashes contain a considerable portion of gypsum, they may be advantageously employed as a top-dressing for cultivated grasses on such soils as contain little or no sulphate of lime. The pure white varieties of granular gypsum are used as ingredients in the composition of earthen-ware and porcelain; and the glaze or enamel with which porcelain is covered, has the purest gypsum,

 or

or even selenite, as one of its ingredients. Its most important use is in the preparation of *stucco:* for this purpose, the gypsum is first exposed to a heat sufficient to drive off its water of crystallization, then finely ground, mixed with a small portion of fine sand and quicklime, and, lastly, a determinate proportion of water is added, which occasions the compound first to swell, and then to contract and harden. Stucco is sometimes used for lining walls and roofs of apartments, in place of common plaster; and occasionally for covering the floors of summer-houses or churches. The finest kind of stucco is used for casts of figures, and statues of various kinds. Artificial coloured marbles are also made of stucco, which are used for covering pillars, walls, altars, pavements of churches, or halls.

Fifth Subspecies.

Sparry Gypsum or Selenite.

Fraueneis, *Werner.*

Gypsum selenites, *Wall.* t. i. p. 165.—Selenite, *Romé de Lisle,* t. i. p. 441.—Fraueneis, *Wern.* Cronst. s. 53.—Broad foliated Gypsum, *Kirw.* vol. i. p. 123.—Fraueneis, *Emm.* b. i. s. 540. —Chaux sulphatée cristallisée, *Hauy,* t. ii. p. 266.—La Selenite, *Broch.* t. i. p. 609.—Spathiger Gyps, *Reuss,* b. ii. 2. s. 406.—Fraueneis, *Lud.* b. i. s. 164. *Id. Suck.* 1r th. s. 675. —Grossblättriger Gyps, *Bert.* s. 109.—Fraueneis, *Mohs,* b. ii. s. 183. *Id. Hab.* s. 84.—Spathiger Gyps, *Leonhard,* Tabel. s. 37.—Chaux sulphatée Selenite, *Brong.* t. i. p. 171. —Spathiger Gyps, *Karsten,* Tabel. s. 52. *Id. Haus,* s. 124. —Selenite, *Kid,* vol. i. p. 66.—Gyps-spath, *Lenz,* b. ii. s. 840. —Durchschtiger Gyps, *Oken,* b. i. s. 400.

External Characters.

Its colours are greyish and yellowish white: from yellowish-

yellowish-white, it passes into wax, honey, and ochre yellow, and clove and yellowish brown; and it is sometimes brownish-black.

It occurs massive, coarsely disseminated; and crystallised in the following figures:

1. Six-sided prism , generally broad, and oblique angular, with two opposite broad, and four smaller lateral planes; or with two opposite very small, and four broader planes; or with alternate broader and narrower lateral planes: the terminal planes or faces are conical or spherical convex, or obtusely bevelled, and the bevelling planes set on obliquely, but parallelly on the broader lateral planes †; or acuminated with four planes, which are set on the smaller lateral planes ‡.
2. Lens.
3. Twin-crystals. These are either formed by two lenses, which are attached by their faces, or by two six-sided prisms pushed into each other in the direction of their breadth, in such a manner that the united summits at one extremity form a re-entering angle, but at the other a salient angle, or four-planed acumination. When two such twin-crystals are pushed into each other in the direction of their length, a
4. Quadruple crystal is formed.

 The

---

* The primitive figure, according to Hauy, is a four-sided prism, whose bases are oblique parallelograms, with angles of $113^{\circ}\ 7'\ 48''$ and $66^{\circ}\ 52'\ 12''$.

† Chaux sulphatée trapezienne, Hauy.

‡ Chaux sulphatée equivalente, Hauy.

The crystals occur of all degrees of magnitude, and are sometimes acicular.

The lateral planes of the prism are sometimes smooth, sometimes longitudinally streaked, and shining; the convex terminal faces, and the lens, are rough and dull.

Internally the lustre is splendent and pearly.

The fracture is generally straight, seldom curved foliated, with a threefold cleavage, of which one of the cleavages is perfect, and very distinct, and the two others imperfect, and at right angles to the former.

The fragments are rhomboidal, in which two of the sides are smooth and splendent, and four are streaked.

It sometimes occurs in granular concretions, which are large, coarse, and small, and their surfaces are generally uneven.

It alternates from semi-transparent to transparent, and in the latter case is observed to refract double.

It is very soft; yields readily to the nail.

It is sectile.

It is very easily frangible.

In thin pieces it is flexible, but not elastic.

Specific gravity, 2.322, *Muschenbröck*. 2.3065, 2.3117, 2.3846, *Kopp*.

Chemical Characters.

It exfoliates before the blowpipe, and, if the flame is directed towards the edge of the folia, it melts into a white enamel, which, after a time, falls into a white powder.

Constituent

Constituent Parts.

Lime, - - -	33.9
Sulphuric Acid, - -	43.9
Water, - - -	21.0
Loss, - - - -	2.1
	100.0

Bucholz, in Gehlen's Journ. b. v. s. 158.

Geognostic Situation.

It occurs principally in the first flœtz gypsum formation, in thin layers; less frequently in rock-salt; also in the third flœtz gypsum formation; more rarely as a constituent part of metalliferous veins; but in considerable quantity in that deposite known in the south of England under the name Blue or London Clay.

Geographic Situation.

It is not unfrequent in the blue clay in the south of England, as at Shotover Hill, near Oxford; Newhaven, Sussex; Isle of Shepey in the Medway; and at Alston in Cumberland. It occurs in the third flœtz gypsum around Paris; in veins of copper-pyrites and grey copper-ore, at Herrengrund, near Newsohl; in a vein of galena or lead-glance at Teschen, in Bohemia; and all over the Continent of Europe, and in other quarters of the globe where foliated granular gypsum occurs.

Uses.

At a very early period, before the discovery of glass, selenite was used for windows; and we are told, that in the time of Seneca, it was imported into Rome from Spain, Cyprus, Cappadocia, and even from Africa. It continued to be used for this purpose until the middle ages; for Albinus informs us, that in his time, the windows of the dome of Merseburg were of this mineral. The first greenhouses, those invented by Tiberius, were covered with selenite. According to Pliny, bee-hives were incased in selenite, in order that the bees might be seen at work. It is used for the finest kind of stucco, and the most delicate pastil-colours. When burnt, and perfectly dry, it is used for cleansing and polishing precious stones, work in gold and silver, and also pearls. It was formerly much used by Roman Catholics for *frosting* the images of the Virgin Mary: hence the names *Glacies Mariæ* or *Frauen-glas* given to it. It has also been named *Lapis specularis*, and *Gypsum speculare* and *glaciale*, from its resemblance to glass or ice.

Observations.

The Gypsum of Montmartre, near Paris, differs from the kinds already described, in containing a considerable portion of carbonate of lime: hence La Methrie, in his lately published work, entitled *Lecons de Mineralogie*, describes it as a particular species, under the name *Montmartrite*. The following description is extracted from the Lecons de Mineralogie, t. ii. p. 380.

Montmartrite.

Chaux sulphatée calcarifère, *Lucas & Hauy*.

Gypsum of Montmartre.

Its colour is yellowish.

It occurs massive, but never crystallised.

It is soft.

It effervesces with nitric acid.

The

The montmartrite is composed of gypsum and carbonate of lime. This carbonate is converted into quicklime in the furnace, and thus a kind of mortar is made: it is on this account, that the *plaster* made of this mineral may be used in work exposed to the weather; while that of pure gypsum, on exposure, soon yields to the action of rain.

The montmartrite contains, about

Sulphate of Lime, -	83
Carbonate of Lime, -	17
	100

2. Anhydrite.

Muriacit, *Werner*.

This species is divided into five subspecies, viz. Compact Anhydrite, Fibrous Anhydrite, Radiated Anhydrite, Sparry Anhydrite, and Scaly Anhydrite.

First Subspecies.

Compact Anhydrite.

Dichter Muriacit, *Werner*.

Dichter Muriacit, *Karsten*, Tabel. s. 52.—Dichter Karstenit, *Haus*. s. 124.—Chaux anhydro-sulphatée concretionnée-contournée, *Hauy*, Tabl. p. 11.—Dichter Anhydrit, *Lenz*, b. ii. s. 847.

External Characters.

Its colours are snow, greyish, and milk white, which passes into pale smalt-blue, and sky-blue; smoke-grey; flesh, brick, and brownish red.

It occurs massive, contorted, and reniform.

It is feebly glimmering, or dull

The fracture is small splintery, passing into even and flat conchoidal.

The fragments are more or less sharp-edged.

It alternates from translucent to translucent on the edges.

It scratches calcareous-spar, but is scratched by fluor-spar.

It is greyish-white in the streak.

It is rather difficultly frangible.

Specific gravity, 2.850, *Klaproth*. 2.906, *Rose*.

Constituent Parts.

Lime, -	41.48	Lime, -	42.00
Sulphuric Acid,	56.28	Sulphuric Acid,	56.50
Water, - -	0.75	Muriate of Soda,	0.25
	100.00	Loss, - -	1.28
			100.00
	Rose, in Karsten's Tabellen.		*Klaproth*, Beit. b. iv. s. 233.

Geognostic and Geographic Situations.

It occurs in beds in the salt-mines of Austria and Salzburg; and also in the first flœtz gypsum, on the eastern foot of the Hartz mountains. The contorted variety has been hitherto found only in the salt-mines of Wieliczka, and those near Bochnia in Poland, and there it is imbedded in clay.

Observations.

The contorted variety, from its resemblance to the convolutions of the intestines, used to be called *Pierre de Tripes*, and was for some time confounded with Heavy-spar.

Second

Second Subspecies.

Fibrous Anhydrite.

Fasriger Muriacit, *Werner.*

Fasriger Muriacit, *Karsten,* Tabel. s. 52.—Fasriger Karstenit, *Haus.* s. 124.—Fasriger Anhydrit, *Lenz,* b. ii. s. 848.

External Characters.

Its colours are hyacinth-red, brick-red, blood-red, and flesh-red.

It occurs massive.

Internally it is glimmering and glistening, and pearly.

The fracture is delicate and parallel fibrous.

The fragments are long splintery.

It is translucent on the edges, or feebly translucent.

It is rather easily frangible.

Geographic Situation.

It is found at Hall in the Tyrol, and at Ischel in Upper Austria.

Third Subspecies.

Radiated Anhydrite.

Strahliger Muriacit, *Karsten.*

Strahliger Karstenit, *Haus.* s. 124.—Strahliger Muriacit, *Karsten,* Tabel. s. 52.—Strahliger Anhydrit, *Lenz,* b. ii. s. 849.

External Characters.

Its colour is sometimes indigo-blue, much mixed with grey,

grey, so that it often passes into a kind of smoke-grey; sometimes the colour is intermediate between Berlin and smalt blue; and rarely, and then only in spots, it is brick-red and aurora-red.

It occurs massive.

Some parts of the fracture-surface (the radiated) are splendent and pearly, others (the splintery) are glistening.

The fracture is partly short, narrow, straight and stellular radiated, and partly splintery.

The fragments are indeterminate angular, and rather blunt-edged.

It is translucent.

It is semi-hard.

Specific gravity, 2.940, *Klaproth.* 2.900, 3.000, *Hausmann.*

Constituent Parts.

Lime, - - -	42.00
Sulphuric Acid, - -	57.00
Oxide of Iron, - -	0.10
Silica, - - -	0.25
	99.35

Klaproth, Beit. b. iv. s. 229.

Geognostic and Geographic Situations.

Along with sparry anhydrite and compact gypsum, it forms a bed in granular gypsum, at the village of Tiede, near Brunswick; and is also found at Hall in the Tyrol, Sulz on the Neckar, Carinthia, and Ischel in Austria.

Uses.

The blue varieties are sometimes cut and polished for ornamental purposes; but it is said that it does not form so good a stucco as common gypsum.

Fourth

Fourth Subspecies.

Sparry Anhydrite or Cube-Spar.

Wurfelspath, *Werner.*

Spathiger Karstenit, *Haus.* s. 124.—Spathiger Muriacit, *Karsten,* Tabel. s. 52.—Chaux sulphatée Anhydre laminaire, *Hauy,* Tabl. p. 10.—Wurfelspath, *Lenz,* b. ii. s. 946.

External Characters.

The principal colour is white, of which it exhibits the following varieties, viz. greyish, reddish, and yellowish white: besides these colours, it also occasionally exhibits the following, viz. dark ash-grey, brick, and pale rose red, pale honey-yellow, and colours intermediate between bluish-grey and violet-blue, and between brick-red and aurora-red.

It occurs massive; and crystallised in the following figures:

1. Rectangular four-sided prism *: it is sometimes so low as to form a four-sided table.
2. Broad six-sided prism.
3. Eight-sided prism; or it may be viewed as the rectangular four-sided prism, truncated on the lateral edges.
4. Broad rectangular four-sided prism, acuminated on the extremities with four planes, which are set on the lateral edges, and the apex of the acumination deeply truncated.

Externally

* According to Bournon, its primitive form is a rectangular four-sided prism, with square bases.

Externally it is shining, or splendent and pearly: internally splendent and pearly.

The fracture is foliated, with a twofold cleavage, in which the cleavages are parallel with the lateral planes of the four-sided prism.

The fragments are cubical.

It sometimes occurs in thick and straight lamellar concretions; also large granular concretions.

It alternates from transparent to strongly translucent, and refracts double.

It is semi-hard; it scratches calcareous-spar, but does not affect fluor-spar.

It is brittle.

It is very easily frangible.

Specific gravity, 2.957, *Bournon.* 2.964, *Klaproth.*

Chemical Characters.

When exposed to the blowpipe, it does not exfoliate, and melt like gypsum, but becomes glazed over with a white friable enamel.

Constituent Parts.

According to Klaproth, it is a Sulphate of Lime, without water, and with a slight admixture of Muriate of Soda.

Geognostic and Geographic Situations.

It is sometimes met with in the gypsum of Nottinghamshire *. In the salt-mines of Hall in the Tyrol; in those of Bex in Switzerland; in quartz, along with talc, sulphur, and iron-pyrites, in the mine of Pesay, also in Switzerland.

Fifth

* Greenough.

Fifth Subspecies.

Scaly Anhydrite.

Anhydrit, *Werner.*

Schupiger Muriacit, *Karsten.*

Schuphiger Karstenit, *Haus.* s. 124.—Schuphiger Muriacit, *Karsten,* Tabel. s. 52.—Chaux sulphatée Anhydre lamellaire, *Hauy,* Tabl. p. 10.—Schuppiger Anhydrit, *Lenz,* b. ii. s. 849.

External Characters.

Its colours are snow, greyish, and milk white, which latter passes into smalt-blue, and rarely into grey.

It occurs massive.

The lustre is splendent and pearly.

The fracture is confused foliated.

It occurs in fine granular scaly concretions.

The fragments are not particularly blunt-edged.

It is translucent on the edges.

It is easily frangible.

It is soft?

Specific gravity, 2.957, *Hauy.*

Constituent Parts.

Lime, - -	41.75
Sulphuric Acid, -	55.00
Muriate of Soda, - -	1.00
	97.75

Klaproth, Beit. b. iv. s. 235.

Geognostic and Geographic Situations.

It is found in the salt-mines of Hall in the Tyrol, 5088 feet above the level of the sea.

3. Vulpinite.

3. Vulpinite.

Vulpinite, *De La Methrie*, Tableaux.—Chaux sulphatée quartzifere, *Hauy*, t. iv. p. 355.—Pierre de Vulpino, dans le Bergamasce, *Fleuriau de Bellevue*.—Chaux anhydro-sulphatée quartzifere, *Hauy*, Tabl. p. 11.—Vulpinit, *Lenz*, b. ii. s. 851.

External Characters.

Its colour is greyish-white, and veined with bluish-grey.

It occurs massive.

Internally it is splendent.

The fracture is foliated, and it is said to exhibit a threefold slightly oblique cleavage.

The fragments are rhomboidal.

It occurs in granular distinct concretions.

It is translucent on the edges.

It is soft.

It is brittle.

It is easily frangible.

Specific gravity, 2.878, *Hauy*.

Chemical Characters.

It melts easily before the blowpipe into a white opaque enamel; and becomes feebly phosphorescent when thrown on glowing coals.

Constituent

Constituent Parts.

Sulphate of Lime, -	92.0
Silica, - - -	8.0
	100.0

Vauquelin, in Bulletin des Sciences de la Société Philomatique, N. 9.; Journal de Physique, t. xlvii. p. 101.; Journal des Mines, N. xxxiv.

Geognostic and Geographic Situations.

It occurs along with granular foliated limestone, and is sometimes associated with quartz, and occasionally with sulphur. It is found at Vulpino in Italy.

Uses.

1. It takes a very fine polish, and is employed by the statuaries of Bergamo and Milan for making slabs, chimney-pieces, &c. It is known to artists by the name *Marmo bardiglio di Bergamo*.

2. It was first particularly noticed by Fleuriau.

4. Glauberite.

Glauberite, *Brongniart.*

Glauberite, *Brong.* Journal des Mines, t. xxiii. p. 5. *Id. Hauy,* Tabl. p. 23. *Id. Lenz,* b. ii. s. 950. *Id. Aikin,* p. 139.

External Characters.

Its colours are greyish-white, and wine-yellow.

It occurs crystallised, in very low oblique four-sided prisms, the lateral edges of which are 104° 28′ and

75°

75° 32, and in which the terminal planes are set on obliquely.

The crystals occur singly, or in groups.

The lateral planes are transversely streaked; the terminal planes are smooth.

It is shining.

The fracture parallel with the terminal planes and edges is foliated; in other directions it is conchoidal.

It is softer than calcareous-spar.

It is transparent.

It is brittle.

Specific gravity, 2.700.

Chemical Characters.

It decrepitates before the blowpipe, and melts into a white enamel. In water, it becomes opaque, and is partly soluble.

Constituent Parts.

Dry Sulphate of Lime,	49.0
Dry Sulphate of Soda,	51.0
	100.0

Brongniart, J. des Mines, t. xxiii. p. 17.

Geognostic and Geographic Situations.

It is found imbedded in rock-salt at Villaruba, near Ocana in New Castile, in Spain.

Observations.

It was brought from Spain to Paris by M. Dumeril, and first analysed and described by Brongniart.

XXV. BORACITE

XXV. BORACITE FAMILY.

This Family contains the following species, viz. Datolite, Botryolite, and Boracite.

1. Datolite.

Datholit, *Werner*.

Chaux Datholite, *Brong*. t. ii. p. 397. *Id. Haus.* s. 123. *Id. Karst* Tabel. s. 17.—Chaux boratée silicieuse *Hauy*, Tabl. p. 17.—Datolit, *Lenz*, b. ii. s. 859. *Id. Aikin*, p. 127.

External Characters.

Its colours are greyish-white milk-white, greenish-white, greenish-grey; which latter inclines to celandine-green, and rarely to muddy honey-yellow.

It occurs massive; and crystallised in the following figures.

1. Oblique four-sided prism *, which is very rarely perfect, generally truncated on the angles and lateral edges; sometimes also bevelled on the lateral edges. When the truncations on the angles becomes so large that they meet, there is formed
2. An oblique four-sided prism, flatly acuminated with four planes, which are set on the lateral edges.

The crystals are small, seldom middle-sized, and form druses: they are sometimes rough, sometimes smooth, and splendent.

 Internally

* According to Hausmann, the primitive figure is an oblique four-sided prism, having lateral edges of 77° 30′ and 102° 30.

Internally it is shining, inclining to glistening, and the lustre is resinous.

The fracture is small grained uneven, which sometimes approaches to splintery, and small conchoidal *.

The fragments are indeterminate angular, and rather sharp-edged.

It occurs in large, coarse, and small angulo-granular distinct concretions, which are easily separable, and sometimes approach to the prismatic and wedge shape.

It is generally translucent, and sometimes transparent.

It is semi-hard; it scratches glass with difficulty.

It is very brittle.

It is difficultly frangible.

Specific gravity, 2.980, *Klaproth.*

Chemical Characters.

When exposed to the flame of a candle, it becomes opaque, and may then be easily rubbed down between the fingers. Before the blowpipe, it intumesces into a milk-white coloured mass, and then melts into a globule of a pale rose colour.

Constituent Parts.

Silica, -	36.50	Silica, -	37.66
Lime, -	35.50	Lime, -	34.00
Boracic Acid,	24.00	Boracic Acid,	21.67
Water, -	4.00	Water, -	5.50
Trace of Iron and Manganese.		Loss, -	1.17
	100.00		100.00
	Klaproth, Beit. b. iv. s. 359.		*Vauquelin,* in Lucas's Tab. Meth. II. 71.

Geognostic

* Hausmann is of opinion, that a double cleavage occurs, in which the cleavages are parallel with the sides of the oblique four-sided prism.

Geognostic and Geographic Situations.

It is associated with large foliated granular calcareous spar, more rarely with violet-blue fluor-spar, and sometimes with apple-green prehnite, with which it occurs in beds, in a variety of mica-slate, subordinate to gneiss, at Arendal in Norway. It has also been found in small veins in greenstone on the Geisalpe, near Sonthofen.

Observations.

1. It was first observed and described by Esmark, who met with it at Arendal.

2. It resembles Prehnite, both in external aspect and in geognostic situation.

2. Botryolite.

Botryolith, *Hausmann.*

This species is divided into two subspecies, viz. Fibrous Botryolite, and Earthy Botryolite.

First Subspecies.

Fibrous Botryolite.

Fasriger Botryolith, *Hausmann.*

Fasriger Botryolith, *Haus.* s. 122.—Chaux boratée silice concretionnée-mamellonné, *Hauy,* Tabl. p. 17.—Fasriger Botryolit, *Lenz,* b. ii. s. 858.—Botryolite, *Aikin,* p. 127.

External Characters.

Externally it is pearl-grey, and yellowish-grey; internally greyish, milk, and reddish white, which passes into pale rose-red. The colours are in concentric stripes.

 It

It occurs generally small, seldom large botryoidal, sometimes approaching to reniform.

The surface is granulated or rough, and dull.

Internally it is glimmering and pearly.

The fracture is very delicate and stellular fibrous, and sometimes passes into splintery.

It occurs in thin and concentric lamellar distinct concretions.

It is translucent on the edges.

It is semi-hard, approaching to soft.

It is brittle.

It is not particularly heavy.

Specific gravity, 2.885, *Klaproth.*

Chemical Characters.

It intumesces and melts into a white glass before the blowpipe.

Constituent Parts.

Silica, - - -	36.00
Lime, - - -	39.50
Boracic Acid, - -	13.50
Oxide of Iron, - -	1.00
Water, - - -	6.50
	96.50

Klaproth, Beit. b. v. s. 125.

Geognostic and Geographic Situations.

It occurs in the Kjenlie mine, near Arendal in Norway, along with common quartz, schorl, calcareous-spar, iron-pyrites, and magnetic ironstone, in a bed in gneiss.

Observations.

Observations.

1. It is distinguished by its colours, botryoidal shape, concentric curved lamellar concretions, delicate and stellular fibrous fracture, its pearly lustre, and specific gravity.

2. It was first described by Abildgaard of Copenhagen, under the name *Semi-globular Zeolite.* Its chemical properties were first noticed by Esmark, who, from the effects produced on it by the blowpipe, conjectured that it contained boracic acid. This conjecture was confirmed by the experiments of Gahn and Hausmann, who discovered, that, like datolite, it contained boracic acid, lime, and silica; and Hausmann, on account of its botryoidal shape, gave it the name it now bears.

Second Subspecies.

Earthy Botryolite.

Erdiger Botryolith, *Hausmann.*

Erdiger Botryolith, *Haus.* s. 122. *Id. Lenz,* b. ii. s. 859.

External Characters.

Its colour is snow-white.
It is small botryoidal.
It is dull.
The fracture is earthy.

Geognostic and Geographic Situation.

It occurs along with the fibrous subspecies.

 3. Boracite.

3. Boracite.

Boracit, *Werner*.

Boracit, *Wid.* s. 533. *Id. Kirw.* vol. i. p. 172. *Id. Estner,* b. ii. s. 1061. *Id. Emm.* b. i. s. 509. *Id. Nap* p. 370. *Id. Broch.* t. i. p. 589. *Id. Hauy,* t. ii. p. 337. *Id Reuss,* b. ii. 2. s. 372. *Id. Lud.* b. i. s. 160. *Id. Suck.* 1r th. s. 578. *Id. Bert.* s. 137. *Id. Mohs,* b. ii. s. 232.—Magnesie boratée, *Lucas,* p. 20.—Borazit, *Leonhard,* Tabel. s. 41.—Magnesie boratée, *Brong.* t. i. p. 167. *Id. Brard,* p. 68.—Boracit, *Karsten,* Tabel. s. 48. *Id. Haus.* s. 120. *Id Kid,* vol. i. p. 118.—Magnesie boratée, *Hauy,* Tabl. p. 16.—Borazit, *Lenz,* b. ii. s. 855. *Id. Oken,* b. i. s. 399. *Id. Aikin,* p. 127.

External Characters.

Its colours are greyish and yellowish white, which passes into yellowish-grey, ash-grey, smoke-grey, and pale greenish-grey.

It has been hitherto found only in a regularly crystallised state, in the following figures:

1. Cube.
   - *a.* Perfect.
   - *b.* More or less deeply truncated on all the angles and edges *.
   - *c.* The edges only truncated.
   - *d.* The angles only truncated.
   - *e.* Truncated on all the edges; but only the alternate angles truncated †.

The crystals are seldom middle-sized, more generally small, and very small, and very rarely two cubes are grown together.

The

* Magnesie boratée surabondante, Hauy.

† Magnesie boratée defective, Hauy.

The external surface is generally rough, seldom smooth, when it is shining or splendent, and adamantine.

Internally the lustre is shining, inclining to glistening, and is adamantine, inclining to resinous.

The fracture is intermediate between small and fine grained uneven and imperfect conchoidal, sometimes concealed foliated, with a threefold cleavage, the folia parallel with the sides of the cube.

The fragments are indeterminate angular, and not particularly sharp-edged.

It is sometimes translucent, sometimes transparent.

It is semi-hard in a high degree; it scratches glass.

It is brittle.

It is easily frangible.

Specific gravity, 2.5662, *Westrumb.* 2.911, *Karsten.*

Physical Characters.

It is pyro-electric on all the angles, those that are diagonally opposite being one positive, and the other negative.

Chemical Characters.

Before the blowpipe, it is fusible, with ebullition, into a yellowish enamel.

Constituent Parts.

	From Luneburg.	From Segeberg.	From Luneburg.
Lime, -	11.00		
Magnesia,	13.50	36.3	16.6
Alumina,	1.00		
Silica, -	2.00		
Oxide of Iron,	0.75		
Boracic Acid,	68.00	63.7	83.4
	96.25	100	100.0
	Westrumb.	*Pfaff.*	*Vauquelin.*

Vauquelin found no lime in the transparent crystals, but only magnesia: hence he is of opinion, that boracite is a simple borate of magnesia. Nearly the same result was obtained by Pfaff, in his analysis of the boracite of the Segeberg.

Geognostic and Geographic Situations.

This curious mineral has been hitherto found only in rocks of gypsum in the Kalkberg, at Luneburg in Hanover; and also in the same formation, in the Segeberg, near Kiel in Holstein.

Observations.

1. Lazius, who first attended to this mineral, named it *Cubic Quartz:* Westrumb found by chemical analysis that it contained boracic acid and earths, and named it *Sedative-spar;* and Werner gave it its present denomination.
2. Fluor-spar is sometimes cut into the form of boracite crystals, and sold as such to the ignorant.

XXVI. BARYTE

XXVI. BARYTE FAMILY.

This Family contains the following species, viz. Witherite, Heavy-Spar, Hepatite, Strontian, and Celestine.

1. Witherite.

Witherit, *Werner.*

Baryt aërée, *De Born,* t. i. p. 267.—Witherit, *Wid.* s. 554.—Barolite, *Kirw.* vol. i. p. 134.—Luft oder Kohlensaurer Baryt, *Estner,* b. ii. s. 1124.—Witerite, *Nap.* p. 387. *Id. Lam.* t. ii. p. 20.—Baryte carbonatée, *Hauy,* t. ii. p. 309.—La Witherite, *Broch.* t. i. p. 613.—Witherit, *Reuss,* b. ii. 2. s. 430 *Id. Lud.* b. i. s. 167. *Id. Suck.* 1r th. s. 693. *Id. Bert.* s. 120. *Id. Mohs,* b. ii. s. 200. *Id. Hab.* s. 93.—Baryte carbonatée, *Lucas,* p. 16.—Witherite, *Leonhard,* Tabel. s. 39.—Baryte carbonatée, *Brong.* t. i. p. 255. *Id. Brard,* p. 60.—Witherit, *Karsten,* Tabel. s. 54. *Id. Haus.* s. 132. *Id. Kid,* vol i. p. 86.—Baryt carbonatée, *Hauy,* Tabl. p. 13.—Witherit, *Lenz,* b. ii. s. 881. *Id. Oken,* b. i. s. 412. *Id. Aikin,* p. 144.

External Characters.

Its colours are greyish and yellowish white, also pale bluish-grey, and greenish-grey, yellowish-grey, pale wine-yellow. and pale flesh-red.

It occurs massive, disseminated, in crusts, cellular, corroded, and crystallised in the following figures:

1. Six-sided prism *, acuminated with six planes, which are set on the lateral planes.

2. Preceding

* Its primitive figure is a rhomboid of 88° 6′ and 91° 54′.

2. Preceding figure, in which the summit of the acumination is more or less deeply truncated.
3. N° 1. in which the edges between the lateral and acuminating planes are truncated.
4. Double six-sided pyramid.

The crystals are middle-sized, small, and very small.

Externally it is glistening; internally it is shining on the principal fracture, and glistening on the cross fracture, and the lustre is resinous.

The principal fracture is intermediate between floriform foliated, and narrow scopiform radiated; the cross fracture uneven, inclining to splintery.

The fragments are wedge-shaped, or indeterminate angular.

The massive varieties occur in thin wedge-shaped distinct concretions, which are often very much grown together, and occasionally pass into coarse granular.

It is translucent, rarely semi-transparent.

It is soft, inclining to semi-hard; scratches calcareous-spar, but is scratched by fluor-spar.

It is brittle.

It is easily frangible.

It is heavy.

Specific gravity, 4.271, *Lichtenberg*. 4.361, *Karsten*.

Chemical Characters.

Before the blowpipe it decrepitates slightly, and melts readily into a white enamel; it is soluble, with effervescence, in diluted muriatic or nitric acid.

Constituent

Constituent Parts.

Carbonate of Barytes,	98.246	Barytes,	74.5	Barytes,	79.66
Carbonate of Strontian,	1.703	Carbonic acid,	22.5	Carbonic acid,	20.00
Alumina, with Iron,	0.043		100.0	Water, -	0.33
Carbonate of Copper,	0.008	*Vauquelin.*			99.99
	100.000			*Buchols*, in Beitr. z. Chem. I. iv.	
Klaproth, Beit. b. ii. s. 86.					

Geognostic and Geographic Situations.

It occurs in Cumberland and Durham, in lead-veins that traverse a flœtz limestone, which rests on red sandstone, and in these it is associated with coralloidal arragonite, brown-spar, earthy fluor-spar, heavy-spar, and galena or leadglance, white lead-ore, green lead-ore, copper-pyrites, azure copper-ore, malachite, iron-pyrites, sparry iron-ore, calamine, and blende. In these counties, it is met with at Aldstone in Cumberland; Arkendale, Welhope, and Dufton in Durham. It also occurs at Merton Fell in Westmoreland; Snailback mine in Shropshire; and at Anglesark in Lancashire, in a vein of galena, along with heavy-spar. It is associated with sparry ironstone near Steinbauer, not far from Neuberg and Mariazell in Stiria: in the Leogang in Salzburg, along with sparry ironstone, and copper-pyrites.

Asia.—At Schlangenberg and Zincof, in the Altain Mountains.

Uses.

It is a very active poison, and in some districts, as in Cumberland, it is employed for the purpose of destreying rats. When dissolved by muriatic acid, the solution of muriate of barytes thus obtained, is said to prove serviceable in scrofula.

Observations.

Observations.

It is distinguished from *Heavy-spar*, by its dissolving in nitrous acid; from *Celestine*, by the same property.

2. Heavy-Spar.

Schwerspath, *Werner.*

This species is divided into eight subspecies, viz. Earthy Heavy-Spar or Heavy-Spar Earth, Compact Heavy-Spar, Granular Heavy-Spar, Lamellar Heavy-Spar, Fibrous Heavy-Spar, Radiated Heavy-Spar or Bolognese Spar, and Prismatic Heavy-Spar.

First Subspecies.

Earthy Heavy-Spar, or Heavy-Spar Earth.

Schwerspath Erde, *Werner.*

Baryta vitriolée terreuse, *De Born,* t. i. p. 268.—Schwerspatherde, *Wid.* s. 558.—Earthy Baroselenite, *Kirw.* vol. i. p. 138.—Schwerspath-erde, *Estner,* b. ii. s. 1143. *Id. Emm.* b. i. s. 550.—Baryta vitriolata terrea, *Nap.* p. 402.—Le spath pesant terreux, *Broch.* t. i. p. 617.—Erdiger Baryt, *Reuss,* b. ii. 2. s. 437. *Id. Lud.* b. i. s. 168. *Id. Suck.* 1r th. s. 697. *Id. Bert.* s. 122.—Baryterde. *Mohs,* b. ii. s. 106.—Erdiger Baryt, *Leonhard,* Tabel. s. 39.—Baryte sulphatée terreux, *Brong.* t. i. p. 252.—Erdiger Baryt, *Haus.* s. 134. *Id. Karsten,* Tabel. s. 54.—Earthy Sulphate of Baryt, *Kid,* vol. i. p. 87.—Baryterde, *Lenz,* b. ii. s. 389.

External Characters.

Its colours are snow, greyish, yellowish, and reddish white;

white; and sometimes incline to pale yellowish-grey, and straw-yellow.

It occurs sometimes loose, sometimes cohering.

It is composed of dull or glimmering dusty particles, which feel meagre, and are heavy.

Constituent Parts.

It is Sulphate of Barytes.

Geognostic and Geographic Situations.

It occurs in drusy cavities in veins of heavy-spar, in Staffordshire and Derbyshire; at Freyberg in Saxony, and Mies in Bohemia; and it occurs in veins, or in large nests, in marl, at Canstein in Westphalia.

Observations.

1. It is distinguished from all other earthy minerals, by its great specific gravity.

2. Some mineralogists are of opinion, that is disintegrated compact Heavy-spar; and others maintain that it is an original formation; which latter opinion is countenanced in those instances where the earthy heavy-spar occurs in close cavities.

Second Subspecies.

Compact Heavy-Spar.

Dichter Schwerspath, *Werner*.

Baryte vitriolata compacte, *De Born,* t. i. p. 268.—Dichter Schwerspath, *Wid.* s. 559.—Compact Baroselenite, *Kirn.* vol. i. p. 138.—Dichter Schwerspath, *Estner,* b. ii. s. 1146. *Id. Emm.* b. i. s. 552.—Barite vitriolata compatta, *Nap.* p. 400.—Le Spath pesant compacte, *Broch.* t. i. p. 618.—Dichter Schwerspath, *Reuss,* b. ii. 2. s. 438. *Id. Lud.* b. i. s. 169. *Id. Suck.* 1r th. s. 698. *Id. Bert.* s. 123. *Id. Mohs,* b. ii. s. 206. *Id. Leonhard,* Tabel. s. 39.—Baryte sulphatée compacte, *Brong.* t. i. p. 252.—Dichter Baryte, *Karsten,* Tabel. s. 54. *Id. Haus.* s. 133.—Baryte sulphatée compacte, *Hauy,* Tabl. p. 13.—Dichter Baryte, *Lenz,* b. ii. s. 890.

External Characters.

Its colours are yellowish and reddish white; yellowish and smoke grey; cream and ochre yellow; and pale flesh-red.

It occurs massive, disseminated, reniform, semi-globular, tuberose, with cubic impressions, and often marked with dendritic delineations.

Internally it is glimmering.

The fracture is coarse earthy, passing into fine-grained uneven: sometimes it is imperfect foliated.

The fragments are indeterminate angular, and blunt-edged.

It occurs in thick curved lamellar concretions.

It is opaque or translucent on the edges.

It is soft.

It is rather sectile.

It is easily frangible.

Specific gravity, 4.484.

Constituent

Constituent Parts.

Sulphate of Barytes, -	83.0
Silica, - - -	6.0
Alumina, - - -	1.0
Water, - - -	2.0
Oxide of Iron, - -	4.0
	96.0

Westrumb, in Bergbaukunde, II. s. 47.

Geognostic and Geographic Situations.

It is found in the mines of Staffordshire and Derbyshire, where it is named *Cawk*. It also occurs at Meis in Bohemia, Freyberg in Saxony, in the Hartz, and the Breisgau: also in clay-slate, near Servos in Savoy; and in Austria and Stiria.

Third Subspecies.

Granular Heavy-Spar.

Körniger Schwerspath, *Werner*.

Blättriger Schwerspath, *Wid.* s. 561.—Körniger Schwerspath, *Emm.* b. i. s. 556.—Le Spath pesant grenue, *Broch.* t. i. p. 620.—Körniger Baryt, *Reuss*, b. ii. 2. s. 441. *Id. Lud.* b. i. s. 169. *Id. Suck.* 1r th. s. 701. *Id. Bert.* s. 124. *Id. Mohs*, b. ii. s. 206. *Id. Leonhard*, Tabel. s. 39.—Baryte sulphatée grenue, *Brong.* t. i. p. 253.—Schuppiger Baryt, *Haus.* s. 133.—Körniger Baryt, *Karsten*, Tabel. s. 54.—Baryte sulphatée granulaire, *Hauy*, Tabl. p. 13.—Körniger Baryt, *Lenz*, b. ii. s. 891.—Granular Heavy-Spar, *Aikin*, p. 145.

External Characters.

The colours are snow, yellowish, greyish, and reddish white,

white, and sometimes dark ash-grey. It is occasionally spotted brown and yellow on the surface.

It occurs massive.

Internally it is glistening, approaching to shining, and is pearly.

The fracture is foliated, sometimes passing into splintery.

The fragments are indeterminate angular, and blunt-edged.

It occurs in distinct concretions, which are generally fine, seldom small granular; sometimes so minute as scarcely to be discernible.

It is feebly translucent.

It is semi-hard, inclining to soft?

It is rather brittle, and easily frangible.

It is heavy.

Specific gravity, 4.380, *Klaproth.*

Constituent Parts.

Barytes, - - -	60
Dry Sulphuric Acid, -	30
Silica, - - -	10
	100

Klaproth, Beit. b. ii. s. 72.

Geognostic Situation.

It occurs principally in beds and lying masses, along with galena, blende, copper-pyrites, and iron-pyrites.

Geographic Situation.

It occurs in beds, along with galena, blende, copper-pyrites, and iron pyrites, at Peggau in Stiria; also in the Hartz, in beds, along with copper and iron pyrites, ga-

lena,

lena, and blende; and at Schlangenberg in Siberia, where it is associated with copper-green, and native copper.

Observations.

1. It bears a striking resemblance to Foliated Granular Limestone, from which, however, it is distinguished by the following characters:

1st, It has a lower degree of lustre.

2d, When the distinct concretions are of the same size as in the foliated granular limestone, they are not so well defined.

3d, It is more easily broken in pieces than the limestone, owing to the concretions being less intimately connected together.

4th, It is much heavier.

2. Foliated Granular Limestone, Granular Heavy-spar, and Granular Gypsum, may be distinguished from each other by the relative distinctness of the concretions: in the Foliated Granular Limestone they are well defined; in Granular Heavy-spar less so; and in Granular Gypsum still more indistinct.

Fourth Subspecies.

Lamellar Heavy-Spar.

This subspecies is divided into three kinds, viz. Curved Lamellar Heavy-Spar, Straight Lamellar Heavy-Spar, and Disintegrated Heavy-Spar.

First Kind.

Curved Lamellar Heavy-Spar.

Krummschaaliger Schwerspath, *Werner.*

Le Spath pesant testacé courbe, ou le Spath lamelleux, *Broch.* t. i. p. 621.—Krummschaaliger Baryt, *Reuss,* b. ii. 2. s. 443. *Id. Lud.* b. i. s. 170. *Id. Suck.* 1r th. s. 700. *Id. Bert.* s. 124. *Id. Mohs,* b. ii. s. 207. *Id. Leonhard,* Tabel. s. 40. —Blättriger Baryt, *Karsten,* Tabel. s. 54.—Baryt sulphatée cretée, *Hauy,* Tabl. p. 13.—Krummschaaliger Baryt, *Lenz,* b. ii. s. 892.

External Characters.

Its principal colours are white, grey, and red: the white varieties are yellowish, greyish, and reddish white; the grey varieties are smoke and pearl grey, and there is a transition from pearl-grey into flesh-red and blood-red, and from yellowish-grey into yellowish-brown.

Sometimes several colours occur together, and are arranged in broad stripes.

It generally occurs massive, frequently also reniform, cellular, and globular, with a drusy surface: the drusy surface is formed of very small, thin, and longish four-sided tables.

Internally it is intermediate between shining and glistening, and the lustre is pearly.

The fracture is curved foliated, which sometimes inclines to splintery, and thus approaches to the compact subspecies.

The fragments are indeterminate angular, and rather blunt-edged.

It occurs in distinct concretions, which are curved and thick lamellar.

It is translucent on the edges.

It

It is soft.
It is brittle.
It is easily frangible.
It is heavy.

Geognostic and Geographic Situations.

It is one of the most common subspecies of heavy-spar. In Scotland, it occurs in trap and sandstone rocks: in Derbyshire, it occurs in flœtz limestone: it characterises a particular venigenous formation at Freyberg in Saxony, where it is associated with radiated pyrites, argentiferous galena, brown blende, calcareous-spar, and fluor-spar. It occurs in Sweden, Carinthia, &c.

Second Kind.

Straight Lamellar Heavy-Spar.

Geradschaaliger Schwerspath, *Werner.*

Gypsum spathosum, *Wall.* t. i. p. 168.—Spath pesant ou seleniteux, *Romé de Lisle,* t. i. p. 577.—Baryte vitriolée spathique, *De Born,* t. i. p. 270.—Var. of Blättriger Schwerspath, *Wid.* s. 561.—Gemeiner Schwerspath, *Emm.* b. i. s. 557.—Baryta vitriolata lamellare, *Nap.* p. 395.—Foliated Baroselenite, *Kirw.* vol. i. p. 140.—Le Spath pesant testacé à lames droites, ou Le Spath pesant commun, *Broch.* t. i. p. 624.—Geradschaaliger Baryt, *Reuss,* b. ii. 2. s. 445.—Frischer Geradschaaliger Baryt, *Lud.* b. i. s. 170. *Id. Suck.* 1r th. s. 702. *Id. Bert.* s. 125. *Id. Mohs,* b. ii. s. 209. *Id. Leonhard,* Tabel. s. 40.—Baryte sulphatée pure crystallisée, *Brong.* t. i. p. 250.—Gemeiner Baryt, *Karst.* Tabel. s. 54. *Id. Haus.* s. 133.—Baryte sulphatée en formes determinables, *Hauy,* Tabl. p. 12.—Geradschaaliger Baryt, *Lenz,* b. ii. s. 894.

External Characters.

Its colours are snow, milk, reddish, yellowish, and

 greenish

greenish white; greyish, ash, smoke, and bluish-grey; greyish-black; smalt-blue, pale sky-blue, and muddy indigo-blue; verdigris-green and olive-green; cream, honey, wax, and wine yellow; brick-red, blood-red, and brownish-red; and yellowish-brown.

It occurs generally massive; and also crystallised in the following figures:

1. Rectangular four-sided table *.
   a. Perfect.
   b. In which the terminal planes are bevelled †, fig. 140.
   c. In which the angles of the bevelment are truncated ‡, fig. 141.
2. Oblique four-sided table.
   a. Perfect.
   b. Truncated on the lateral edges.
   c. Truncated on the acute terminal edges, and sometimes also on the acute angles.
   d. Truncated on the obtuse angles ||.
   e. Truncated on the obtuse angles and terminal edges.
   f. Bevelled on the obtuse terminal edges.
3. Longish six-sided table.
   a. Perfect §, fig. 142.
   b. Bevelled on the terminal planes.
   c. Bevelled on the terminal and lateral edges.

d. With

---

* The primitive form, according to Hauy, is a rectangular prism, whose bases are rhombs, with angles of 101° 30′ and 78° 30′.

† Baryte sulphatée trapezienne, Hauy.

‡ Baryte sulphatée epointée, Hauy.

|| Baryte sulphatée apophane, Hauy.

§ Baryte sulphatée retrecée, Hauy.

d. Bevelled on the lateral planes, and truncated on the bevelling edges.

e. The lateral edges acutely bevelled, and the acute angles truncated.

4. Eight-sided table.

a. Perfect.

b. Bevelled on the terminal planes.

c. Slightly truncated on the lateral and terminal edges.

d. Bevelled on the lateral and terminal planes.

The crystals vary in size, from large to small; and rest on one another, or intersect one another.

Externally they are smooth and splendent; internally shining and splendent, and the lustre intermediate between resinous and pearly.

The fracture is perfect foliated, with a threefold cleavage.

The fragments are intermediate between cubical and rhomboidal.

The massive variety occurs in straight and thin lamellar distinct concretions, which present in profile a radiated appearance.

It is translucent, or transparent, and refracts double.

It scratches calcareous-spar, but is scratched by fluor-spar.

It is brittle.

It is easily frangible.

Specific gravity, 4.300 to 4.700.

Chemical Characters.

It decrepitates briskly before the blowpipe, and, by continuance of the heat, melts into a hard white enamel. A piece exposed for a short time to the blowpipe, and then laid on the tongue, gives the flavour of sulphureted hydrogen, or rotten eggs. When pounded, and thrown on glowing coals, it phosphoresces with a yellow light.

Constituent Parts.

Sulphate of Barytes, - -	97.60
Sulphate of Strontian, - -	0.85
Water, - - - -	0.10
Oxide of Iron, - -	0.80
Alumina, - - -	0.05

Klaproth, Beit. b. ii. s. 78.

Geognostic Situation.

It is found almost always in veins, which occur in granite, gneiss, mica-slate, clay-slate, grey-wacke, limestone, and sandstone. It is often accompanied with ores, particularly the flesh-red variety, and these are, native silver, silver-glance or sulphureted silver, copper-pyrites, lead-glance, white cobalt ore, light red silver-ore, native arsenic, earthy cobalt-ore, cobalt-bloom or red cobalt, antimony, and manganese. It occurs sometimes in beds, and encrusting the walls of drusy cavities.

Geographic Situation.

It occurs in veins in the different primitive rocks in this island; and in a similar repository in transition rocks, and also in flœtz limestone, sandstone, and trap. Beautiful crystallised varieties are found in the lead-mines of Cumberland, Durham, and Westmoreland. It is very frequent on the Continent of Europe, and also in America, particularly in mining districts.

Uses.

Uses.

It is said to form a good manure for clover fields: when burnt, and finely ground, is used in place of bone-ashes for cupels; the white varieties are employed as white colours in painting, and for pastil-pencils; and it is sometimes used as a flux for ores of particular kinds.

Third Kind.

Disintegrated Heavy-Spar.

Mulmicher oder mürber geradschaaliger Schwerspath, *Werner*.

Mulmiger Baryt, *Reuss*, b. ii. 2. s. 455. *Id. Leonhard*, Tabel. s. 40. *Id. Karsten*, Tabel. s. 54.—Aufgelöster Baryt, *Lenz*, b. ii. s. 899.

External Characters.

Its colours are greyish, greenish, yellowish, and reddish white.

It occurs massive.

It is glistening.

The fracture and concretions the same as in the preceding species.

It is opaque, or faintly translucent on the edges.

It is soft, passing into very soft.

It is very easily frangible.

It is pearly.

Geognostic and Geographic Situations.

It was formerly met with in considerable quantity at Freyberg in Saxony, in a mixture of galena, blende, and iron-pyrites.

Fifth Subspecies.

Fibrous Heavy-Spar.

Fasriger Schwerspath, *Werner.*

Fasriger Baryt, *Leonhard,* Tabel. s. 40. *Id. Karsten,* Tabel. s. 54. *Id. Haus* s. 133.—Baryt sulphatée concretionnée-fibreuse, *Hauy,* Tabl. p. 13.—Fasriger Baryt, *Lenz,* b. ii. s. 900.

External Characters.

Its colour is chesnut-brown.

It is of a form intermediate between reniform and botryoidal.

Internally it is shining, and the lustre is resinous.

The fracture is coarse, and plumiform or scopiform fibrous.

The fragments are indeterminate angular, and not particularly sharp-edged.

It is translucent on the edges.

It is soft.

It is brittle.

It is easily frangible.

It is heavy.

Specific gravity, 4.080, *Klaproth.* 4.239, *Noeggerath.*

Constituent Parts.

Sulphate of Barytes,	-	99.0
Trace of Iron.		
		99.0

Klaproth, Beit. b. iii. s. 288.

Geognostic

Geognostic and Geographic Situations.

It is found at Neu-Leinengen in the Palatinate; also in an ironstone mine in clay-slate, at Chaud-Fontaine, near Lüttich, in the Ourthe department.

Observations.

It was first described by Karsten, and analysed by Klaproth. It was sent to Klaproth as a rare variety of ca lamine.

Sixth Subspecies.

Radiated Heavy-Spar or Bolognese Spar.

Bologneser Spath, *Werner.*

Gypsum spathosum, opacum, semi-pellucidum, *Wall.* t. i. p. 169. —Var. of Blættriger Schwerspath, *Wid.* s. 561.—Bologneserstein, *Emm.* b. iv. s. 572.—Litheosphore, *Lam.* t. i. p. 24. —Baryte sulphatée rayonnée, *Hauy,* t. ii. p. 302.—La Spath de Bologne, ou la pierre de Bologne, *Broch.* t. i. p. 633.—Strahliger Baryt, *Reuss,* b. ii. 2 s. 460.—Bologneser Spath, *Lud.* b. i. s. 172. *Id. Suck.* 1r th. s. 712.—Kieselerdiger schwefelsaurer Baryt, *Bert.* s. 130.—Bologneser spath, *Mohs,* b. ii. s. 227.—Strahliger Baryt, *Leonhard,* Tabel. s. 40.—Baryt sulphatée pure radiée, *Brong.* t. i. p. 251.—Strahliger Baryt, *Haus.* s. 133. *Id. Karsten,* Tabel. s. 54.—Bologna Stone, *Kid,* vol. i. p. 89.—Baryt sulphatée radiée, *Hauy,* Tabl. p. 13.—Strahliger Baryt, *Lenz,* b. ii. s. 901.

External Characters.

It principal colour is smoke-grey, which passes into ash-grey and yellowish-grey.

It

It occurs in original roundish and compressed blunt angular pieces, which are always covered with clay or marl.

Internally it is shining or glistening, and the lustre is resinous.

The cross fracture is foliated; the longitudinal fracture radiated, sometimes approaching to fibrous, and is parallel and scopiform radiated.

The fragments are rhomboidal, or wedge-shaped.

It generally occurs in granular concretions; more rarely the concretions are wedge-shaped.

It is translucent.

In other characters it agrees with the preceding.

Chemical Characters.

It is remarkably phosphorescent when heated. This property was first observed in the year 1630, by a shoemaker named Vincenzo Casciarolo, during his search after the philosopher's stone. When the mineral is calcined, pulverised, and made into cakes, it acquires a strong phosphorescent property by exposure to light; the phosphorescence is visible upon simply taking it into a dark place.

Constituent Parts.

Sulphate of Barytes, -	62.00
Lime, - - -	2.00
Silica, - - -	16.00
Alumina, - -	14.75
Oxide of Iron, - -	0.25
Water, - - -	2.00
	97.00 *Afzelius.*

Geognostic

Geognostic and Geographic Situations.

It occurs imbedded in marl in Monte Paterno, near Bologna; also at Rimini; and in Jutland.

Observations.

1. Its uneven surface, shews that the rounded pieces are of cotemporaneous formation with the marl in which they are contained, and not rolled pieces.

2. When rendered phosphorescent, it is known under the name *Bologaian Phosphorus*.

Seventh Subspecies.

Columnar Heavy-Spar.

Stängenspath, *Werner.*

Var. Blættriger Schwerspath, *Wid.* s. 561.—Stangenspath, *Emm.* b. i. s. 569.—Le Spath pesant en barres, *Broch.* t. i. p. 631. —Baryte sulphatée bacillaire, *Hauy,* t. ii. p. 302.—Stänglicher Baryt, *Reuss,* b. ii. 2. s. 458.—Stangenspath, *Lud.* b. i. s. 172. *Id. Suck.* 1r th. s. 711. *Id. Bert.* s. 130. *Id. Mohs,* b. ii. s. 225.—Stänglicher Baryt, *Leonhard,* Tabel. s. 40.—Baryte sulphatée pure baccillaire, *Brong.* t. i. p. 251. —Stänglicher Baryt, *Karsten,* Tabel. s. 54.—Stangenspath, *Haus.* s. 133.—Baryte sulphatée bacillaire, *Hauy,* Tabl. p. 13. —Stänglicher Baryt, *Lenz,* b. ii. s. 903.—Columnar Heavy-spar, *Aikin,* p. 145.

External Characters.

Its colours are yellowish, greyish, and greenish white.

It occurs crystallised, in acicular oblique four-sided prisms, which are always columnarly aggregated, and intersect each other.

Externally it is frequently invested with iron-ochre, but

but when unsoiled, it is shining and pearly; internally it is shining and pearly.

The fracture is foliated, with a threefold cleavage.

The fragments are indeterminate angular, and rather sharp-edged.

It is translucent.

It is soft.

It is brittle, and easily frangible.

It is heavy.

Specific gravity, 4.500.

Constituent Parts.

Barytes, - - -	63.00
Sulphuric Acid, -	33.00
Strontian Earth, -	3.10
Oxide of Iron, - -	1.50
Water, - - -	1.20

Lampadius.

Geognostic and Geographic Situations.

It was formerly found in the vein of Lorenzgegentrum, near Freyberg in Saxony, along with ores of different kinds, and also fluor-spar, quartz, and straight and curved lamellar heavy-spar. It is also mentioned as occurring in Derbyshire.

Observations.

1. It has been sometimes confounded with White Lead-ore, but is distinguished from that mineral by the following characters: White Lead-ore has an adamantine lustre, its fracture is small conchoidal, and its specific gravity is 6.558; whereas Columnar Heavy-spar has a pearly lustre, foliated fracture, and a specific gravity of 4.500.

Eighth

Eighth Subspecies.

Prismatic Heavy-Spar.

Saulenspath, *Werner.*

Sauliger Baryt, *Reuss,* b. ii. 2. s. 455.—Saulenspath, *Lud.* b. i. s. 172. *Id. Mohs,* b. ii. s. 226.—Sauliger Baryt, *Leonhard,* Tabel. s. 40.—Saulenspath, *Lenz,* b. ii, s. 905.

External Characters.

Its colours are greenish, yellowish, ash, smoke, and pearl grey; the pearl-grey passes into flesh-red; the yellowish-grey into wax, wine, and honey yellow: it occurs also pale indigo and sky blue.

It occurs sometimes massive; but more frequently crystallised in the following figures:

1. Oblique four-sided prism, acutely bevelled on the terminal planes, the bevelling planes set on the acute lateral edges, fig. 143. Sometimes one of the bevelling planes is so large as to cause the other to disappear.
2. The preceding figure, in which the bevelment is again flatly and deeply bevelled, and sometimes the angles on the obtuse lateral edges bevelled, and these bevelling planes set on the terminal edges.
3. Oblique four-sided prism, acuminated with four planes, which are set on the lateral edges. Sometimes the two opposite obtuse edges are truncated, and thus there is formed a
4. Broad six-sided prism, acuminated with four planes, two of which are on the acute lateral edges, and two on the opposite broader lateral planes

5. Six-

5. Six-sided prism, bevelled on the terminal planes, the bevelling planes set on the two opposite lateral planes.
6. Six-sided prism, bevelled on the terminal planes, the bevelling planes set on the acute lateral edges: the acute lateral edges truncated.

The crystals are middle-sized and small, and are generally promiscuously aggregated.

The surface of the crystals is splendent, and the lateral planes are transversely streaked.

Internally it is shining, or splendent and resinous.

The fracture is more or less perfect foliated.

The fragments approach to the rhomboidal in shape.

It alternates from translucent to semi-transparent.

It is soft.

It is brittle.

It is easily frangible.

It is heavy.

Specific gravity, 4.500.

Geognostic Situation.

It occurs in veins, along with fluor-spar, and ores of silver and cobalt.

Geographic Situation.

It occurs at Kongsberg in Norway; Mies in Bohemia; and Freyberg, Marienberg, and Ehrenfriedersdorf in Saxony.

3. Hepatite.

3. Hepatite.

Gypsum, lapis hepaticus, *Wall.* Syst. i. s. 165.—Baryte sulphatée fetide, *Hauy*, t. ii. p. 304.—Hepatit, *Reuss*, b. ii. 2. s. 463. *Id. Lud.* b. ii. s. 157. *Id. Suck.* 1r th. s. 714. *Id. Bert.* s. 131. *Id. Mohs*, b. ii. s. 228.—Baryte sulphatée fetide, *Lucas*, p. 15.—Hepatit, *Leonhard*, Tabel. s. 40.—Baryte sulphatée fetide, *Brong.* t. i. p. 253. *Id. Brard*, p. 59.—Hepatit, *Karsten*, Tabel. s. 54. *Id. Haus.* s. 134.—Baryte sulphatée fetide, *Hauy*, Tabl. p. 13.—Hepatit, *Lenz*, b. ii. s. 908. *Id. Oken*, b. i. s. 404. *Id. Aikin*, p. 171. 2d edit.

External Characters.

Its colours are greyish-white, yellowish and smoke grey, greyish and brownish black.

It occurs massive, disseminated, and in globular or elliptical pieces, from an inch to a foot and upwards in diameter.

Externally it is feebly glimmering; internally shining, and intermediate between pearly and resinous.

The fracture is foliated, and either straight, curved, or floriform foliated, which latter inclines to radiated.

The fragments are indeterminate angular, and blunt-edged.

It occurs in curved lamellar concretions; sometimes there is a tendency to wedge-shaped concretions.

It is opaque, or translucent on the edges.

It is soft.

It affords a greyish-white coloured streak.

It is heavy.

Chemical

Chemical Characters.

It burns white before the blowpipe; and when rubbed or heated, gives out a fetid sulphureous odour.

Constituent Parts.

Sulphate of Barytes, with a trace of Sulphate of Strontian, 93.58 Sulphate of Lime, 3.58 Oxide of Iron, 0.87 Water, Carbonaceous Matter, Sulphur, and Alumina, 2.00 100.00 *John*, Chem. Unt. b. ii. s. 73.	Sulphate of Barytes, - 92.75 Carbon and Bitumen, - 2.00 Sulphate of Lime, 2.00 Oxide of Iron, 1.50 Water, - 1.25 Sulphur, Oxide of Manganese, Chromic Acid, and Alumina, very small quantity. 99.05 *John*, Chem. Unt. b. ii. s. 69.	Sulphate of Barytes, . 85.25 Carbon, - 0.50 Sulphate of Lime, 6.00 Oxide of Iron, 5.00 Alumina, - 1.00 Loss, including Moisture and Sulphur, - 2.25 100.00 *Klaproth*, Beit. b. v. s. 121.

Geognostic and Geographic Situations.

It occurs at Buxton in Derbyshire; at Kongsberg in Norway, in veins that traverse mica-slate and hornblende-slate, along with native silver, heavy-spar, coal-blende, and iron-pyrites; and at Andrarum in Schonen, in transition alum-slate, in the form of balls. These balls are sometimes impregnated with iron-pyrites, or the iron-pyrites forms the central part.

Observations.

1. It is named *Hepatite*, from the disagreeable sulphureous odour it exhales when rubbed or exposed to heat.

2. Marggraf, Linnæus, and Cronstedt, arrange this mineral with Limestone.

4. Strontianite.

4. Strontianite.

Strontian, *Werner*.

Strontian, *Wid.* s. 571. *Id. Kirw.* vol. i. p. 332. *Id. Estner,* b. ii. s. 48 —Kohlensaurer Strontianit, *Emm.* b. i. s. 310.—Strontianite, *Nap.* p. 391.—Strontites, *Hope,* Edin. Trans. for 1790. *Id. Lam.* t. ii. p. 130.—Strontiane carbonatée, *Hauy,* t. ii. p. 327.—La Strontianite, *Broch.* t. i. p. 637.—Strontianit, *Reuss,* b. ii. 2. s. 416. *Id. Lud.* b. i. s. 174. *Id. Suck.* 1r th. s. 684. *Id. Bert.* s, 133. *Id. Mohs,* b. ii. s. 198. Strontiane carbonatée, *Lucas,* p. 18.—Kohlensaurer Strontianit, *Leonhard,* Tabel. s. 41.—Strontiane carbonatée, *Brong.* t. i. p. 259. *Id. Brard,* p. 64.—Strontian, *Karsten,* Tabel. s. 54.—Strontianite, *Haus.* s. 131. *Id. Kid,* vol. i. p. 82.—Strontiane carbonatée, *Hauy,* Tabl. p. 15.—Strontianite, *Lenz,* b. ii. s. 915. *Id. Oken,* b. i. s. 411.—Strontian, *Aikin,* p. 145.

External Characters.

Its colour is pale asparagus-green, which sometimes inclines to apple-green, sometimes to yellowish-white and greenish-grey. The greenish-grey variety sometimes passes into milk and yellowish white, and pale straw-yellow.

It occurs massive, and crystallised in acicular six-sided prisms, acuminated with six planes, which are set on the lateral planes.

The crystals are sometimes scopiformly and manipularly aggregated.

The lustre of the principal fracture is shining or glistening; of the cross fracture glistening, and is pearly.

The principal fracture is narrow and scopiform radiated,

ed, which sometimes inclines to fibrous, sometimes to floriform foliated *; the cross fracture is uneven.

The fragments are generally wedge-shaped.

It sometimes occurs in imperfect longish angulo-granular concretions.

It is more or less translucent, and sometimes semitransparent.

It yields readily to the knife.

It is brittle.

It is easily frangible.

Specific gravity, 3.675, *Klaproth*. 3.644, *Kirwan*. 3.6583, 3.675, *Hauy*.

Chemical Characters.

It is infusible before the blowpipe, but becomes white and opaque, and tinges the flame of a dark purple colour. It is soluble, with effervescence, in muriatic or nitric acid; and paper dipped in the solutions thus produced, burns with a purple flame.

Constituent Parts.

Strontian,	61.21	69.5	62.0	74.0
Carbonic Acid,	30.20	30.0	30.0	25.0
Water, -	8.50	0.5	8.0	0.5
	100.00	100.0	100.0	99.5
	Hope, Edin. Trans. for 1790.	*Klaproth*, Beit. b. i. s. 270.	*Pelletier*, Jour. des Mines, N. 21. p. 46.	*Bucholz*, in Lenz Min. b. ii. s. 916.

Geognostic and Geographic Situations.

It occurs at Strontian in Argyleshire, in veins that traverse gneiss, along with galena or leadglance, heavy-spar, and calcareous-spar; also at Braunsdorf in Saxony, along

* Towards the centre of the radiation, the colour is generally pale.

along with calcareous-spar, iron and copper pyrites; and at Pisope, near Popayan in Peru.

Observations.

The peculiar earth which characterises this mineral was discovered by Dr Hope, and its various properties were made known to the public in his excellent memoir on Strontites, inserted in the Transactions of the Royal Society of Edinburgh for the year 1790.

5. Celestine.

Celestin, *Werner.*

This species is divided into four subspecies, viz. Foliated Celestine, Radiated Celestine, Fibrous Celestine, and Compact Celestine.

First Subspecies.

Foliated Celestine.

Blättricher Celestin, *Karsten.*

Schaaliger Celestin, *Werner.*

Strontiane sulphatée, *Hauy,* t. ii. p. 313.—Blättricher Schützit, *Reuss,* b. ii. 2. s. 423.—Blättricher Celestin, *Suck.* 1r th. s. 688. *Id. Bert.* s. 134. *Id. Mohs,* b. ii. s. 230.—Blättriger schwefelsaurer Strontianit, *Leonhard,* Tabel. s. 41.—Blättriger Celestin, *Karsten,* Tabel. s. 54.—Strontian sulphatée laminaire, *Hauy,* Tabl. p. 14.—Blättriger Celestin, *Lenz,* b. ii. s. 923.

External Characters.

Its colours are milk-white, bluish-grey, pale indigo-blue; also reddish-white, and pale flesh-red.

It occurs massive; and crystallised in the following figures:

1. Irregular six-sided table, with two lateral edges of 77° 2′, and four of 128° 31 *.
2. Irregular eight-sided table, having four lateral edges of 128° 31′, and four of 142° 24′.
3. Rectangular four-sided table, which is bevelled on two or four terminal planes, with bevelling edges of 75° 12′, and 101° 32′; and sometimes the edges of the bevelment are truncated.

The crystals are middle-sized and small.

The surface of the massive varieties is streaked, and the same is the case with the lateral planes of the tables.

Externally it is shining, and splendent and pearly; internally it is shining and pearly, inclining to vitreous.

The fracture is straight foliated.

The fragments are rhomboidal.

It occurs in lamellar distinct concretions, which are generally straight, or slightly curved, and in which the surfaces are smooth and shining.

It is translucent or semi-transparent.

It scratches calcareous-spar, and heavy-spar, but is scratched by fluor-spar.

It is rather sectile.

It is very easily frangible.

Specific gravity, 3.960, *Clayfield.* 3.967, *Karsten.*

Chemical Characters.

It melts before the blowpipe into a white friable enamel, without very sensibly tinging the flame: after a short

* The primitive form, according to Hauy, is a right rhomboidal four-sided prism, with lateral edges of 104° 48′ and 75° 12′.

short exposure to heat, it becomes opaque, and has then acquired a somewhat caustic acrid flavour, very different from that of sulphureted hydrogen, which heavy-spar acquires in similar circumstances.—*Aikin.*

These characters apply also to the other subspecies.

Constituent Parts.

Strontian,	57.64	Strontian, and Sulphuric Acid,	97.208	Strontian, and Sulphuric Acid,	97.601
Sulphuric Acid,	43.00	Sulphate of Barytes, -	2.222	Sulphate of Barytes, -	00.975
	100.64	Silica, -	0.254	Silica, -	00.107
Rose, in Karsten's Tabellen, s. 55.		Oxide of Iron,	0.116	Oxide of Iron, and intermixed Hydrate of Iron,	00.646
		Water, -	0.190	Water, -	00.248
		Petroleum, a minute portion.			
			99.099		99.577
		Stromeyer, in Gött. Gell. Anz. 1811, 188.		*Stromeyer*, in Gött. Gel. Anz. 1812, 12. 114.	

Geognostic and Geographic Situations.

It occurs in red sandstone at Inverness. It is frequent along with some of the other subspecies at Aust Passage, and elsewhere in the neighbourhood of Bristol, and in the islands in the Bristol Channel, particularly in Barry Island, on the coast of Glamorganshire; and it has been found on the banks of the Nidd, near Knaresborough, Yorkshire. It forms a bed, about one-fourth of a fathom thick, in a coal-mine, which appears to be connected with the shell limestone, at Süntel in Hanover; and also near Karlshütte, on the road from Göttingen to Hanover; in the Canton of Aargau in Switzerland; Monte Chio Maggiore in the Vicentine, in vesicular cavities in basaltic amygdaloid; and Montmartre, near Paris.

Second

Second Subspecies.

Radiated Celestine.

Strahliger Celestin, *Karsten.*

Sáulenformiger Celestin, *Werner.*

Strahliger Celestin, *Karsten,* Tabel. s. 54.—Strontiane sulphatée fibro-lamellaire, *Hauy,* Tabl. p. 14.—Strahlicher Cœlestin, *Lenz,* b. ii. s. 927.

External Characters.

Its colours are yellowish-white, sky-blue, and indigo-blue.

It occurs massive, in long adhering tubes; and also crystallised in the following figures:

1. Very oblique four-sided prism, bevelled on the extremities, the bevelling planes set on the obtuse lateral edges *, fig. 144. Sometimes the acute edges are truncated †, fig. 145.
2. Sometimes the angles between the bevelling and lateral planes are more or less deeply truncated, and thus form a four-planed acumination, in which the acuminating planes are set on the lateral edges ‡, fig. 146.
3. Sometimes the acute edges of the preceding figure are truncated, and thus a six-sided prism is formed ||, fig. 147.

The

* Strontiane sulphatée unitaire, Hauy.

† Strontiane sulphatée emousée, Hauy.

‡ Strontiane sulphatée dodecaedre, Hauy.

|| Strontiane sulphatée epointée, Hauy.

The crystals are middle-sized, and scopiformly aggregated.

Externally it is smooth, splendent, and vitreous.

Internally it is glistening and pearly.

The principal fracture is broad or narrow, and scopiformly radiated; the cross fracture is uneven, passing into concealed foliated. The fracture of the crystals is foliated; the cleavages parallel to the planes of the rhomboidal prism: of these folia, that parallel to the base of the prism, is the only one which is very distinct. The cross fracture is uneven.

It occurs in distinct concretions, which are thin prismatic, and wedge-shaped.

It is translucent or transparent.

In other characters, it agrees with the preceding subspecies.

Geognostic and Geographic Situations.

It occurs in strata of gypsum, along with sulphur, in the valleys of Noto and Mazzara, in Sicily; in the Canton of Aargaw in Switzerland; in amygdaloid, along with calcareous-spar, in the neighbourhood of Greden, in the circle of the Inn; and in gypsum, near Cadiz.

Observations.

It has much the appearance of Prismatic Heavy-spar, but is distinguished from that mineral by its more distinct foliated fracture, inferior lustre, less regular concretions, greater sectility, easier frangibility, and inferior weight.

Third Subspecies.

Fibrous Celestine.

Fasriger Cœlestin, *Werner*.

Fasriger Cœlestin, *Karsten*, Tabel. s. 54 —Strontiane sulphatée fibreuse-conjointe, *Hauy*, Tabl. p. 14.—Fasriger Cœlestin, *Lenz*, b. ii. s. 931.

External Characters.

Its colour is indigo-blue, which passes on the one side into bluish-grey, on the other into milk-white.

It occurs massive.

Internally it is glistening and pearly.

The fracture is parallel fibrous, or narrow radiated; the longitudinal fracture is concealed foliated.

The fragments are splintery.

It is translucent.

Specific gravity, 3.721, *Karsten*. 3.830, *Klaproth*.

In other characters it agrees with the preceding subspecies.

Constituent Parts.

Strontian, - -	56.0
Sulphuric Acid, -	42.0
Trace of Oxide of Iron.	
	98.0

Klaproth, Beit. b. ii. s. 97.

Geognostic and Geographic Situations.

It occurs in the red sandstone formation near Bristol; imbedded in marl, which is probably connected with gypsum,

gypsum, at Frankstown in Pennsylvania; and at Bouveron, near Toul, in the department of Meurthe in France.

Fourth Subspecies.

Compact Celestin.

Dichter Cœlestin, *Karsten.*

Dichter Cœlestin, *Karsten,* Tabel. s. 54.—Strontiane sulphatée calcarifère, *Hauy,* Tabl. p. 14.—Dichter Cœlestin, *Lenz,* b. ii. s. 921.

External Characters.

Its colours are snow-white, and yellowish-white, yellowish-grey, ochre-yellow, and yellowish-brown.

It occurs massive, or in spheroidal or reniform masses, which are often traversed by fissures which divide its surface into quadrangular pieces, which are sometimes lined with minute crystals of celestine.

Internally it is dull or glimmering.

The fracture is fine splintery, passing into uneven, or minutely foliated.

The fragments are blunt-edged.

It is opaque, or translucent on the edges.

It is soft.

Specific gravity, 3.592, *Hauy.*

In other characters it agrees with the preceding subspecies.

Constituent

Constituent Parts.

Sulphate of Strontian,	91.42
Carbonate of Lime, -	8.33
Oxide of Iron, - -	0.25
	100.00

Vauquelin, in Brogniart's Mineralogie, t. i. p. 258.

Geognostic and Geographic Situations.

It occurs imbedded in marly-clay, with gypsum, at Montmartre, near Paris; and it is said to form a whole bed in Champagne

XXVII. HALLITE

XXVII. HALLITE FAMILY.

This Family contains but one species, viz. Cryolite.

Cryolite.

Kryolith, *Werner*.

Alumine fluatée alkaline, *Hauy*, t. ii. p. 398.—Chryolith, *Reuss*, b. ii. 2. s. 556. *Id. Lud.* b. ii. s. 148. *Id. Suck.* 1r th. s. 532. *Id. Bert.* s. 278. *Id. Mohs*, b. ii. s. 237.—Alumine fluatée alkaline, *Lucas*, p. 27.—Kryolith, *Leonhard*, Tabel. s. 42.—Alumine fluatée, *Brong.* t. i. p. 164. *Id. Brard*, p. 87.—Kryolith, *Karsten*, Tabel. s. 48. *Id. Haus.* s. 121.—Alumine fluatée alkaline, *Hauy*, Tabl. p. 22.—Kryolith, *Lenz*, b. ii. s. 943. *Id. Oken*, b. i. s. 399. *Id. Aikin*, p. 126.

External Characters.

Its colour is pale greyish-white, snow-white, and yellowish-brown.

It occurs massive, and disseminated.

It is shining, inclining to glistening, and the lustre is vitreous, inclining to pearly.

The principal fracture is foliated, with a threefold cleavage, of which the folia are parallel to the planes of a rectangular parallelopiped; the cross fracture is uneven.

The fragments are cubical or tabular.

It occurs in straight and thick lamellar distinct concretions.

It is translucent.

It is softer than fluor-spar.

It

It is brittle.

It is easily frangible.

Specific gravity, 2.949, *Hauy*. 2.953, *Karsten*.

Chemical Characters.

It becomes more translucent in water, but does not dissolve in it. It melts before it reaches a red heat, and when simply exposed to the flame of a candle. Before the blowpipe, it at first runs into a very liquid fusion, then hardens, and at length assumes the appearance of a slag.

Constituent Parts.

Alumina, - -	24.0	21.0
Soda, - - -	36.0	32.0
Fluoric Acid, and Water,	40.0	47.0
	100.0	100.0
	Klaproth, Beit. b. iii. s. 214.	*Vauquelin*, Hauy, Traité, t. ii. p. 400.

Geognostic and Geographic Situations.

This curious and rare mineral has been hitherto found only in West Greenland, and but in one place of that dreary and remote region, viz. the Fiord or arm of the sea named Arksut, situated about thirty leagues from the colony of Juliana Hope. It occurs in two thin layers in gneiss: one of these contains the greyish and snow white cryolite, and is not intermixed with other minerals; the other is wholly composed of the yellowish-brown coloured variety, mixed with galena, iron-pyrites, sparry iron-ore, quartz, and felspar. They are situated very near each other: the first is washed at high water by the tide, and a considerable portion of it is exposed, the

the superincumbent gneiss being removed. It varies in thickness from one foot to two feet and a half in thickness *.

Observations.

1. As this mineral, when exposed to a very low heat, melts almost like ice, it was named *Cryolith*, from κρυος, *ice*, and λιθος, *stone*.

2. It has been confounded with Heavy-spar, from which it is distinguished, by inferior specific gravity, and its easy fusibility before the blowpipe: it might also be mistaken for some varieties of Gypsum, but is distinguished from these by superior specific gravity, and its not exfoliating when exposed to the blowpipe.

---

* Allan and Giesecké, in Thomson's Annals, vol. ii. p. 389.

CLASS II.

CLASS II.

SALINE MINERALS.

FOSSIL SALTS.

The saline substances included under this Class, are those only which are found in a natural state: the numerous artificial salts described in some systems of mineralogy, are consequently excluded. The greater number of these natural salts appear to have been formed by the agency of air and water, and are therefore more properly atmospherical than terrestrial products. Natural Rock-salt is perhaps the only exception, it being found in the interior of the earth.

The characters by which the substances of this class are distinguished from those of the other classes, are principally their *taste*, and *easy solubility*.

They may be conveniently arranged in the following order:

Order I. Earthy Salts.—*Genera*, Alumina, Magnesia.
II. Alkaline Salts.—*Genera*, Soda, Potash, Ammonia.
III. Metallic Salts.—*Genera*, Iron, Copper, Zinc, Cobalt.

ORDER

ORDER I.—EARTHY SALTS.

This Order contains all the Salts having an Earthy basis. They are easily soluble in water, and then have either a sweetish astringent, or saline bitter taste.

Genus I.—Salts of Alumina.

This Genus contains but two species, viz. Alum, and Rock-Butter.

1. Alum.

Natürlicher Alaun, *Werner.*

Alumen nativum, *Wall.* t. i. p. 31.—Naturlicher Alaun, *Wid.* s. 593.—Alum, *Kirw.* vol. ii. p. 13.—Naturlicher Alaun, *Estner,* b. iii. s. 39. *Id. Emm.* b. ii. s. 9.—L'Alumen Natif, *Broch.* t. ii. p. 6.—Alumine sulphatée alkaline, *Hauy,* t. ii. p. 387. 398.—Alaun, *Reuss,* b. iii. s. 58. *Id. Suck.* 2ter th. s. 23. *Id. Bert.* s. 322. *Id. Mohs,* b. ii. s. 272. *Id. Leonhard,* Tabel. s. 46.—Alumine sulphatée, *Brong.* t. i. p. 155. —Alaun, *Karsten,* Tabel. s. 56. *Id. Haus.* s. 119.—Sulphate of Alumine, *Kid,* vol. ii. p. 13.—Alaun, *Lenz,* b. ii. s. 1007.

External Characters.

Its colour is yellowish and greyish white.

It occurs as a mealy efflorescence, stalactitic, or in delicate capillary crystals *; and some varieties have a shining pearly lustre, and curved and parallel fibrous fracture,

* The crystallizations of purified alum, are the octahedron, and cube, and the intermediate figures.

fracture, and others are glistening and vitreous, and imperfect conchoidal.

Its taste is sweetish astringent.

Chemical Characters.

It is soluble in from sixteen to twenty times its weight of water. It melts easily by means of its water of crystallization; and by continuance of the heat, it is converted into a white spongy mass.

Constituent Parts.

Natural Alum of Freinwald.

Alumina, - -	15.25
Potash, - - -	0.25
Oxide of Iron, - -	7.50
Sulphuric Acid, and Water,	77.00
	100.00

Klaproth, Beit. b. iii. s. 103.

Geognostic Situation.

It generally occurs as an efflorescence on aluminous minerals, as alum-slate, alum-earth, alum-stone, aluminous-coal, aluminous-slate-clay and bituminous shale, and also encrusting lavas.

Geographic Situation.

Europe.—It occurs as an efflorescence on the surface of bituminous-shale and slate-clay at Hurlet, near Paisley; also encrusting alum-slate near Moffat, in Dumfriesshire; Ferrytown of Cree, in Galloway; and at Whitby, in Yorkshire. On the Continent of Europe, it is met with in many places, as in the alum-slate rocks near Christiania in Norway; in coal-mines in Bohemia; also in Austria,

tria, Bavaria, Hungary, Italy, and the islands of Stromboli, Milo, &c. in the Mediterranean Sea.

Africa.—In Egypt.

America.—In Real del Monte in Mexico, on a porphyritic stone.

Uses.

It is employed as a mordant in dyeing; also in the manufacture of leather and paper; as a medicine; for preserving animal substances from putrefaction; and it is sometimes mixed with bread, in order to give it a whiter colour.

Observations.

1. The minerals that afford alum, either contain it ready formed, or only its constituent parts, which are disposed to unite, and form alum, when placed in favourable circumstances. This latter is the most frequent case.

2. The Romans and Grecians appear to have been unacquainted with alum: the *alumen* of the Romans, and the συπτηρια of the Greeks, being vitriol of iron.

2. Rock-Butter.

Bergbutter, *Werner*.

Bergbutter, *Wid.* s. 589. *Id. Emm.* b. ii. s. 13.—Le Beurré de Montagne, ou Le Bergbutter, *Broch.* t. ii. p. 10.—Bergbutter, *Reuss,* b. iii. s. 66. *Id. Lud.* b. i. s. 182. *Id. Suck.* 2ter th. s. 26. *Id. Bert.* s. 323. *Id. Leonhard,* Tabel. s. 46. *Id. Karsten,* Tabel. s. 56. *Id. Haus.* s. 119. *Id. Lenz,* b. ii. s. 1009.

External Characters.

The colours are yellowish-white, yellowish-grey, cream-yellow, straw-yellow, and pale sulphur-yellow.

It

It occurs massive, and tuberose.
Internally it is strongly glimmering, and resinous.
The fracture is straight foliated.
The fragments are blunt-edged.
It is translucent on the edges.
It feels rather greasy.
It is easily frangible.

Constituent Parts.

It is Alum, mixed with Alumina and Oxide of Iron.

Geognostic Situation.

It oozes out of rocks that contain alum, or its constituents, as alum-slate, bituminous-shale impregnated with iron-pyrites, or alum-earth.

Geographic Situation.

It occurs at the Hurlet Alum-work, near Paisley; oozing out of rocks of alum-slate in the island of Bornholm, in the Baltic; at Muskau in Upper Lusatia; Saalfeld in Thuringia; and, according to Pallas, in aluminous rocks on the banks of the river Jenisei, in Siberia.

Genus II.—Salts of Magnesia.

This Genus contains but one species, viz. Epsom Salt.

Epsom Salt.

Naturlicher Bittersalz, *Werner.*

Sal neutrum acidulare, *Wall.* t. ii. p. 71.—Sal d'Epsom; Sel de Sedlitz; Sel d'Angleterre; Vitriol de Magnesia, *Romé de Lisle,* t. i. p. 306.—Naturlicher Bittersalz, *Wid.* s. 595.—Epsom Salt, *Kirw.* vol. i. p. 12.—Naturlicher Bittersalz, *Estner,* b. iii. s. 44. *Id. Emm.* b. ii. s. 14.—Le Sel amere natif, ou Le Sel d'Epsom natif, *Broch.* t. ii. p. 11.—Magnesie sulphatée, *Hauy,* t. ii. p. 331.-336.—Bittersalz, *Reuss,* b. iii. s. 53. *Id. Lud* b. i. s. 182. *Id. Suck.* 2ter th. s. 21. *Id. Bert.* s. 324. *Id. Mohs,* b ii. s. 271. *Id. Leonhard,* Tabel. s. 46.—Magnesie sulphatée, *Brong.* t. i. p. 165.—Haarsalz, *Karst.* Tabel. s. 56 —Bittersalz-fasriges, haarförmiges, & mehliges, *Haus.* s. 120. *Id. Lenz,* b. ii. s. 1015.—Sulphat of Magnesia, *Aikin,* 2d edit. p. 251.

External Characters.

Its colours are snow-white, greyish-white, and yellowish-white, and sometimes ash-grey, and smoke-grey.

It occurs in farinaceous crusts, in flakes, small botryoidal, reniform, and in acicular and capillary prismatic crystals *.

The farinaceous variety is dull, the others shining, or glistening, and pearly.

It varies from transparent to opaque.

It is soft.

It

* The crystallizations of Artificial Epsom Salt are the following:

1. Rectangular four-sided prism, either bevelled on the extremities, or acuminated with four planes.
2. Six-sided prism, acuminated on the extremities with four planes, and the edges of the acumination truncated.

It is brittle, and easily frangible.

Its taste is bitter and saline.

Chemical Characters.

Before the blowpipe, it dissolves very easily by the assistance of its water of crystallization, but it is difficultly fusible. Its solution gives a precipitate with lime-water.

Constituent Parts.

The constituent parts of purified Epsom Salt, the Sulphate of Magnesia of chemists, are, according to

	Bergmann,	*Kirwan.*
Sulphuric Acid, -	33.0	29.46
Magnesia, - -	19.0	17.00
Water of Crystallization,	48.0	53 54
	100.0	100.00

Geognostic and Geographic Situations.

It occurs as an efflorescence at Hurlet, near Paisley, along with natural alum; and sometimes effloresces on old walls: at Jena it encrusts rocks of gypsum; and half burnt clay at Witschiz in Bohemia; on porphyry-slate, also in Bohemia; at Solfatera, on decomposing lava; at Gran, in Hungary, it effloresces on sandstone, clay, and compact limestone.

Uses.

When purified, it is used as a purgative medicine; and it is valued by chemists on account of the magnesia which can be obtained from it.

ORDER II.—ALKALINE SALTS.

This Order contains all the Natural Salts having an earthy basis. They are more or less soluble in water: their tastes are alkaline, saline, or bitter; the predominating colour is white; and they are light.

Genus I.—Salts of Soda.

Of this genus there are five species, viz. Natron, Glauber Salt, Reussite, Rock-Salt, and Borax.

1. Natron, or Soda.

Natürliches Mineral Alkali, *Werner*.

This species is divided into two subspecies, viz. Common Natron, and Radiated Natron.

First

First Subspecies.

Common Natron.

Gemeines Natron, *Werner.*

Nitrum, *Plin.* Hist. Nat. xxxi. 10. p. 46.—Alkali minerale natron, *Wall.* t. ii. p. 61.—Alkali fixe mineral, *Romé de Lisle,* t. i. p. 146.—Natürliches mineral alkali, *Wid.* s. 579.—Natron, *Kirw.* vol. ii. p. 6. *Id. Estner,* b. iii. s. 18.—Natürliches mineral alkali, *Emm.* b. ii. s. 31.—Carbonate de Natron, *Lam.* t. i. p. 462.—Soude carbonatée, *Hauy,* t. ii. p. 373. 379.—L'Alkali mineral, ou le Carbonate de Soude, *Broch.* t. ii. p. 30.—Natron, *Reuss,* b. iii. s. 4.—Natürliches Mineral Alkali, *Lud.* b. i. s. 176. *Id. Suck.* 2r th. s. 2. *Id. Bert.* s. 331. *Id. Mohs,* b. ii. s. 254.—Gemeines Natron, *Leonhard,* Tabel. s. 44.—Soude carbonatée, *Brong.* t. i. p. 149.—Gemeines Natron, *Karsten,* Tabel. s. 56.—Soda, *Haus.* s. 120.—Carbonate of Soda, *Kid,* vol. ii. p. 4.—Soude carbonatée, *Hauy,* Tabl. p. 21.—Gemeines Natron, *Lenz,* b. ii. s. 960.—Natron, *Aikin,* p. 153.

External Characters.

Its colours are yellowish and greyish white; also smoke-grey and cream-yellow.

When fresh, it is compact, sometimes granular, sometimes radiated *, vitreous and glistening, and more or less translucent: when weathered, it is in loose, dull, opaque parts.

It has an alkaline taste.

* The crystallization of Artificial Soda,—for that of Natural Soda has not been met with,—is an oblique octahedron, which is either perfect, or truncated on the angles or edges of the common basis.

Chemical Characters.

It effervesces with acids. Is easily soluble in water, and its solution colours blue vegetable tinctures green. It is very fusible before the blowpipe.

Constituent Parts.

Egyptian Natron.		Bohemian Natron.		Natron of Hungary.	
Dry sub-carbonate of Soda,	32.6	Carbonate of Soda,	89.18	Carbonate of Soda,	14.2
Dry Sulphate of Soda,	20.8	Carbonate of Lime,	7.44	Muriate of Soda,	22.4
Dry Muriate of Soda,	15.0	Carbonate of Magnesia,	1.35	Sulphate of Soda,	9.2
Water,	31.6	Extractive matter,	2.03	Earthy residuum,	9.2
				Water,	45.0
	100.0		100.00		100.0
Klaproth, Beit. b. iii. s. 80.		*Reuss*, Min. b. iii. s. 5.		According to *Lampadius*.	

Geognostic Situation.

It occurs as an efflorescence on the surface of soil—on decomposing rocks of particular kinds,—on the sides and bottoms of lakes that become dry during the summer season,—also on the walls and bottoms of caves,—and dissolved in the water of lakes and springs. In Hungary, according to Ruckert and Pazmand, there are so many natron lakes, that 50,000 quintals of soda could be obtained from them annually. In some places of the same country, it effloresces on the surface of the soil, heath, &c. According to Dr Reuss, it is observed efflorescing on meadows near Priesen and Sebnitz in Bohemia, and on decomposing gneiss in the vicinity of Bilin, where it is renewed every spring. To the west of the Delta of Egypt are several lakes, some of which hold carbonate of soda, or natron, in solution, others muriate of soda, or common salt. In some of these lakes, both these salts are contained,

contained, and are deposited alternately on the sides of the lake, in consequence of the evaporation of the water that held them in solution, this alternate deposition depending on the different degrees of the solubility of these salts: the common salt being the least soluble, is first crystallised; and when that has been separated, to such an extent as to leave a considerable excess of carbonate of soda or natron in solution, then this latter substance begins to crystallise. Berthollet is of opinion, that the natron is formed by the decomposition of the common salt by means of carbonate of lime, during which process the lime unites with the muriatic acid of the common salt, forming with it muriate of lime, while the carbonic acid thus disengaged from the lime, unites with the soda, and thus forms the natron. In some of the lakes, the eastern part contains only common salt, and the western natron, and these two solutions never mix together. It effloresces on the surface of lava in Italy, and in Switzerland, and France, on the walls and roof of caves.

Geographic Situation.

Europe.—It occurs in Bohemia, at Bilin, Carlsbad and Eger; in Hungary, in the neighbourhood of Debrezin, district of Bahar, &c.; in Switzerland, at Schwartzberg, in the canton of Berne; in the Phlegrean Fields, Monte Nuovo, near Naples; and Mount Ætna in Sicily.

Africa.—It occurs in considerable quantity in Egypt at the town of Nitria; in the valley of the Natron Lakes, Nubia; and the island of Teneriffe.

Asia.—In the vicinity of Smyrna, and the ancient city of Ephesus; in Bengal; near Bombay; near Tegapatnam, on the western coast of India; Sina, in the neighbourhood of Pekin; in Thibetian Tartary; Persia; Natolia,

tolia; district of Ochotsk, in the government of Irkutsk; in the neighbourhood of Nertschinsk in Siberia; and in the Crimea.

America.—Dissolved in the lakes of Mexico.

Second Subspecies.

Radiated Natron.

Strahliches Natron, *Klaproth.*

Strahliges Natron, *Reuss*, b. ii. 3. s. 3. 9. *Id. Leonhard*, Tabel, s. 44. *Id. Karsten*, Tabel. s. 56.—Trona, *Haus*. s. 120.—Soude carbonatée aciculaire, *Hauy*, Tabl. p. 21.—Strahliches Natron, *Lenz*, b. ii. s. 961.

External Characters.

Its colours are greyish and yellowish white.

Occurs in crusts, and crystallised in capillary or acicular crystals, which are aggregated on one another.

The lustre is glistening and vitreous.

The fracture is radiated, inclining to foliated.

It is translucent.

It has an alkaline taste.

Chemical Characters.

Same as those of Common Natron.

Constituent Parts.

Water of Crystallization, -	22.50
Carbonic Acid, - -	38.00
Pure Soda, - - -	37.00
Sulphate of Soda, - -	2.50
	100.00

Klaproth, Beit. b. iii. s. 87.

Geognostic

Geognostic and Geographic Situations.

Mr Bagge, Swedish consul at Tripoli, has given the following information respecting this interesting subspecies of natron. " The native country of this natron, which is there called *Trona*, is the province Sukena, two days journey from Fezzan. It is found at the bottom of a rocky mountain, forming crusts, usually the thickness of a knife, and sometimes, although rarely, of an inch, on the surface of the earth. It is always crystalline: in the fracture, it consists of cohering, longish, parallel, frequently radiated crystals, having the aspect of unburnt gypsum. Besides the great quantity of trona which is carried to the country of the Negroes and to Egypt, fifty tons are annually carried to Tripoli. It is not adulterated with salt. The salt-mines are situated on the sea-shore; but the trona occurs twenty-eight days journey up the country *." According to the accounts of Mr Barrow, it would appear also to occur in the district of Tarka in Boshieman's Land, in Southern Africa.

Uses of Natron.

It is principally employed in the manufacture of glass, in the manufacture of soap, in dyeing, and for the washing of linen. It is sometimes purified before it is used, but more frequently (particularly that from Egypt) it is used in its natural state. In Hungary, particularly at Debrezin, it is used in great quantity in the manufacture of soap: it has been also employed in considerable quantity in Scotland and England for the same purpose. In Siberia a fine white glass is manufactured with it. In the Levant, the natron of Suckena is mixed with tobacco, in order to give it

* Bagge, in the Abhandl. d. Schwed. Acad. v. j. 1773, b. xxxv. s. 131.

it a sharper taste. The ancient Egyptians are said to have macerated dead bodies in it for several months previous to preparing them as mummies. It is sometimes also purified for the alkali it contains, and is then used as a flux, &c.

Observations.

1. Klaproth restored to this species the old name *Natron*, a word said to be derived from Nitria in Upper Egypt, where, as already mentioned, it occurs in considerable quantity.

2. The terms *Natrum* and *Nitrum*, which are used indiscriminately by ancient writers, are most commonly applicable to this mineral: but they are sometimes applicable to saltpetre or nitre, and sometimes to sal-ammoniac.

2. Glauber Salt, or Sulphate of Soda.

Natürliches Glaubersalz, *Werner*.

Sal mirabile, *Wall.* t. ii. p. 70.—Sel de Glauber, *Romé de L.* t. i. p. 301. *Id. Born*, t. ii. p. 26.—Natürliches Wundersalz, *Wid.* s. 597.—Glauber Salt, *Kirw.* vol. ii. p. 9.—Natürliches Glaubersalz, *Estner*, b. iii. s. 50 *Id. Emm.* b. iii. s. 401.—Le Sel de Glauber natif, *Broch.* t. ii. p. 14.—Glaubersalz, *Reuss*, b. iii. s. 49. *Id. Lud.* b. i. s. 183. *Id. Suck.* 2r th. s. 18. *Id. Mohs*, b. ii. s. 273. *Id. Leonhard*, Tabel. s. 46. —Soude sulphatée, *Brong.* t. i. p. 113.—Glaubersalz, *Karsten*, Tabel. s. 56.—Glauberite, *Haus.* s. 120.—Soude sulphatée, *Hauy*, Tabl. p. 19.—Glaubersalz, *Lenz*, b. ii. s. 1027.

External Characters.

Its colours are greyish and yellowish white; seldom snow or milk white.

It

It occurs in the form of mealy efflorescences; in crusts, seldom stalactitic, small botryoidal, reniform; and crystallised,

1. In acicular crystals *.
2. In six-sided prisms, more or less flatly acuminated by three planes, which are set on the alternate edges or planes.

The acicular crystals are small, the prisms middle-sized, and so grown together, that they are with difficulty distinguishable.

Internally it is shining, and the lustre is vitreous.

The principal fracture of the crystallised varieties is foliated, with a threefold cleavage: the cross fracture is small conchoidal; that of others fine-grained uneven. When decomposed, the fracture is earthy.

The fragments are indeterminately angular, and blunt-edged.

It occurs sometimes in small and fine granular distinct concretions.

It is soft; the earthy varieties are friable.

It is brittle.

It is easily frangible.

Its taste is first cooling, and then saline and bitter.

Chemical Characters.

Before the blowpipe, it is affected in the same manner as Epsom salt; but its solution does not, like that of Epsom salt, afford a precipitate with an alkali.

Constituent

* The primitive form of Glauber salt is an octahedron.

Constituent Parts.

Natural Glauber Salt of Eger, according to Reuss, (Chemisch-medicinische Beschreibung des Kaiser Franzens Bades, Dresden, 1794), contains,

Sulphate of Soda, - -	67.024
Carbonate of Soda, -	16.333
Muriate of Soda, - -	11.000
Carbonate of Lime, -	5.643
	100

Geognostic Situation.

It occurs, along with rock-salt and Epsom salt, on the borders of salt-lakes, and dissolved in the waters of lakes, in efflorescences on muirish ground, sandstone, marl-slate, and on old and newly built walls.

Geographic Situation.

Europe.—At Eger in Bohemia, it occurs efflorescent on meadow-ground; as an efflorescence on the walls of old galleries in mines, at Grenoble in France; in old salt-mines at Aussee, Ischel and Hallstadt, in Upper Austria; at Altenberg in Stiria; Felsobanya in Hungary; Hildesheim; Durrenberg, near Hallein in Salzburg; Schwartzburg in Switzerland; near Aranjuez in Spain; Solfatara in Italy.

Asia.—It occurs on the banks, and in the water of many Siberian salt-lakes; neighbourhood of the Lake Baikal; the desert plains of Iset, Ischim and Barebyn.

Africa.—Egypt.

Uses.

It is used as a purgative medicine; and in some countries as a substitute for soda, in the manufacture of white glass.

3. Reussite.

3. Reussite.

Reussin, *Karsten.*

Reussin, *Karsten,* Tabel. (1. Ausg. 46. 75.) *Id. Leonhard,* Tabel. s. 46. *Id. Reuss,* Min. ii. 3. 46.—*Karsten,* Tabel. 2. Ausg. s. 56. *Id. Lenz,* b. ii. s. 1019.

External Characters.

Its colours are snow-white and yellowish-white, which latter inclines to wine-yellow.

It occurs as a mealy efflorescence, in loose, earthy, dull particles; and crystallised in

1. Flat six-sided prisms, with two broad, and four narrow lateral planes, and bevelled on the extremities.
2. Acicular crystals.

The first are small and middle-sized, the latter loose, or scopiformly aggregated.

Internally it is shining and vitreous, and the fracture of the crystals small conchoidal.

It is soft.

Constituent Parts.

Sulphate of Soda, -	66.04
Sulphate of Magnesia,	31.35
Muriate of Magnesia,	2.19
Sulphate of Lime, -	0.42
	100.0

Reuss, in Crell's Annalen, 1791, 11. 18.

Geognostic

Geognostic and Geographic Situations.

It is found in the country around Sedlitz and Saidschutz, where it effloresces on the surface during the spring of the year: also at Pilln, near Brüx.

4. Rock-Salt.

Natürliches Kochsalz, *Werner.*

This species is divided into two subspecies, viz. Rock-Salt, and Lake Salt.

First Subspecies.

Rock-Salt.

Steinsalz, *Werner.*

This subspecies is divided into two kinds, viz. Foliated Rock-Salt, and Fibrous Rock-Salt.

First Kind.

Foliated Rock-Salt.

Blättriches Steinsalz, *Werner.*

Muria fossilis pura; Sal gemmæ, *Wall.* t. ii. p. 53.—Sel marin et Sel gemme, *Romé de L.* t. i. p. 374.—Blättriches Steinsalz, *Wern.* Pabst. b. i. s. 361.—Lamellar Sal Gem, *Kirw.* vol. ii. p. 32.—Blättriches Steinsalz, *Estner,* b. iii. s. 63. *Id. Emm.* b. i. s. 19.—Soude muriatée cristallisé, et Soude muriatée amorphe, *Hauy,* t. ii. p. 356. 365.—Le Sel Gem lamelleux, *Broch.* t. ii. p. 21.—Blättriches Steinsalz, *Reuss,* b. iii. s. 30. *Id. Leonhard,* Tabel. s. 44. *Id. Karsten,* Tabel. s. 56. *Id. Haus.* s. 121.—Soude muriatée laminaire, *Hauy,* Tabl. p. 20. —Blättriches Steinsalz, *Lenz,* b. ii. s. 975.

External Characters.

Its most common colours are white and grey. Of white

white, it occurs greyish, yellowish, and milk white; but it seldom approaches to snow-white Of grey, ash, smoke, and pearl grey. From pearl-grey it passes, though rarely, into flesh, blood, and brick red. Still seldomer do we observe the white varieties marked with Berlin, azure, violet, or lavender blue spots or patches.

It is said also to occur ochre-yellow, wine-yellow, and emerald-green.

It occurs massive, disseminated, in minute veins, in crusts, plates, reniform, stalactitic, tuberose, corroded, cellular, with impressions, and also crystallised in cubes *.

On the fresh fracture it is splendent, and the lustre is resinous, inclining to vitreous.

The fracture is straight foliated, with a complete three-fold rectangular cleavage, the folia parallel with the planes of the cube.

The fragments are cubic.

It occurs in large, coarse, and small granular distinct concretions; also in globular and columnar concretions.

In general, it is strongly translucent, sometimes semi-transparent and transparent.

It is soft.

It feels rather greasy.

It is brittle.

It is easily frangible.

It has a saline and sweetish taste.

Specific gravity, 2.143, 2.2, *Hassenfratz*.

* The primitive form, according to Hauy, is the cube.

Second Kind.

Fibrous Rock-Salt.

Fasriges Steinsalz, *Werner*.

Fasriges Steinsalz, *Wern.* Pabst. b. ii. s. 363.—Fibrous Sal Gem, *Kirw.* vol. ii. p. 25.—Fasriges Steinsalz, *Estner*, b. iii. s. 71. *Id. Emm.* b. ii. s. 23.—Le Sel Gem fibreuse, *Broch.* t. ii. p. 25.—Soude muriatée fibreuse, *Hauy*, t. ii. p. 356. 365.—Fasriges Steinsalz, *Reuss*, b. iii. s. 27. *Id. Leonhard*, Tabel. s. 45. *Id. Karsten*, Tabel. s. 56. *Id. Haus.* s. 121.—Soude muriatée fibreuse-conjointe, *Hauy*, Tabl. p. 20.—Fasriges Steinsalz, *Lenz*, b. ii. s. 979.

External Characters.

Its colour is greyish, yellowish and snow white; seldom marked with stripes of ash-grey, Berlin and violet blue.

It occurs massive.

Internally, it is shining and glistening, and the lustre is pearly.

The fracture is parallel, sometimes delicate, sometimes coarse, and usually curved, fibrous.

The fragments are splintery.

It is strongly translucent, verging on semi-transparent.

In other characters it resembles the preceding kind.

Chemical Characters.

It decrepitates briskly when exposed to the action of the blowpipe, or when laid on burning coals.

Constituent

Constituent Parts.

Cheshire Rock-Salt.

Muriate of Soda,	-	$983\frac{1}{4}$
Sulphate of Lime,	-	$6\frac{1}{2}$
Muriate of Magnesia,	-	$0\frac{5}{16}$
Muriate of Lime,	-	$0\frac{1}{16}$
Insoluble matter,	-	10
		1000.0

Henry, in Philosophical Transactions for 1810, part i. p. 97.

Geognostic Situation.

It is sometimes found in transition gypsum in the transition-slate of the Allee Blanche, and between the grey-wacke and black transition limestone near Bex, below the Dent de Chamossaire; in alpine limestone, at Hall in the Tyrol, which is now considered as one of the members of the transition class; but the greatest formation is that in muriatiferous clay, in which the salt occurs, either disseminated, or in beds, alternating with the clay, or in vast irregular masses included in it. In this formation, the salt is occasionally associated with thin layers of anhydrite, stinkstone, limestone, and sandstone. It appears to lie upon, or even to alternate with, the first flœtz gypsum formation. Humboldt mentions a formation of rock-salt in muriatiferous clay, lying on a very new sandstone, at Punta Araya in America *; and small quantities of it occur in the third flœtz gypsum in the Segeberg, near Kiel in Holstein.

* Humboldt's Personal Narrative, vol. ii. p. 269.

Geographic Situation.

Europe.—The principal deposite of salt in this island is that in Cheshire, where there are several beds that vary in thickness from four feet to upwards of one hundred and thirty feet, and alternate with clay and marl, which contain compact, foliated, granular and radiated gypsum. Rock-salt also occurs at Droitwich in Worcestershire.

In the north of Germany, rocks of the salt formation and salt springs occur,—an evidence of the existence of salt beds, as all salt springs issue from salt beds, or rocks richly impregnated with salt; but no salt beds appear at the surface until we come to the Circle of Austria, and the neighbouring countries. The range of salt beds commences at Hall in the Tyrol *, passes through Reichenthal in Bavaria, continues to Hallein in Salzburg, Hallstadt, Ischel, and Ebensee in Austria, and terminates at Aussee in Stiria. The further continuation of the salt deposition is found at a considerable distance, that is, in Hungary, at Marmoros, Rhona, Szek, and Speries: then again in the great inclosed circular valley of Transylvania; from thence it extends through Wallachia, Moldavia, Buckovina, Gallicia, to Upper Silesia.

The salt repository of Marmoros is well known. In Transylvania, which is a vast circular valley, having its bottom covered with salt, there are many extensive salt-works: in Moldavia there also numerous salt-mines; and, what is worthy of remark, the rock-salt itself there forms hills. The salt-mines of Wieliczka are situated about two leagues south-east of Cracau in West Gallicia, and about

* At Sulzbach, on the Necker, in Swabia, there is a great bed of clay richly impregnated with salt, and sometimes even containing great masses of it. This appears to be the farthest limit of the salt-beds on that side of Germany.

about nine miles to the north-east of the Carpathian Mountains. They have been worked since the year 1251, and their depth and extent is very great; by some said to be 900 feet, and having an extent of more than a league from east to west. According to Abbé Estner, the salt of Hungary, Transylvania, and West Gallicia, occurs only of a grey colour. The party-coloured is found principally in Upper Austria, Salzburg, Stiria, and the Tyrol *. The beautiful blue foliated variety was formerly found at Ischel in Upper Austria; and the very rare green variety is at present found at Berchtesgaden and Hallein in Salzburg, where the fibrous blue variety which occurs in the Tyrol is also met with.

There are, besides, immense deposites of salt in Old and New Castile in Spain: thus, at Cordova, it is said to form a hill between 300 and 400 feet high; and in France there are salt-springs, but no salt-beds have hitherto been discovered.

Africa.—Besides the great beds of this mineral found in Europe, it is also very extensively distributed in the other quarters of the globe. In the northern part of Africa, on both sides of the Atlas Mountains, vast quantities of rock-salt occur. Mr Horneman, on his journey from Cairo to Ummosogeir, discovered a plain on the summit of the chain of limestone mountains that bound the Desart of Lybia to the north, consisting of a mass of rock-salt, spread over so large a tract of surface, that in one direction no eye could reach its termination, and its width he computed at several miles. It also occurs on the western and southern borders of the Great Desart of Sahra; in the country of Congo; and in Abyssinia †.

* The salt-mines in the Tyrol are 5000 feet above the level of the sea.

† Bruce mentions, that in some parts of Abyssinia, cubic pieces of rock-salt pass as current coin.

Asia—There is a considerable mine of rock-salt twenty versts from Jena-Tayerska, in the desart between the Volga and the Uralian Mountains; another, named Iletzki, near Astracan; and there are several others in Siberia *. Salt-mines are worked in that part of China which borders on Tartary. At Teflis, Tauris, and other places in Persia, there are great masses of rock-salt; and we are informed, that in the Desart of Caramania, and also in Arabia, rock-salt is so abundant, and the atmosphere is so dry, that the inhabitants use it for building houses. The Island of Ormuz, situated in the mouth of the Persian Gulf, is principally composed of rock-salt. Rock-salt is one of the mineral productions of the valley of Caschmere; and in the province of Lahore in India, there is a hill of rock-salt, equal to that of Cordova; the salt of this hill is cut into dishes, plates, and stands for lamps †.

America.—Rock-salt is found in vast quantity on the elevated Desarts of Peru, where it is very hard, and has usually a violet colour; also in the Cordilleras of New Granada, at the height of 2000 toises ‡. It occurs in considerable quantity in Upper Louisiana; and great masses of it have been found at the junction of the stream of Atha-pus-caou with the Atha-pus-caou Lake; and in California.

New Holland.—According to Governor Hunter, it is found in considerable quantity on the east coast of that country.

Uses.

* Pallas speaks of rock-salt in the neighbourhood of the river Jaik, which is sometimes so hard as to snap the pick-axes made use of in quarrying it.

† Pennant's Outlines of the Globe, vol. i. p. 42.

‡ Humboldt's Personal Narrative, vol. ii. p. 268.

Uses.

Its uses are very various and important. We employ it daily as a seasoning for our food: vast quantities are employed for the preservation of animal flesh, butter, &c.; it is also used as a manure, in the manufacture of earthenware, soap-making, and in many metallurgic operations. It affords muriatic acid and soda, by certain chemical processes. It is sometimes employed in its crude state, but is more commonly purified.

Second Subspecies.

Lake-Salt.

Seesalz, *Werner.*

Seesalz, *Reuss,* b. iii. s. 36.

External Characters.

Its colour is white.

Internally it is shining or glistening, and intermediate between vitreous and resinous.

It occurs either in thin plates, which are formed on the surface of salt lakes or inland seas, or in grains on their bottoms.

It is translucent.

Geognostic and Geographic Situations.

It is found on the bottoms and sides of salt-lakes.

Europe.—It is collected in the islands of Cyprus and Milo, in the Mediterranean sea. Nearly the half of the peninsula of the Crimea is filled with salt-lakes, which afford a great quantity of lake-salt.

Asia.—Lake-salt is collected in the neighbourhood of the Caspian.

Africa.—At Manzelach, near Alexandria, there are two salt-lakes, which afford a great quantity of fine white salt. The bottoms of the salt-lakes in the land of the Hottentots and the Caffres, are so compactly covered with salt, that it appears like ice, and the grains or distinct concretions adhere so closely together, that the mass is as hard as stone. Many extensive districts are supplied with salt from the lake of Dombu, which is situated in the great Desart of Bilma, in the kingdom of Bornu.

America.—Lake-salt is collected in several of the salt-lakes in North America, as in Mexico.

5. Borax.

Borax Tincal, *Wall.* t. ii. p. 82.—Tinkal, *Leonhard*, Tabel. s. 44. *Id. Karsten*, Tabel. s. 56.—Soude boratée, *Hauy*, Tabl. p. 20.—Borax, *Lenz*, b. ii. s. 1024.

External Characters.

Its colours are greyish, yellowish, and greenish-white: also greenish grey, and mountain-green.

It occurs crystallised in the following figures:

1. Oblique six-sided prism, with alternate broad and narrow lateral planes, and oblique terminal planes.
2. Oblique six-sided prism, with two opposite broader lateral planes: sometimes bevelled on the extremities, the bevelling planes set on the smaller lateral planes.

3. Flat

3. Flat double four-sided pyramid.

The surface of the crystals is sometimes smooth, sometimes rough, and with a white, grey, or brown crust.

The crystals occur loose, and of various sizes.

Internally it is shining and resinous.

The fracture is partly foliated, partly flat conchoidal.

The fragments are blunt-edged.

It is semi-transparent.

It refracts double in a high degree.

It is soft and very soft.

It is brittle.

It is easily frangible.

Specific gravity, 1.569, *Karsten*. 1.705, *Klaproth*.

Its taste is alkaline and sweet.

Chemical Characters.

It intumesces before the blowpipe, and melts into a transparent glass.

Constituent Parts.

Boracic Acid, - -	37.00
Soda, - - -	14.50
Water, - - -	47.00
	98.50

Klaproth, Beit. b. iv. s. 353.

Geognostic and Geographic Situations.

It occurs dissolved in the water of many springs in Persia; also in the soil of different parts of Persia; and in Thibet, it is found in the soil, or in the water of lakes.

It is said also to occur in China, and in Peru.

Uses.

Uses.

It is used as a flux for metals, and as an ingredient in artificial gems; but its great use is to facilitate the soldering of the more precious metals. It is employed as a flux by mineralogists in examining the properties of minerals before the blowpipe; and is sometimes used in medicine as a refringerant.

Observations.

1. The name *Borax* occurs in Geber, who wrote in the ninth century; and is derived from the word Baurach, in use among the Arabians. It has been confounded with the Chrysocolla of Pliny, in consequence of the use that jewellers make of it in soldering gold. It is brought from India in an impure state, under the name *tinkal*, enveloped in a kind of fatty matter, which is soap, with soda for its base. When purified in Europe, it takes the name of *borax*. This purification is performed by the Dutch; but the process which they follow is unknown.

2. Karsten describes, under the name *Sassolin*, a mineral principally composed of boracic acid. The following account contains the principal information we possess in regard to it:

Sassoline, or Native Boracic Acid.

Natürliches Sedativsalz, *Estner*, b. iii. s. 84.—Sassolin, *Reuss*, b. ii. 3. s. 12. *Id. Leonhard*, Tabel. s. 44. *Id. Karsten*, Tabel. s. 56.—Acide boracique, *Hauy*, Tabl. p. 56.

Its colours are greyish and yellowish white, and cream-yellow.

It occurs in grains, crusts, very small corroded pieces, which

which appear to be composed of crystalline grains, and acicular crystals.

Externally it is uneven.

It is dull or glimmering, and the lustre is resinous.

The fracture passes from uneven into small foliated.

The fragments are blunt-edged.

It is feebly translucent.

It becomes resinous in the streak.

It is soft, and friable.

Chemical Characters.

It melts easily before the blowpipe into a transparent globule.

Constituent Parts.

Boracic Acid, - -	86.0
Ferruginous Sulphate of Manganese, - - - -	11.0
Sulphate of Lime, - -	3.0
	100.0

Klaproth, Beit. b. iii. s. 99.

Geognostic and Geographic Situations.

It is found on the edges of hot-springs near Sasso, in the territory of Florence.

GENUS

Genus II.—Salts of Potash.

This Genus contains but one species, viz. Nitre.

Nitre.

Natürlicher Salpeter, *Werner*.

Nitrum terra mineralisatum, *Wall.* t. ii. p. 45.—Nitrate de Potasse, *De Born*, t. ii. p. 57.—Natürlicher Salpeter, *Wid.* s. 602.—Nitre, *Kirw.* vol. ii. p. 25.—Natürlicher Salpeter, *Estner*, b. iii. s. 55. *Id. Emm.* b. ii. s. 16.—Nitrate de Potasse, *Lam.* t. i. p. 468.—Potasse nitratée, *Hauy*, t. ii. p. 346. 355.—Le Nitre natif, *Broch.* t. ii. p. 17.—Salpeter, *Reuss*, b. iii s. 21.—Natürlicher Salpeter, *Lud.* b. i. s. 177. *Id. Suck.* 2ter th. s. 9. *Id. Bert.* s. 325.—Potasse nitratée, *Lucas*, p. 21.—Salpeter, *Leonhard*, Tabel. s. 44.—Potasse nitratée, *Brong.* t. i. p. 112.—Salpeter, *Karsten*, Tabel. s. 56. *Id. Haus.* s. 121.—Nitrate of Potash, *Kid*, vol. ii. p. 3.—Salpeter, *Lenz*, b. ii. s. 965.

External Characters.

Its colours are greyish-white, yellowish-white, and snow-white.

It occurs in flakes, crusts, and in capillary crystals, seldom in six-sided prisms, acutely acuminated with six planes, which are set on the lateral planes *.

It is dull, glimmering, or shining, and the lustre vitreous.

The fracture is delicate and straight fibrous, or small conchoidal.

It

* The crystals of artificial Nitre are the octahedron, or double four-sided pyramid, and the six-sided prism, acuminated with six planes, which are set on the lateral planes.

It alternates from translucent to transparent.
It is soft, or friable.
It is brittle, and easily frangible.
Its taste is cooling and saline.
Specific gravity, 1.9369, *Hassenfratz*.

Chemical Characters.

It deflagrates when thrown on hot coal.

Constituent Parts.

The Natural Nitre of Molfetta, according to Klaproth:

Nitrate of Potash, -	42.55
Sulphate of Lime, -	25.45
Carbonate of Lime, -	30.40
Muriate of Potash? -	0.20
Loss, - - -	1.40
	100.00

Klaproth, Beit. b. i. s. 320.

Geognostic Situation.

It is usually found in thin crusts on the surface of soil, and sometimes also covering the surface of compact limestone, chalk, and calc-tuff. In many countries it germinates in certain seasons out of the earth, and when this earth is accumulated in heaps, so as to expose a large surface to the atmosphere, it is found to produce it annually.

Geographic Situation.

Europe.—It is found in great quantities in many plains in Spain; very abundantly in the plains of Hungary, the Ukraine, and Podolia; in France, on the walls and floors of chalk caves; in the county of Bamberg, on a species of of limestone marl; on marly sandstone in the neighbourhood of Göttingen; at Hornberg, near Wurzburg, encrusting calc-tuff; but the most remarkable repository of natural nitre in Europe, is that discovered by Abbé Fortis, near Molfetta, in the kingdom of Naples. It there occurs encrusting a yellowish-grey coloured compact limestone. It is never found in beds with the limestone, as mentioned by some mineralogists, but always on its surface.

Asia.—Nitre is very abundant in India; also in Persia; and in the valley between Mount Sinai and Suez, in Arabia.

Africa.—This salt is abundant in Egypt; also at Ludamar, in the interior of Africa; and in the Karoo Desart, to the east of the Cape of Good Hope.

America.—The nitre used for the manufacture of gunpowder in the United States of America, is obtained from an earth collected in the limestone caves of Kentucky. It effloresces in considerable abundance on the soil near Lima; and in Tucuman, in South America.

Uses.

In Hungary, Spain, Molfetta, and the East Indies, considerable quantities of natural nitre are collected; but the greatest proportion of that used in commerce, is obtained by working artificial nitre beds. These consist of the refuse of animal and vegetable bodies, undergoing putrefaction, mixed with calcareous and other earths. Its principal use is in the fabrication of gunpowder: it is also used in medicine, and many of the arts.

Genus

Genus III.—Salts of Ammonia.

This Genus contains two species, viz. Sal Ammoniac, and Mascagnine.

1. Sal Ammoniac, or Muriate of Ammonia.

Natürlicher Salmiack, *Werner*.

Sal ammoniacum, *Wall.* t. ii. p. 77.—Sel Ammoniac, *Romé de L.* t. i. p. 382. *Id. Born,* t. ii. p. 54.—Natürlicher Salmiack, *Wid.* s. 610.—Sal Ammoniac, *Kirw.* vol. ii. p. 33.—Natürlicher Salmiack, *Estner,* b. iii. s. 78. *Id. Emm.* b. ii. s. 24. —Muriate d'Ammoniac, *Lam.* t. i. p. 473.—Le Sel Ammoniaque, *Broch.* t. ii. p. 27.—Ammoniaque muriatée, *Hauy,* t. ii. p. 380. 386.—Salmiack, *Reuss,* b. iii. s. 38. *Id. Lud.* b. i. s. 180. *Id. Suck.* 2ter th. s. 180. *Id. Bert.* s. 328. *Id. Mohs,* b. ii. s. 267.—Ammoniaque muriatée, *Lucas,* p. 25.—Salmiack, *Leonhard,* Tabel. s. 45.—Ammoniac muriatée, *Brong.* t. i. p. 109.—Salmiack, *Karsten,* Tabel. s. 56. *Id. Haus.* s. 122.—Muriate of Ammonia, *Kid,* vol. ii. p. 12.—Salmiak, *Lenz,* b. ii. s. 984.—Sal Ammoniac, *Aikin,* p. 251.

This species is divided into two subspecies, viz. Volcanic Sal Ammoniac, and Conchoidal Sal Ammoniac.

First Subspecies.

Volcanic Sal Ammoniac.

Vulcanischer Salmiak, *Karsten.*

External Characters.

The colours are yellowish and greyish-white; pearl-grey

grey and smoke-grey; wine-yellow; sometimes apple-green, sulphur-yellow. and brownish-black.

It occurs in efflorescences, crusts, stalactitic, small botryoidal, tuberose, corroded; and crystallised in the following figures:

1. Octahedron.
2. Cube, more or less deeply truncated on the edges.
3. Rectangular four-sided prism, acuminated with four planes, which are set on the lateral planes, or on the lateral edges.
4. Garnet or rhomboidal dodecahedron.
5. Leucite crystallization, or the double eight-sided pyramid, acuminated with four planes.

The crystals are small and very small; and their lateral planes are usually smooth.

Externally it is dull or glistening; internally shining and vitreous.

The fracture is even or uneven, sometimes inclining to fibrous.

It sometimes occurs in granular distinct concretions.

It alternates from transparent to opaque.

It is soft.

It is slightly ductile, and elastic.

Specific gravity, 1.5442, *Hassenfratz*.

Its taste is pungent and saline.

Chemical Characters.

When moistened, and rubbed with quicklime, it gives out a pungent ammoniacal odour.

Constituent

Constituent Parts.

Sal Ammoniac of Vesuvius.

Muriate of Ammonia,	-	99.5
Muriate of Soda,	- -	0.5
		100.0

Klaproth, Beit. b. iii. s. 91.

Geognostic Situation.

As its name implies, it is a volcanic production, occurring in the fissures, or on the surface of volcanic or pseudo-volcanic rocks.

Geographic Situation.

Europe.—It occurs in the vicinity of burning beds of coal, both in Scotland and England. It is found in the Island of Iceland. On the Continent, it is met with at Solfatara, Vesuvius, Ætna, the Lipari Islands, and Tuscany.

Asia.—Thibet, Persia, and the Isle of Bourbon.

America.—In volcanic districts both in North and South America.

Second Subspecies.

Conchoidal Sal Ammoniac.

Muschlicher Salmiak, *Karsten.*

External Characters.

Its colour is greyish-white.

It occurs in angular pieces.

Its surface is uneven.

 Externally

Externally it is glimmering; internally it is shining and vitreous.

The fracture is nearly perfect conchoidal.

The fragments are indeterminate angular.

It is semi-transparent or transparent.

It is malleable.

It is soft.

It is light.

Its taste is pungent and urinous.

Constituent Parts.

Muriate of Ammonia,	-	97.50
Sulphate of Ammonia,	-	2.50
		100

Klaproth, Beit. b. iii. s. 94.

Geognostic and Geographic Situations.

This mineral is said to occur, along with sulphur, in rocks of indurated clay or clay-slate, in the country of Bucharia.

Uses.

This salt is used for a variety of purposes. Great quantities of artificial sal ammoniac are annually exported from this country to Russia, where it appears to be used by the dyers. It is employed by coppersmiths, to prevent the oxidation of the surface of the metals they are covering with tin. It has the property of rendering many metallic oxides volatile, and is frequently used to separate metals from each other. Dissolved in nitric acid, it forms the fluid named *aqua regia*, employed in the solution of gold; and the pure ammonia is also obtained from this salt.

Observations.

Observations.

It is an opinion entertained by many, that this salt is the same with the sal ammoniac (ἁλς ἀμμωνιακως) of the ancients; but the accounts of Pliny, Dioscorides, Collumella, Synesius, Herodotus, Strabo, and Arrian, prove that they understood by sal ammoniac rock-salt, and even the ancient Arabian physicians Avicenna and Serapion, who flourished during the eleventh century, describe rock-salt under the name sal ammoniac. The first account we have of sal ammoniac is in a treatise of Geber's, the date of which is uncertain.—Vid. Beckman, Beiträge zur Geschichte der Erfindungen, b. v. s. 254,—285.

2. Mascagnine *, or Sulphate of Ammonia.

Mascagnin, *Karsten.*

Mascagnin, *Reuss,* b. ii. 3. s. 45. *Id. Karsten,* Tabel. s. 56.—Ammoniaque sulphatée, *Hauy,* Tabl. p. 21.—Mascagnin, *Lenz,* b. ii. s. 985.

External Characters.

Its colours are yellowish-grey and lemon-yellow.
It occurs in mealy crusts, or stalactitic.
Internally it is dull or glistening.
The fracture is uneven or earthy.
It is semi-transparent or opaque.
Its taste is sharp and bitter.

* It is named after the discoverer M. Mascagni.

Chemical Characters.

It is easily soluble in water; partly volatilised by heat; and becomes moist on exposure to the air.

Constituent Parts.

It is a compound of Ammonia, Sulphuric Acid, and Water.

Geognostic and Geographic Situations.

It occurs among the lavas of Ætna, and Vesuvius; in the Solfatara by Puzzæolo; in the lagunes, near Siena in Tuscany; and on the bottom of a hot spring in Dauphiny.

ORDER III.

ORDER III.—METALLIC SALTS.

All the Salts included under this Order, have a metallic basis. They are easily soluble in water, and taste bitter, sweetish, or metallic. They are soft, light, and either white, green, blue, or red.

Genus I.—Salts of Iron.

Of this genus but one species has hitherto been met with in the mineral kingdom, viz. Iron-Vitriol or Sulphate of Iron.

Iron-Vitriol, or Sulphate of Iron.

Eisen-Vitriol, *Werner.*

Στυπτηρια of the Greeks.—Alumen, *Plin.* xxxv. 15. p. 52.—Chalcanthum, *Plin.* ?—Vitriolum ferri, *Waller,* Syst. Min. ii. p. 22.—Fer sulphatée, *Hauy,* t. iv. p. 122.—Eisen Vitriol, *Reuss,* b. ii. 3. s. 68. *Id. Karsten,* Tabel. s. 56. *Id. Haus.* s. 137.—Sulphat of Iron, *Kid,* vol. ii. p. 20.—Eisenvitriol, *Lenz,* b. ii. s. 989.—Green Vitriol, *Aikin,* p. 250.

External Characters.

Its colours are emerald, apple, and verdigris green, and sometimes grass-green: on exposure to the weather, it becomes straw-yellow, cream-yellow, ochre-yellow, and yellowish-brown.

It occurs pulverulent, massive, disseminated, stalactitic, tuberose, botryoidal, reniform; and crystallised in the following figures:

 1. Oblique

1. Oblique double four-sided pyramid.
2. Rhomb, variously modified by truncations on the angles and edges.
3. Capillary crystals.

It is shining, both externally and internally, and the lustre is vitreous, with exception of the capillary varieties, which are pearly.

The fracture is flat conchoidal or foliated, and occasionally fibrous.

It alternates from semi-transparent to opaque.

It refracts double.

It is soft.

It is friable, brittle, and easily frangible.

Specific gravity, 1.972, *Brisson.*

It tastes sweetish, styptic, and metallic.

Chemical Characters.

Before the blowpipe, on charcoal, it becomes magnetic, and colours glass of borax green.

Constituent Parts.

Oxide of Iron, - -	25.7
Sulphuric Acid, - -	28.9
Water, - - -	45.4
	100.0 *Berzelius,*

Geognostic and Geographic Situations.

It is always associated with iron-pyrites, by the decomposition of which it is formed. It occurs in several coal-mines in this country, and in many iron and coal mines on the Continent of Europe; and also in America, and Asia.

Uses.

Uses.

It is employed to dye linen yellow, and wool and silk black; in the preparation of ink; of Berlin-blue; for the precipitation of gold from its solution; and sulphuric acid can be obtained from it by distillation. The residue of the latter process (colcothar of iron) is used as a red paint, and, when washed, for polishing steel.

Observations.

Vitriol of iron, by exposure, becomes yellow, and at last brown: in this state, it appears to answer to the *Misy*, Plin. Hist. Nat. xxxiv. 31. The Melanteria, or Inkstone of Pliny, the Lapis atramentarius flavus of Wallerius, appear to be varieties of this mineral.

GENUS II.—SALTS OF COPPER.

Of this Genus but one species has been hitherto enumerated by mineralogists, viz. Blue Vitriol or Sulphate of Copper.

Blue Vitriol, or Sulphate of Copper.

Kupfervitriol, *Werner*.

Χαλκανθον of the Greeks.—Chalcanthum atramentum sutorium, *Plin.* Hist. Nat. xxxiv. 12. p. 32.—Vitriolum cupri, *Waller.* Syst. Min. t. ii. p. 20.—Cuivre sulphatée, *Hauy*, t. iii. p. 580. —Kupfervitriol, *Reuss*, b. ii. 3, s. 73. *Id. Karst.* Tabel. s. 56. —Sulphate of Copper, *Kid*, vol. ii. p. 23.—Kupfervitriol, *Lenz*, b. ii. s. 993.—Blue Vitriol, *Aikin*, p. 249.

External Characters.

The common colour is dark sky-blue, which sometimes

 approaches

approaches to verdigris-green. By exposure to the air, it becomes yellow.

It occurs massive, disseminated, stalactitic, dentiform; and crystallised in the following figures :

1. Cube, either truncated on the edges or angles.
2. Capillary crystals.

Externally and internally it is shining and vitreous.

The fracture is conchoidal.

The fragments are rather sharp-edged.

It is translucent.

It is soft, brittle, and easily frangible.

Specific gravity, 2.1943, *Hassenfratz.*

Its taste is nauseous, bitter, and metallic.

Chemical Characters.

When a portion of it is dissolved in water, and spread on the surface of iron, it immediately covers it with a film of copper.

Constituent Parts.

Oxide of Copper,	-	32.13
Sulphuric Acid,	-	31 57
Water,	- -	36.30
		100.00 *Berzelius.*

Geognostic and Geographic Situations.

It occurs, along with copper-pyrites, in Pary's mine in Anglesea; and also in the copper-mines in the county of Wicklow, in Ireland It is found in considerable quantity in several copper-mines on the Continent of Europe; and also in Siberia.

Uses.

It is used in cotton and linen printing, and the oxide separated from it is used by painters.

GENUS

GENUS III.—SALTS OF ZINC.

This Genus contains but one species, viz. White Vitriol or Sulphate of Zinc.

White Vitriol or Sulphate of Zinc.

Vitriolum zinci, *Waller*. Syst. Min. t. ii. p. 24.—Zinc sulphatée, *Hauy*, t. iv. p. 180.—Zinkvitriol, *Leonhard*, Tabel. s. 45. *Id. Karsten*, Tabel. s. 56.—White Vitriol or Sulphate of Zinc, *Kid*, vol. ii. p. 24.—Zinkvitriol, *Lenz*, b. ii. s. 996.—White Vitriol, *Aikin*, p. 250.

External Characters.

Its colours are greyish, yellowish, reddish, and greenish-white.

It occurs massive, stalactitic, reniform, botryoidal, in crusts; and crystallised in the following figures:

1. Rectangular four-sided prism, acuminated with four planes, which are set on the lateral planes.
2. Acicular crystals, which are promiscuously aggregated.

It is shining.

The fracture is fibrous and radiated.

It occurs in granular and prismatic concretions,

It is translucent.

It is soft, brittle, and easily frangible.

Specific gravity, 2.00, *Born*.

Its taste is nauseous metallic.

Chemical Characters.

It intumesces before the blowpipe, but does not phosphoresce: it dissolves in 2.285 parts of boiling water.

Constituent

Constituent Parts.

From Rammelberg.		Ditto.
Oxide of Zinc, -	27.5	21.739
Oxide of Manganese,	0.5	6.522
Sulphuric Acid, -	22.0	71.739
Water, - - -	50.0	
	100.0	100
	Klaproth, Beit. b. v. s. 196.	*Herz.* Archiv. b. iii. s. 537.

Geognostic and Geographic Situations.

It occurs in mineral repositories that contain blende, and appears to be formed by the decomposition of that ore. It occurs at Holywell in Flintshire; and it is said also in Cornwall: in the Rammelsberg in the Hartz; at Spitz in Austria; at Schemnitz in Hungary; and Sahlberg in Sweden.

Uses.

It is used as a medicine; is employed in great quantities by varnishers to make oil drying; and a fine white colour named *Zinc-white*, which is more durable than white-lead, is prepared from it. To prepare this colour, the salt is dissolved in water, and the white oxide, which is the zinc-white, is precipitated from it by means of potash or chalk.

GENUS IV.

GENUS IV.—SALTS OF COBALT.

This Genus contains one species, viz. Red Vitriol or Sulphate of Cobalt.

Red Vitriol or Sulphate of Cobalt.

Kobaltvitriol, *Werner*.

Kobaltvitriol, *Kopp*, in Leonhard's Taschenbuch, b. i. s. 111. *Id. Leonhard*, Tabel. s. 45. *Id. Karsten*, Tabel. s. 56. *Id. Lenz*, b. ii. s. 1003.—Red Vitriol, *Aikin*, p. 250.

External Characters.

Its colour is flesh-red, inclining to rose-red.

It occurs coralloidal, stalactitic, also in crusts.

The surface is rough and longitudinally furrowed.

It is dull, and seldom shining on the surfaces of the distinct concretions, and the lustre is pearly.

The fracture is earthy.

The fragments are blunt-edged.

It occurs in granular distinct concretions.

It is opaque.

It affords a yellowish-white streak.

It is easily friable, and brittle.

It is light.

It tastes styptic.

Chemical Characters.

Its solution affords, with carbonate of potash, a pale bluish precipitate, which tinges borax of a pure blue colour.

Constituent

Constituent Parts.

Oxide of Cobalt, -	38.71
Sulphuric Acid, -	19.74
Water, - - -	41.55
	100.00

Koppe, in Journal fur die Chemie, Physik et Mineralogie, b. vi. Heft 1. 1808, s. 157.

Geognostic and Geographic Situations.

It occurs in mining-heaps in Biber, along with lamellar heavy-spar, earthy cobalt-ore, and grey cobalt-ore; and it has been also found in the Leogang.

CLASS III.

CLASS III.

INFLAMMABLE MINERALS.

The Minerals belonging to this class, do not vary much in colour, and almost the only tints are of black, brown, and yellow: they are very rarely crystallised; the hardest yield easily to the knife; their specific gravity seldom exceeds 2.0; and they feel warmer than earthy or saline minerals. Some of them are very combustible, and the rest burn more or less easily before the blowpipe. They are more nearly allied to the Metallic than to the Saline or Earthy classes of Minerals.

I. SULPHUR FAMILY.

This Family contains but one species, viz. Sulphur.

Sulphur.

Natürlicher Schwefel, *Werner.*

This species is divided into two subspecies, viz. Common Sulphur, and Volcanic Sulphur.

First

First Subspecies.

Common Sulphur.

Gemeiner natürlicher Schwefel, *Werner.*

Sulphur nativum purum flavum, *Wall.* t. ii. p. 123.—Soufre, *Romé de Lisle,* t. i p. 28. *Id. De Born,* t. ii. p. 91.—Natürlicher Schwefel, *Wid* s 646. *Id. Wern.* Pabst. b. i. s. 368.—Native Sulphur, *Kirw.* vol. ii. p. 69.—Natürlicher Schwefel, *Estner,* b. iii. s. 178. *Id. Emm.* b. ii. s. 189.—Soufre, *Lam.* t. i. p. 68.—Le Soufre natif, *Broch.* t. ii. p. 37.—Soufre, *Hauy,* t. ii p. 277.–287 —Schwefel, *Reuss,* b. iii. s. 84. *Id Lud* b. i. s. 184. *Id. Suck.* 2ter th. s. 38. *Id. Bert.* s. 338. *Id Mohs,* b. ii s. 277. *Id. Leonhard,* Tabel. s. 47.—Soufre, *Brong.* t. ii. p. 68.—Schwefel, *Karsten,* Tabel. s. 58. *Id. Haus.* s. 67.—Sulphur, *Kid,* vol. ii. p. 30.—Gemeiner natürlicher Schwefel, *Lenz,* b. ii. s. 1033.—Sulphur, *Aikin,* p. 5.

External Characters.

The principal colour is sulphur-yellow, of different degrees of intensity: it occurs also honey-yellow, lemon-yellow, and wax-yellow, with an intermixture of grey and brown.

It occurs massive, disseminated, in crusts, in dusty particles; and crystallised in the following figures:

1. Very acute octahedron, or double four-sided pyramid, in which the base is oblique *, fig. 148. The pyramid sometimes terminates in a line †, fig 149.; sometimes appears truncated on the apices ‡, fig. 150. or two opposite angles ||, fig.

---

* Soufre primitif, Hauy.

† Soufre cunciforme, Hauy.

‡ Soufre basé, Hauy.

|| Soufre unitaire, Hauy.

fig. 151.; the edge of the common basis is occasionally truncated *, fig. 152.

2. Octahedron, acuminated with four planes, which are set on the lateral planes †, fig. 153. The apices of the acumination are sometimes truncated ‡, fig. 154.

3. Double six-sided pyramid, with two opposite broad, and four smaller, lateral planes, and which end in a line ||, fig. 155.

4. The preceding figure acuminated on the extremities with four planes, and the acuminating planes set on the smaller lateral planes §, fig. 156.

5. In delicate acicular crystals.

The crystals are middle-sized, small, and very small, seldom large.

The surface of the crystals is generally smooth, seldom drusy.

Internally it is intermediate between shining and splendent, and the lustre is adamantine, approaching to resinous.

The fracture is uneven, inclining sometimes to splintery, sometimes to conchoidal.

The fragments are angular, and blunt-edged.

It is translucent; the crystals are semi-transparent and transparent, and they refract double.

It is soft.

It is brittle, and easily frangible.

When

* Soufre prisme, Hauy.

† Soufre diodecaedre, Hauy.

‡ Soufre octodecimal, Hauy.

|| Soufre emoussé, Hauy.

§ Soufre unibinaire, Hauy.

When rubbed, it exhales a faint sulphureous smell, and becomes resino-electric.

It is light.

Specific gravity, 1.990 to 2.033.

Chemical Characters.

It is easily inflammable, burning with a lambent bluish flame, and a suffocating odour.

Geognostic Situation.

It occurs in primitive and flœtz rocks, but most abundantly in the flœtz formations. It occurs in veins of copper-pyrites, that traverse granite at Schwartzwald in Suabia; in mica-slate at Glasshütte, near Schemnitz in Hungary; in mica-slate, in the form of a bed, along with quartz, between Alausi and Ticsan, in Quito in Peru; and also in primitive porphyry in the same country. In the flœtz rocks, it occurs either in imbedded masses, of various magnitudes, or lining the walls of cavities, along with radiated celestine, in the beds or strata of gypsum, marl, or limestone, in which it is contained. One of the newest formations of this subspecies, is that which occurs in superimposed crystals on bituminous wood, or earth-coal, in Thuringia.

Geographic Situation.

Europe.—In the island of Iceland, where it occurs in considerable quantity, it is associated with gypsum; in Ireland, it occurs in small nests in limestone; in Sicily, where it is abundant, it is frequently beautifully crystallised, and is contained in rocks of gypsum; in Spain, it occurs principally in Arragon and Murcia, and it is so abundant, as to answer the consumption of the whole country; the sulphur of the mines of Swarzowicé in Poland,

land, is distributed in kidneys through a kind of marl; magnificent crystals are found at Conila, near Cadiz; and beautiful crystals at Bevieux, in the Canton of Berne, in Switzerland; massive varieties are collected in gypsum, near the glaciers of Pesay and Gebrulartz, in Switzerland; and in that of Lauenstein in Westphalia; an earthy variety is found in drusy cavities in flint, in the neighbourhood of the Abbey of La Charité, and the village of Neuville, in the department of Doubs in France; it is disseminated through sandstone at Buodoshegy, in Transylvania; it is intermixed with red manganese ore at Kapnic; and with red orpiment at Felsobanya.

Asia.—In the gold mines of Catharinenburg; and the galena veins in the Altain Mountains, in Siberia.

America.—In California, and Peru.

Africa.—Imbedded in basalt in the Isle de Bourbon; and in considerable abundance in Mount Atlas, in northern Africa.

Second Subspecies.

Volcanic Sulphur.

Vulcanischer Natürlicher Schwefel, *Werner*.

Le Soufre natif volcanique, *Broch.* t. ii. p. 42.—Vulcanischer Schwefel, *Reuss,* b. iii. s. 90. *Id. Mohs,* b. ii. s. 282. *Id. Leonhard,* Tabel. s. 47. *Id. Karsten,* Tabel. s. 38. *Id. Lenz,* b. ii. s. 1038.—Volcanic Sulphur, *Aikin,* p. 6.

External Characters.

The principal colour is sulphur-yellow, which passes into orange-yellow. It is sometimes yellowish-grey, and brick and cochineal red.

 It

It occurs massive, in blunt-edged pieces, in crusts, stalactitic *, fused-like, cellular, corroded, amorphous; and crystallised in the following figures:

1. Acute single three-sided pyramid.
2. Rhomboid †.
3. Acicular crystals.

It is glistening, and the lustre is resinous, inclining to adamantine.

The fracture is coarse grained uneven.

The fragments are indeterminate angular, and blunt-edged.

It is slightly translucent.

In other characters it agrees with the preceding subspecies.

Geognostic and Geographic Situations.

Europe.—It occurs only in volcanic countries, where it is found more or less abundantly amongst lavas. Solfatara, in the vicinity of Vesuvius, is one of the most famous repositories of volcanic sulphur, and it is there collected in considerable quantities for the purposes of commerce. It is also found in the island of Iceland; on Ætna; and in the Lipari Islands.

Africa—Island of Teneriffe, and island of Bourbon.

America.—In the islands of St Lucia, St Domingo, Martinique, and Guadaloupe.

Asia.—Island of Java.

Uses.

---

* Spallanzani observed stalactites of volcanic sulphur, three feet long and two inches thick, in a grotto formed in the walls of the crater of Vulcano.

† This variety, which is intermixed with calcareous earth, is found on Mount Ætna.

Uses.

It is used as a medicine; and it enters as a principal ingredient into the composition of gunpowder; its vapour is used for whitening wool and silk, and also for fumigations *; and sulphuric acid, which is so useful in the arts, is prepared from it.

* Sulphur was much employed by the ancients, particularly in their lustrations: it is mentioned by Homer, Pliny, Theocritus, Ovid, and Juvenal, as being used for that purpose.

II. BITUMINOUS FAMILY.

This Family contains the following species, viz. Naphtha, Mineral Oil or Petroleum, Mineral Pitch or Bitumen, Brown Coal, and Black Coal.

1. Naphtha.

Ναφθα of the ancient Greeks. Vid. *Plin.* Hist. Nat. t. ii. (ed. Bip. 1. p. 198.)—Bitumen candidum? *Plin.* Hist. Nat. xxxv. (ed. Bip. v. p. 324.)—Bitumen Naphtha, *Wall.* Syst. Min. t. ii. p. 98.—Naphte, *Romé de L.* t. ii. p. 192. *Id. De Born*, t. ii. p. 75. *Id. Wid* s. 617. *Id. Kirw.* vol. ii. p. 42.—Bitumine liquide blanchatre, *Hauy*, t. iii. p. 312.—Le Naphte, *Broch.* t. ii. p. 59.—Naphtha, *Reuss*, b. iii. 3. s. 96.—Bitume Naphte, *Brong.* t. ii. p. 19.—Naphtha, *Leonhard*, Tabel. s. 48.—Liquides Bergöl, *Karsten*, Tabel. s. 58.—Naphtha, *Hauy*, s. 117. *Id. Kid*, vol. ii. p. 61. *Id. Lenz*, b. ii. s. 1045. *Id. Aikin*, p. 3.

External Characters.

Its colours are yellowish-white, yellowish-grey, and wine-yellow.

It is perfectly fluid.

It is shining and resinous.

It feels greasy.

It exhales an agreeable bituminous smell.

Specific gravity, 0.7.

Chemical Characters.

It takes fire on the approach of flame, affording a bright white light.

Constituent

Constituent Parts.

It is a compound of Carbon, Hydrogen, and a little Oxygen.

Geognostic and Geographic Situations.

This mineral is seldom found in a pure state. It is said to occur in considerable springs on the shores of the Caspian Sea; in the Caucasus; Japan; Persia; and France; also in Sicily; and some districts in Italy, as Calabria, Modena, and Parma. These springs issue from rocks of different kinds, as limestone, marl, and sandstone.

Uses.

In Persia, Japan, and some parts of Italy, where it occurs in considerable quantity, it is used in lamps, in place of oil, for lighting streets, churches, &c. When mixed with certain vegetable oils, it forms an excellent varnish; and formerly it was employed as a vermifuge medicine.

2. Mineral Oil, or Petroleum.

Erdöl, *Werner*.

Bitumen liquidum, *Plin.* Hist. Nat. xxxv.—Maltha tarde fluens, *Wall* t. ii. p 92.—Petrole, *Romé de Lisle,* t. ii. p. 591. *Id. De Born,* t. ii. p. 75 —Petrol, *Kirw.* vol. ii. p. 43. *Id. Estner,* b. iii. s 97 *Id. Emm.* b. ii. s. 43.—Bitumine liquide noiratre, *Hauy,* t. iii. p. 312.—L'Huile minerale commune, ou le Petrol, *Broch.* t. ii. p. 59, 60.—Gemeines Bergöl, *Reuss,* b. iii. s. 96.–101. *Id. Mohs,* b. ii. s. 302. *Id. Leonhard,* Tabel. s. 48.—Bitumen Petrole, *Brong* t. ii. p. 24.—Verdicktes Bergöl, *Karsten,* Tabel. s. 58.—Steinöl, *Haus.* s. 117. —Petroleum, *Kid,* vol. ii. p. 62.—Bergöl, *Lenz,* b. ii. s. 1047. Petroleum, *Aikin,* p. 3.

External Characters.

Its colours are reddish and blackish brown, and brownish-black, sometimes inclining to green.

It is fluid, but approaches more or less to the viscid state.

It is shining and resinous.

It feels greasy.

It is semi-transparent, translucent, and opaque.

It exhales a strong bituminous odour.

It is so light as to swim on water.

Chemical Characters.

It inflames easily, emits a bluish flame, and yields a smoke more or less opaque, according to the density of the oil, and sometimes leaves a very small earthy residue.

Constituent

Constituent Parts.

It is composed of Carbon, Hydrogen, and a little Oxygen.

Geognostic Situation.

It generally flows from rocks of the coal formation, and usually from the immediate vicinity of beds of coal; also from limestone rocks. It occurs in marshes, on the surface of spring water; or it flows or trickles unmixed from its repository.

Geographic Situation.

Europe.—Oozing from flœtz rocks at St Catherine's Well, near Edinburgh, and in the island of Pomona, one of the Orkneys. Filling cavities and veins in shell-limestone at Pitchford and Madeley in Shropshire. Several springs of this mineral occur in France, as at Gabian in Languedoc, and in Auvergne. It is also found on the Lake Tegern, in Bavaria; near Neufschatel in Switzerland; at Amiano, twelve leagues from Parma; and in Mount Zibio, near Modena. It is also met with in Sicily; in the salt-mines in Transylvania; in Gallicia; and Moldavia.

Asia.—On the shores of the Caspian Sea; at Semenowa, in Siberia; and the stream of Taliza, in the Altain Mountains. There are very productive mines of mineral oil in the kingdom of Ava: about five hundred shafts or pits are sunk through soil, sandstone, slate-clay, and coal, and it is from the coal that the oil issues. When drawn from the pit, it is much mixed with water, from which it is separated by decantation. Mineral oil is also met with in Japan.

America.—On the banks of the Ohio; and, according to

to travellers, many springs in Pennsylvania carry along with them immense quantities of this mineral. It is also a production of Newfoundland; and the island of Trinidad.

Uses.

In Piedmont, Persia, Japan, and other countries, it is used in lamps, in place of oil, for lighting streets and churches. It is also used for warming rooms, when mixed with earth, and inflamed. It is occasionally employed instead of common tar, to preserve wood from decay, and from worms: also as a varnish; and in the composition of fire-works.

2. Mineral Pitch or Bitumen.

This species is divided into three subspecies, viz. Earthy Mineral Pitch, Slaggy Mineral Pitch, and Elastic Mineral Pitch.

First Subspecies.

Earthy Mineral Pitch.

Erdiges Erdpech, *Werner.*

Semi-compact Mineral Pitch, or Maltha, *Kirw.* vol. ii. p. 46.—Le Poix mineral terreuse, *Broch.* t. ii. p. 65.—Erdiges Bergpech, *Reuss,* b. iii. s. 107. *Id. Lud.* 1r th. s. 193. *Id. Suck.* 2ter th. s. 45. *Id. Bert.* s. 342. *Id. Mohs,* b. ii. s. 307. *Id. Leonhard,* Tabel. s. 48.—Thonartiges Erdpech, *Karst.* Tabel. s. 58.—Erdiges Erdpech, *Haus.* s. 117. *Id. Lenz,* b. ii. s. 1051.—Cohesive Mineral Pitch, *Aikin,* p. 3.

External Characters.

Its colour is blackish-brown.

It

It occurs massive.

It is faintly glimmering, inclining to dull.

The fracture is earthy, or small grained uneven, sometimes inclining to splintery.

The fragments are blunt-edged.

The streak is shining and resinous.

It is very soft.

It is sectile.

It feels greasy.

It is so light as almost to swim on water.

It smells strongly bituminous.

Chemical Characters.

It burns with a clear and brisk flame, emits an agreeable bituminous smell, and deposites much soot.

Constituent Parts.

Inflammable Matter, -	50.50
Silica, - - -	28.50
Alumina, - - -	15.50
Lime, - - -	4.25
Oxide of Iron, - -	1.19

Lenz, Min. b. ii. s. 1052.

Geognostic and Geographic Situations.

It occurs in the Iberg in the Hartz, along with slaggy mineral pitch, in veins that traverse grey-wacke; at Prague, in calcareous-spar veins that traverse transition greenstone.

Second

Second Subspecies.

Slaggy Mineral Pitch.

Schlackiges Erdpech, *Werner.*

Ασφαλτος of the Greeks, *Aristotelis,* lib. de Min. Ascult. expl. a J. Beckman, Gott. 1786, 4to, p. 280.—Bitumen, *Plin.* Hist. Nat. xxxv.—Bitumen solidum coagulatum friabile, Asphaltum, *Wall.* t. ii. p. 93.—Asphalte, ou Bitume de Judée, *Romé de L.* t. ii. p. 592. *Id. De Born,* t. ii. p. 78 —Bergpech, ou Judenpech, *Wid.* s. 624.—Asphaltum, or Compact Mineral Pitch, *Kirw.* vol. ii. p. 46.—Asphaltum, *Hatchett,* Lin. Trans. vol. iv. p. 132. —Schlackiges Erdpech, *Estner,* b. iii. s. 110. *Id. Emm.* b. ii. s. 50.—Asphalte, *Lam.* t. ii. p. 533. 635.—Bitume solide, *Hauy,* t. iii. p. 313.—La Poix minerale scoriacée, *Broch.* t. ii. p. 66.—Schlackiges Bergpech, *Reuss,* b. iii. s. 113. *Id. Lud.* b. i. s. 193. *Id. Suck.* 2ter th. s. 48. *Id. Bert.* s. 343. *Id. Mohs,* b. ii. s. 307.—Bitume asphalte, *Brong.* t. ii. p. 25.—Schlackiges Erdpech, *Leonhard,* Tabel. s. 48. *Id. Karsten,* Tabel. s. 58. *Id. Lenz,* b. ii. s. 1052.—Compact Mineral Pitch, *Aikin,* p. 3.

External Characters.

Its colour is velvet-black, which sometimes approaches to brownish-black.

It occurs massive, disseminated, and globular.

Internally it is splendent and shining; often glistening, and the lustre is resinous.

The fracture is either imperfect, or very perfect conchoidal.

The fragments are pretty sharp-edged.

It is very soft, passing into soft.

It is opaque.

It

It is sectile.

It retains its lustre in the streak.

It is not very difficultly frangible.

It feels greasy.

Specific gravity, 1.0—1.6.

When held between the fingers emits a bituminous smell.

Constituent Parts.

Slaggy Mineral Pitch from Avlona in Albania.

Carbonated Hydrogen, -	36	cubic inches.
Bituminous Oil, - -	32	grains.
Ammoniacal Water, -	6	
Carbon, - - -	30	
Silica, - - - -	7.50	
Alumina, - - -	4.50	
Lime, - - - -	0.75	
Oxide of Iron, - -	1.25	
Oxide of Manganese, -	0.50	

Klaproth, Beit. b. iii. s. 318.

The above quantities were obtained from 100 grains, and are partly products, partly educts.

Geognostic and Geographic Situations.

Europe.—It occurs in veins and imbedded masses in flœtz limestone in Fifeshire; in clay ironstone in East Lothian; in veins, at Haughmond Hill in Shropshire; and in mineral veins in Cornwall. At Iberg, near Grund in the Hartz, along with sparry ironstone, brown ironstone, and heavy-spar; in veins, along with calcareous-spar and brown ironstone, at Kamsdorf in Saxony: at Violenberg, near Grund, in pieces the size of a hen's egg, mixed with slaty glance-coal, in veins composed of compact

pact brown ironstone, cellular quartz, and straight lamellar heavy-spar. It is also met with at Nordberg and Dannemora in Sweden; Morsfeldt in the Palatinate; in the quicksilver mines of Deux Ponts; in Salzburg; Switzerland; Avlona in Albania; and Semenowa in Russia.

Asia.—In the Mountains of Caucasus; in Lake Asphaltes in Judea; Uralian Mountains.

America.—Mexico. In the Island of Trinidad, there is a lake three miles in circumference, covered with a bituminous substance, of the nature of slaggy mineral pitch.

Uses.

The Egyptians employed it in the process of embalming bodies *. The Arabians still use a solution of it in oil to besmear their horse harness, to preserve it from insects. The ancients also used it as an ingredient in mortar; and it is said that the walls of the famous city of Babylon were built with a mortar of this kind. The German translator of J. Bar. de Vignola's *Civil Baukunst*, observes, "I may here also remark, that we find in the accounts of travellers that buildings are often constructed with pitch; and that Peter de Val mentions, that he examined very old buildings, the stones of which were cemented by means of mineral pitch, and which were still very firm, and in good order." Klaproth says, that the slaggy mineral pitch of Avlona burns with a strong and lively

* Rouelle concludes, from experiments which he made on mummies, that the Egyptians employed slaggy mineral pitch in embalming the dead. This operation was performed in three different ways: the first with slaggy mineral pitch alone; the second with a mixture of this bitumen, and a liquor extracted from cedar, called *Cedria*; and the third with a similar mixture, to which resinous and aromatic substances were added.—*Hauy*, Mineralog. t. ii. p. 315, 316.

lively flame, and is considered as the principal ingredient in the *Greek Fire*, so much employed in former times.

Observations.

1. The substances described under the names *Asphalt*, *Jews Pitch*, *Mumia mineralis*, *Mineral Pitch*, *Bitumen of Judea*, principally belong to this subspecies, although under these names, by some mineralogists, earthy mineral pitch is understood.

2. Gagat or Jet, is a variety of pitch coal, and therefore cannot be arranged under this species.

3. The term *Asphalt*, sometimes applied to this substance, is derived from the name of the Lake of Judea, where this mineral occurs in abundance.

Third Subspecies.

Elastic Mineral Pitch.

Elastiches Erdpech, *Werner.*

Elastic Bitumen, *Hatchett*, Linn. Trans. vol. iv. p. 146. &c.—Mineral Cahoutchou, *Kirw.* vol. ii. p. 48.—Elastiches Erdpech, *Estner*, b. iii. s. 106. *Id. Emm.* b. iii. s. 106.—Cahoutchou fossile, *Lam.* t. ii. p. 540.—La Poix minerale elastique, *Broch.* t. ii. p. 64.—Bitume elastique, *Hauy*, t. iii. p. 313, 314.—Elastiches Bergpech, *Reuss*, b. iii. s. 110. *Id. Lud.* b. i. s. 192. *Id. Suck.* 2ter th. s. 46.—Elastisches Federharz, *Bert.* s. 343.—Elastisches Erdpech, *Mohs*, b. ii. s. 304. *Id. Leonhard*, Tabel. s. 48. *Id. Haus.* s. 117. *Id. Karst.* Tabel. s. 58. *Id. Lenz*, b. ii. s. 1052.—Elastic Mineral Pitch, *Aikin*, p. 4.

External Characters.

Its colours are blackish-brown, sometimes inclining to brownish-black, sometimes to reddish-brown.

It

It occurs massive, disseminated, globular, stalactitic, and with impressions.

Internally it is shining and glistening, and the lustre is resinous.

The principal fracture is curved slaty; the cross fracture is conchoidal

The fragments are indeterminately angular, and also slaty.

It is translucent on the edges.

It is shining in the streak.

It is very soft.

It is perfectly sectile.

It is elastic flexible.

It is light, verging on swimming.

Specific gravity, from 0.9053 to 1.233, *Hatchett.* 0.930, *La Metherie.* 0.9021, *Jordan.*

Constituent Parts.

100 Grains afforded the following products and educts:

Carbonated Hydrogen, -	38	cubic inches.
Carbonic Acid, - -	4	
Bituminous Oil, - -	73	grains.
Acid Water, - - -	1.50	
Carbon, - - - - -	6.25	
Lime, - - - - -	2.0	
Silica, - - - -	1 50	
Oxide of Iron, - -	0 75	
Sulphate of Lime, - -	0.50	
Alumina, - - - -	0.25	

Klaproth, Beit. b. iii. s 112.

Geognostic

Geognostic and Geographic Situations.

It is found in the cavities of a vein in the lead-mine called Odin, which is situated near the base of Mamtor, to the north of Castletown in Derbyshire. The vein traverses limestone, and contains galena or lead-glance, accompanied with fluor-spar, calcareous-spar, heavy-spar, quartz, blende, calamine, selenite, and slaggy mineral pitch.

Observations.

1. According to Hatchett, a transition is to be observed from Mineral Oil, through Slaggy Mineral Pitch, to Elastic Mineral Pitch.

2. Like the elastic gum, called Caoutchouc, it removes the traces of graphite, (black lead), but it at the same time soils the paper a little.

3. The first account of this mineral was published by Dr Lister in the Philosophical Transactions for 1673 *. It was found in an old forsaken mine. He calls it a subterraneous fungus, and is uncertain whether it belongs to the vegetable or mineral kingdom; but rather inclines to the former, and hints that it may have grown out of the old birch props used in the mine. It was first accurately examined by Mr Hatchett.

3. Brown Coal.

This species is divided into six subspecies, viz. Bituminous Wood or Brown Coal, Earth Coal or Earthy Brown Coal, Alum Earth, Common Brown Coal or Conchoidal Brown Coal, Granular Brown Coal, Moor Coal or Trapezoidal Brown Coal.

First

* Vol. viii.

First Subspecies.

Bituminous Wood or Fibrous Brown Coal.

Bituminöses Holz, *Werner.*

Vegetabile fossile bituminosum, *Wall.* t. ii. p. 415.—Bituminöses Holz, *Wid.* s. 631. *Id. Wern.* Pabst. b. i. s. 365.—Carbonated Wood, *Kirw.* vol. ii. p. 60.—Bituminöses Holz, *Estner,* b. iii. s. 166. *Id. Emm.* b. ii. s. 54.—Le Bois bitumineux commun ou parfait, *Broch.* t. ii. p. 44.—Bituminöses Holz, *Reuss,* b. iii. s. 146.—Holzige Braunkohle, *Lud.* b. i. s. 186. *Id. Suck.* 2ter th. s. 60.—Bituminöses Holz, *Bert.* s. 351. *Id. Mohs,* b. ii. s. 311.—Lignite fibreux, *Brong.* t. ii. p. 32.—Bituminöses Holz, *Leonhard,* Tabel. s. 48.—Holzige Afterkohle, *Haus.* s. 116.—Fasrige Braunkohle, *Karsten,* Tabel. s. 59.—Bituminöses Holz, *Lenz,* b. ii. s. 1057.

External Characters.

Its colours are pale and dark blackish brown, clove-brown, wood-brown, hair-brown, and yellowish-brown.

Its external shape resembles exactly that of stems and branches of trees, but is usually compressed.

Its principal fracture is glimmering, sometimes approaching to glistening: the cross fracture is shining. The first is lighter coloured than the second

The fracture is fibrous in the small, slaty in the great, and corresponds with the woody texture: the cross fracture in some varieties is imperfect conchoidal.

The fragments are splintery, or cuneiform, but seldom indeterminately angular.

It is opaque.

The streak is shining.

It

It is soft.
It is sectile.
It is rather elastic-flexible.
It is rather easily frangible.
It is so light as almost to swim on the surface of water.

Chemical Characters.

It burns with a clear flame, and evolves, during combustion, a peculiar bituminous smell, which is very different from that of black coal.

Constituent Parts.

According to Vauquelin, the bituminous wood of Rollo contains the following ingredients:

Vegetable Earth, -	54.0
Sulphate of Iron, -	10.7
Sulphur, - - -	0.8
Oxide of Iron, - -	12.7
Sulphate of Lime, -	0.7
Silica, - - - -	0.2
Loss, - - -	

Geognostic Situation.

It usually occurs in alluvial land, in beds of common brown coal; sometimes also forming whole beds, part of which is converted into common brown coal and earth coal. It sometimes also occurs in fragments, branches, &c. in clay; and in the Prussian amber-mines it is found in considerable quantity, and occasionally with adhering amber. Rocks of the flœtz-trap formation sometimes contain beds or imbedded portions of this mineral; and

 it

it is also met with in imbedded masses in limestone and sandstone, belonging to some of the coal formations.

Geographic Situation.

In England, at Bovey Tracey, near Exeter; at the mouth of the Ouse, in Sussex: in Scotland, in the flœtz-trap formation, accompanied with pitch-coal, in the island of Skye; in separate pieces in trap-tuff, in the island of Canna; in flœtz limestone in the island of Skye; and in the coal formation in the counties of Fife and Mid Lothian. It occurs in considerable beds in the trap rocks of the island of Iceland. On the Continent of Europe, it is met with both in Upper and Lower Saxony; also in Bohemia, Silesia, Moravia, Bavaria, Austria, Stiria, Transylvania, Russia, Poland, and France.

Use.

It is employed as fuel where great heats are not required.

Observations.

1. In Iceland, where it occurs in great quantity, it is called *Suturbrand.*

2. It passes into Common Brown Coal, with which it is often confounded.

Second

Second Subspecies.

Earth Coal or Earthy Brown Coal.

Erdkohle, *Werner.*

Le Bois bitumineux terreux, *Broch.* t. ii. p 45.—Erdkohle, *Reuss,* b. ii 3. s. 159.—Lignite terreux, *Brong.* t. ii, p. 33.—Erdige Braunkohle, *Leonhard,* Tabel. s. 49.—Erdige Afterkohle, *Haus.* s. 116.—Erdige Braunkohle, *Karsten,* Tabel. s. 58 — Erdige bituminöses Holz, *Lenz,* b. ii. s. 1059.

External Characters.

Its colours are yellowish-brown, and blackish-brown.

Its occurs massive. Its consistence is between cohering and loose, but more inclined to the latter.

Its particles are dusty, and soil a little.

Internally it is faintly glimmering, passing into dull.

The fracture in the small is intermediate between uneven and fine earthy.

The fragments are indeterminately angular, and blunt-edged.

The streak is somewhat shining.

It is very soft, passing to friable.

It is light, almost swimming.

Chemical Characters.

It burns easily, and diffuses, during combustion, a smell like that of burning bituminous wood. Alkohol dissolves a brownish-coloured bitter substance, having many of the properties of vegetable extract. By distillation it affords a honey-coloured oil, which is soluble in alkohol, and appears to be intermediate between resin and volatile oil.

oil. When this oil is freed of its watery parts, by exposure to a gentle heat, and then allowed to cool, it acquires the consistence of white cerate *.

Constituent Parts.

Earth Coal of Schraplau.

Carbonated Hydrogen,	59.0	cubic inches.
Carbonic Acid, -	8.5	
Acid Water, - -	12.0	
Empyreumatic Oil, -	30.0	
Coal, - - -	20.25	
Lime, - - -	2.0	
Sulphate of Lime, -	2.5	
Clay, - - -	0.5	
Oxide of Iron, - -	1.0	
Sand, - - -	11.5	

Klaproth, Beit. b. iii. s. 320.—323.

Geognostic and Geographic Situations.

It is found, along with bituminous wood, in Thuringia, in the district of Mansfeldt; and in the circles of Saal and Leipsic, it occurs in beds from twenty to forty feet thick, having an extent of several square miles.

Uses.

It is used as fuel where no great degree of heat is required, as in heating rooms, salt, nitre, and alum works, and in distillation. But to render it fit for these purposes, it must be moistened with water, beat in troughs, then

* This oil, acccording to Klaproth, resembles very much in its properties the substance called Sea or Lake Wax, which is found at Bargusin, on the shores of the Lake Baikal.

then made into bricks, and dried. Sometimes it is intermixed with small black coal, to increase the intensity of the heat. Its ashes are used with advantage as a manure; and a colour resembling umber is prepared from it.

Observations.

1. It passes into Bituminous Wood, from which it differs principally in its state of aggregation.

2. Umber or Cologne Earth is said to be a variety of earth coal.

3. When much iron-pyrites is dispersed through it, alum is prepared from it, as is the case at Muhlbach, and Komothau in Bohemia.

4. Its name is derived from its state of aggregation.

Third Subspecies.

Alum Earth.

Alaunerde, *Werner.*

Terra aluminaris, *Wall.* t. ii. p. 32.—Alaunerde, *Wid.* s. 398.—*Id. Estner,* b. ii. s. 647. *Id. Emm.* b. ii. s. 299.—Aluminite bitumineux, *Lam.* t. ii. p. 116.—La terre alumineuse, *Broch.* t. i. p. 383.—Alaunerde, *Reuss,* b. ii. 3 s. 152. *Id. Lud.* b. i. s. 110. *Id. Suck.* 2ter th. s. 528. *Id. Bert.* s. 218. *Id. Mohs,* b. ii. s. 311. *Id. Leonhard,* Tabel. s. 48. *Id. Karsten,* Tabel. s. 58.—Erdige Afterkohle, *Haus.* s. 116.—Alaunerde, *Lenz,* b. ii. s. 1063.

External Characters.

Its colours are blackish-brown, and brownish-black.

 It

It is massive.

It is dull, sometimes glimmering; but this is owing to an intermixture of mica.

The fracture is earthy, with a tendency to straight slaty.

It breaks into tabular pieces.

The streak is shining.

It feels rather meagre, and sometimes greasy.

It is sectile.

It is very soft, inclining to friable.

It is light.

Chemical Characters.

When exposed to heat, it burns with a flame; and when left some time exposed to a moist atmosphere, it becomes warm, and at length takes fire.

Constituent Parts.

Charcoal, - - -	19.65
Sulphur, - - -	2.85
Silica, - - -	40.00
Alumina, - - -	16.00
Oxide of Iron, - -	6.40
Sulphate of Iron, - -	1.80
Sulphate of Lime, -	1.50
Magnesia, - - -	0.50
Sulphate of Potash, -	1.50
Muriate of Potash, -	0.50
Water, - - -	10.75
	101.45

Klaproth, in Gehlen's Journ. vi. 44.

Geognostic Situation.

It occurs frequently in beds of great magnitude in alluvial land, and sometimes also in the flœtz-trap formation.

Geographic

Geographic Situation.

It is found in Bohemia, Saxony, Austria, Naples, Hungary, and in the Vivarais in France.

Uses.

It is lixiviated, to obtain the alum it contains: it is even sometimes used for fuel.

Fourth Subspecies.

Common Brown Coal, or Conchoidal Brown Coal.

Gemeine Braunkohle, *Werner*.

Braun Kohle, *Estner*, b. iii. s. 126—La Houille brune, *Broch.* t. ii. p. 47.—Gemeine braun Kohle, *Reuss*, b. ii. 3. s. 154. *Id. Lud.* b. s. 187. *Id. Suck.* 2ter th. s. 63. *Id. Bert.* s. 345. *Id. Mohs*, b. ii. s. 311. *Id. Leonhard*, Tabel. s. 48. *Id. Karsten*, Tabel. s. 58. *Id. Haus.* s. 117.—Muschliche Braunkohle, *Lenz*, b. ii. s. 1060.

External Characters.

Its colour is light brownish-black, which passes into blackish-brown.

It occurs massive.

Internally it is shining, and sometimes glistening; and the lustre is resinous.

The fracture is rather imperfect large conchoidal; and sometimes shews the fibrous woody texture.

The fragments are indeterminate angular, and more or less sharp-edged.

The colour is lighter in the streak.

It is soft, and very soft.

It is rather sectile.

 It

It is not very brittle.
It is pretty easily frangible.
It is light.

Chemical Characters.

It burns with a weak blue-coloured flame, and emits a smell like that of burning bituminous wood.

Constituent Parts.

200 Grains of the Bovey brown coal, by distillation, yielded,

	Grains.
1. Water, which soon came over acid, and afterwards turbid, by the mixture of some bitumen, - - - - -	60
2. Thick brown oily bitumen, - -	21
3. Charcoal, - - - - - -	90
4. Mixed gas, consisting of hydrogen, carbonated hydrogen, and carbonic acid,	29
	200

Hatchett, Phil. Trans. 1804.

Geognostic Situation.

It occurs in alluvial land, and in flœtz-trap rocks.

Geographic Situation.

It is found at Bovey near Exeter; in the Leitmeritzer, Saatzer, and Ellbogner circles in Bohemia; in the counties of Mansfeldt, Thuringia, Magdeburg, and the circles of Saal and Leipsic, in Lower Saxony; in Hessia, in the famous hill called the Meissner; at Kaltennordheim, in the district of Eisenach; at Stockhausen and Hoen in Westerward; Island of Bornholm in Denmark; in the Faroe Islands; Greenland.

Use.

Use.

It is used as fuel.

Observations.

1. We find in it, 1. Iron-pyrites: 2. Honeystone: 3. Amber: 4. A substance resembling the Retin-asphaltum of Hatchett.

2. It is to be observed passing into Bituminous Wood and Moor-Coal; sometimes also into Pitch-Coal.

Fifth Subspecies.

Granular Brown Coal.

Körnige Braunkohle, *Karsten.*

Körnige Braunkohle, *Karsten,* Tabel. s. 96. *Id. Lenz,* b. ii. s. 1062.

External Characters.

Its colour is blackish-brown on the longitudinal fracture, and velvet-black on the cross fracture.

It occurs in the ligneous external form.

The longitudinal fracture is glimmering and silky, and delicate and straight fibrous; the cross fracture shining, and small and imperfect conchoidal.

The fragments are trapezoidal.

It exhibits small granular distinct concretions, on the cross fracture.

It is soft.

It yields a blackish-brown streak.

It is light.

Geognostic Situation.

It occurs at Maremma, in the district of Sienna.

Sixth

Sixth Subspecies.

Moor-Coal or Trapezoidal Coal.

Moorkohle, *Werner*.

Moorkohle, *Estner*, b. iii. s. 129.—La Houille limoneuse, *Broch*, t. ii. p. 48.—Moorkohle, *Reuss*, b. iii. s. 157. *Id. Lud.* b. i. s. 187.—Moorbraunkohle, *Suck.* 2ter th. s. 64.—Moorkohle, *Bert.* s. 346. *Id. Mohs*, b. ii. s. 313. *Id. Leonhard*, Tabel. s. 49.—Trapezoidische Braunkohle, *Karsten*, Tabel. s. 58. *Id. Haus.* s. 116.—Moorkohle, *Lenz*, b. ii. s. 1065.

External Characters.

Its colour is dark blackish-brown.

It occurs massive.

Internally it is glistening, and the lustre is resinous.

The principal fracture is imperfect slaty; the cross fracture even, approaching to flat conchoidal.

The fragments are trapezoidal, approaching to cubical.

It is soft and very soft.

It is rather sectile.

The streak is shining.

It is uncommonly easily frangible;—the most frangible species of coal.

It is light.

Chemical Characters.

Nearly the same as those of brown coal.

Geognostic Situation.

It occurs in great beds in alluvial land, and flœtz-trap rocks.

Geographic

Geographic Situation.

It occurs in the Leitmeritzer, Saatzer and Ellbogner circles in Bohemia; at Thalern, near Krems in Austria; also in Transylvania, Moravia, the island of Bornholm in the Baltic Sea, and the Faroe Islands. It occurs more frequently in Bohemia than in any other country.

Observations.

1. When it is for some time exposed to the air, it bursts and falls into pieces.
2. It passes into the preceding subspecies.

4. Black Coal.

Schwartzkohle, *Werner.*

This species contains six subspecies, viz. Pitch-Coal, Slate-Coal, Cannel-Coal, Foliated Coal, Coarse Coal, and Soot-Coal.

First Subspecies.

Pitch-Coal or Jet.

Pechkohle, *Werner.*

Gemma Samothracea, *Plin.* Hist. Nat. xxxvii. ?—Bitumen gagas, *Waller,* Syst. Min. t. ii p. 106.—Pechkohle, *Estner,* b. iii. s. 132.—La Houille piciforme, ou le Pechkohle, *Broch.* t. ii. p. 49.—Pechkohle, *Reuss,* b. iii. s. 142. *Id. Lud.* b. i. s. 188. —Pechsteinkohle, *Suck.* 2ter th. s. 58.—Pechkohle, *Bert.* s. 349. *Id. Mohs,* b. ii. s. 317. *Id. Leonhard,* Tabel. s. 49. *Id. Karsten,* Tabel. s. 58. *Id. Lenz,* b. ii. s. 1066.—Jet, *Aikin,* p. 4.

External Characters.

Its colour is velvet-black, and brownish-black when it is passing into bituminous-wood.

It

It occurs massive, in plates, and sometimes in the shape of branches, with a regular woody internal structure.

Internally it is shining, sometimes splendent, and the lustre is resinous.

The fracture is large and perfect conchoidal.

The fragments are indeterminate angular, and rather sharp-edged.

It is soft.

It affords a brown-coloured streak.

It is rather brittle.

It is easily frangible.

It does not soil.

It is light.

Specific gravity, according to Wiedenman, 1.308.

Chemical Characters.

It burns with a greenish flame, and a bituminous odour.

Geognostic Situation.

It occurs, along with brown coal, in beds in flœtz trap and limestone rocks; also in beds and in imbedded portions in bituminous-shale.

Geographic Situation.

It occurs, along with slate-coal, in the coal fields of Scotland; in bituminous-shale and slate-clay near Whitby. On the Continent, it is met with in the Meissner hill in Hessia; at Irsenberg in Bavaria; in a bed of bituminous wood at Kunnerdorf in Bohemia; in a bed of loam above moor-coal, in the Saxon Erzgebirge; in bituminous-shale in limestone in Stiria.

Uses.

Uses.

It is used as fuel, either in its natural state, or when converted into coaks. According to a report published in the "Journal des Mines," twelve hundred men are employed in the district of Aude in France, in fabricating, with the pitch-coal of that neighbourhood, rosaries, buttons, ear-rings, necklaces, bracelets, snuff-boxes, drinking-vessels, &c. One thousand hundred weight are yearly expended for this purpose; and, to Spain alone, the value of 18,000 livres is sold. In Prussia, the amber-diggers, who name it Black Amber, cut it into various ornamental articles.

Observations.

1. It is the darkest coloured, and its conchoidal fracture is the most perfect of any of the subspecies of coal.

2. According to Voigt, it is to be observed passing on the one side into glance-coal, and on the other into brown coal.

3. Its name is derived from its pitchy aspect. It was formerly known by the name Gagat or Jet, a name derived from the river Gaga, or the city Gagas in Lesser Asia, where it was formerly dug.

4. It is named Black Amber by the Prussian amber-diggers, because it is found accompanying amber, and, when rubbed, becomes faintly electric.

5. Several varieties of slaggy mineral pitch, and cannel coal, are known by the name of Jet.

Second

Second Subspecies.

Slate-Coal.

Schieferkohle, *Werner*.

Schieferkohle, *Estner*, b. iii. s. 147.—La Houille schisteuse, ou le Schieferkohle, *Broch.* t. ii. p. 52.—Schieferkohle, *Reuss*, b. iii. s. 132. *Id. Voigt*, s. 10. *Id. Lud.* b. i. s. 189.—Schiefer Steinkohle, *Suck.* 2r th. s. 53.—Schieferkohle, *Bert.* s. 347. *Id. Mohs*, b. ii. s. 316. *Id. Leonhard*, Tabel. s. 49. *Id. Karsten*, Tabel. s. 58. *Id. Lenz*, b. ii. s. 1068.

External Characters.

Its colour is intermediate between velvet-black and dark greyish-black. Sometimes it presents a pavonine or peacock-tail tarnish, sometimes a rainbow tarnish.

It occurs massive *.

It is shining or glistening, and the lustre is resinous.

The principal fracture is nearly straight, and generally thick slaty; the cross fracture is small grained uneven, passing into even and perfect conchoidal.

It sometimes occurs in ovoidal and columnar distinct concretions.

The fragments are sometimes slaty, sometimes trapezoidal, or indeterminate angular.

It is soft.

It is brittle, inclining to sectile.

It is easily frangible.

It is light.

Specific

* According to Hauy, this coal may be split into right rhomboidal prisms of about 95°.—*Lucas*, t. ii. p. 259.

Specific gravity,—

According to *Kirwan*, 1.250 to 1.370 English.
1.259 From Irvine in Scotland.
Wiedeman, 1.277
Richter, 1.28125 to 1.3730 From Sabrze in Silesia.
1.32132 to 1.3820 Bielschowitz.

Chemical Characters.

It burns longer than cannel or columnar coal; cakes more or less, and after combustion leaves a slag.

Constituent Parts.

Slate-Coal of Waldenburg.		Slate-Coal of Sarbze.		Slate-Coal of Bielschowitz.		Slate-Coal of Whitehaven.	
Bitumen,	36.875	Bitumen,	32.934	Bitumen,	37.890	Carbon,	56.8
Carbon,	57.993	Carbon,	63.312	Carbon,	58.172	Mixture of Asphalt and Maltha, in which the asphalt predominates,	43.0
Earth,	5.823	Earth, and Oxide of Iron,	3.904	Earths, and Oxide of Iron,	3.937		
Iron, and Oxide of Manganese,	1.157						
Richter, Neue Gegenst. d. Chem. vi. 234.		*Richter*, Neue Gegenst. d. Chem. vi. 224.		*Richter*, Neue Gegenst. d. Chem. vi. 224.		*Kirwan.*	

Geognostic and Geographic Situations.

In England it is found in vast quantity, at Newcastle, and in the great expanse of the coal formation in that neighbourhood; in the whole tract of the coal formation which stretches from Bolton, by Allonby, Workington to Whitehaven; in Scotland, in almost every quarter of the great river-district of the Forth; in great quantity in the river-district of the Clyde, at Cannoby, Sanquhar, and Kirkconnel, in Dumfriesshire: it is found also in Thuringia;

Thuringia; electorate of Saxony; Bohemia; Silesia, Hungary; the Tyrol; Stiria; Bamberg; Bavaria; Salzburg; and France.

Observations.

1. It passes sometimes into Cannel and Foliated Coal.
2. It very frequently contains mineral charcoal, and also a variety of pitch-coal, which is named Cherry-coal.
3. Its name is derived from its slaty fracture.

Third Subspecies.

Cannel-Coal.

Kennelkohle, *Werner*.

Cannelkohle, *Estner*, b. iii. s. 151.—La Houille de Kilkenny, ou le Kennelkohle, *Broch.* t. ii. p. 53.—Cannelkohle, *Reuss*, b. iii. s. 130. *Id. Voigt*, s. 172. *Id. Lud.* b. i. s. 189. *Id. Suck.* 2r th. s. 53. *Id. Bert.* s. 348. *Id. Mohs*, b. ii. s. 320.—Houille compacte, *Brong* t. ii. p. 3.—Kannelkohle, *Leonhard*, Tabel. s. 50. *Id. Karsten*, Tabel. s. 58.—Cannel Coal, *Kid*, vol. ii. p. 52.—Kennelkohle, *Lenz*, b. ii. s. 1071.—Candle Coal, *Aikin*, p. 5.

External Characters.

Its colour is dark greyish-black.

It is massive.

Internally it is glistening, and the lustre is resinous.

The fracture is sometimes large and flat conchoidal, sometimes even

The fragments are irregular or cubical.

It soft and semi-hard.

It is brittle.

It

It is easily frangible.
It is light.
Specific gravity,—

According to	*Kirwan*,	1.232
	Watson,	1.237
	La Metheric,	1.270
	Blumenbach,	1.275

Geognostic Situation.

It occurs along with the preceding subspecies in the coal formation.

Geographic Situation.

It is found in England near Whitehaven, Wigan in Lancashire, Brosely in Shropshire, Athercliff near Sheffield; in Scotland, at Gilmerton in the neighbourhood of Edinburgh, and Muirkirk in Clydesdale.

Uses.

On account of its solidity, and the good polish it is capable of receiving when pure, it is cut into drinking-vessels of various kinds, inkholders, snuff-boxes, &c.; but its principal use is as fuel.

Observations.

According to the Bishop of Llandaff, its name is derived from the word *candle*, because in some places the poor people use it in place of lights. In Scotland it is named *Parrot Coal.*

Fourth Subspecies.

Foliated Coal.

Blätterkohle, *Werner.*

Id. Estner, b. iii. s. 155.—Le Charbon lamelleux, *Broch.* t. ii. p. 54.—Blätterkohle, *Reuss,* b. iii. s. 128. *Id. Voigt,* s. 72. *Id. Lud.* b. i. s. 189. *Id. Suck.* 2ter th. s. 52. *Id. Bert.* s. 347. *Id. Mohs,* b. ii. s. 347. *Id. Leonhard,* Tabel. s. 50. *Id. Karsten,* Tabel. s. 58. *Id. Lenz,* b. ii. s. 1069.

External Characters.

Its colour is intermediate between velvet-black and greyish-black; sometimes it has a pavonine tarnish on the fracture.

It occurs massive.

The principal fracture is splendent; the cross fracture is glistening, and the lustre is resinous.

The principal or longitudinal fracture is straight foliated, with a single cleavage, which is more or less perfect; the cross fracture is slaty.

The fragments are indeterminate angular, approaching to cubical.

It is soft and very soft.

It is intermediate between brittle and sectile.

It is very easily frangible.

It is light.

Geognostic Situation.

It occurs in the coal formation.

Geographic Situation.

It is found in the Electorate of Saxony, and in Silesia.

Fifth

Fifth Subspecies.

Coarse Coal.

Grobkohle, *Werner.*

Id. Estner, b. iii. s. 158.—La Houille grossiere, ou la Grobkohle, *Broch.* t. ii. p. 55.—Grobkohle, *Reuss,* b. iii. s. 123. *Id. Lud.* b. i. s. 190. *Id. Suck.* 2ter th. s. 51. *Id. Bert.* s. 346. *Id. Leonhard,* Tabel. s. 50. *Id. Lenz,* b. ii. s. 1073.

External Characters.

Its colour is dark greyish-black.

It occurs massive.

It is glistening.

The cross fracture is coarse-grained uneven ; the longitudinal fracture is generally slaty.

The fragments are sometimes indeterminate angular, sometimes rather blunt-edged.

It is semi-hard ; it is the hardest subspecies of coal.

It is rather brittle.

It is rather easily frangible.

It is light, passing into not particularly heavy.

Geognostic Situation.

It is found in the coal formation.

Geographic Situation.

It occurs in coalworks in the neighbourhood of Dresden ; also near Horzowitz in Bohemia ; near Sabrze in Upper Silesia ; and Amberg in the Upper Palatinate.

Observations.

1. There are several formations of Black or Common Coal. Thus, it occurs in beds in transition limestone, very abundantly, along with slate-clay, bituminous shale, micaceous sandstone, clay ironstone, claystone, porphyry, greenstone, and other trap rocks, either above or below the old red sandstone; and also in beds in shell limestone.

2. Black Coal appears in general to be an original formation or deposit, apparently formed in the same manner as Glance-Coal, and Graphite.

Sixth Subspecies.

Soot-Coal *.

Russ-Kohle, *Voigt*.

Russ-Kohle, *Karsten,* Tabel. s. 58.—Houille fuligineuse, *Hauy.*

External Characters.

Its colour is dark greyish-black.

It occurs massive.

It is dull or glistening, and the lustre inclines to semimetallic.

The fracture is uneven, sometimes inclining to earthy.

The fragments are blunt edged.

It soils.

It is soft.

It

* It is named *Clod Coal* in West Lothian.

It is brittle, and easily frangible.

It is light.

Chemical Characters.

It burns with a bituminous smell, cakes, and leaves a small quantity of ashes.

Geognostic and Geographic Situations.

It occurs, along with slate-coal, in West Lothian, and other parts in the river district of the Forth; and on the Continent, it is met with in Saxony and Silesia.

III. GRAPHITE FAMILY.

This Family contains three species, viz. Glance-Coal, Graphite, and Mineral Charcoal.

1. Glance-Coal.

Glanzkohle, *Werner*.

This species contains three subspecies, viz. Conchoidal Glance-Coal, Slaty Glance-Coal, and Columnar Glance-Coal.

First Subspecies.

Conchoidal Glance-Coal.

Muschliche Glanzkohle, *Werner*.

Id. Estner, b iii. s. 135.—La Houille eclatant, ou le Glanzkohle, *Broch.* t. ii. p. 50.—Glanzkohle, *Reuss,* b. iii. s. 138. *Id. Voigt,* s. 90. *Id. Leonhard,* Tabel. s. 49 —Schlagiger Anthracit, *Karsten,* Tabel. s 58 *Id. Haus.* s. 115.—Muschlicher Anthracite, *Lenz,* b. ii. s. 1077.

External Characters.

Its colour is iron-black, of various degrees of intensity, which rather inclines to brown; and on the surface it has a tempered-steel coloured tarnish.

It occurs massive and vesicular; the interior of the vesicles has a tempered-steel coloured tarnish.

Internally

Internally it is shining, verging on splendent, and the lustre is metallic.

The fracture is large and small conchoidal; yet we can occasionally observe a trace of the ligneous texture.

The fragments are indeterminate and angular, sharp-edged.

It is soft, sometimes bordering on semi-hard.

It is rather brittle.

It is easily frangible.

It is light.

In thin pieces it emits a ringing sound.

Chemical Characters.

It burns without flame or smell, and leaves a white-coloured ash.

Constituent Parts.

Inflammable Matter, -	96.66
Alumina, - - -	2.00
Silica and Iron, - -	1.33

Schraub. Beschr. d. Meissner, s. 146.

Geognostic Situation.

It occurs in beds in the oldest flœtz coal formation, and along with flœtz-trap rocks.

Geographic Situation.

It occurs in beds in the coal formation of Ayrshire, as near Cumnock and Kilmarnock; in the coal districts in the river district of the Forth; and in Staffordshire in England. On the Continent, it is met with in the Meissner in Hessia.

Observations.

1. It appears to pass into Slaty Glance-coal, (coal-blende).

2. On the Meissner, it occurs along with other subspecies of coal, in the following order, beginning with the uppermost: 1. Columnar coal; 2 Conchoidal glance-coal: 3. Pitch-coal: 4. Common brown-coal, passing into pitch-coal: 5. Brown-coal, with inclosed bituminous wood and earth-coal: 6. Bituminous wood.—*Voigt.*

Second Subspecies.

Slaty Glance-Coal.

Schriefege Glanzkohle, *Werner.*

Plombagine charbonneuse, ou Anthracolite, *De Born,* t. ii. p. 296. Kohlenblende, *Wid.* s. 653 —Native mineral Carbon, *Kirw.* vol. ii p. 49.—Kohlenblende, *Estner,* b. iii. s. 197. *Id. Emm.* b. ii. s. 77.—Anthracite de Dolomieu, *Lam.* t. i. p. 76.—La Blende charbonneuse, ou la Kohlenblende, *Broch.* t ii. p. 79.—Anthracite, *Hauy,* t. iii. p. 307.—Kohlenblende, *Reuss,* b. iii. s. 183.—Anthracite, *Brong.* t. ii. p. 55.—Kohlenblende, *Leonhard,* Tabel. s. 50.—Gemeiner Anthracit, *Karsten,* Tabel. s. 58. *Id. Haus.* s. 115.—Schiefriger Anthrazit, *Lenz,* b. ii, s. 1078.

External Characters.

Its colour is dark iron-black, seldom inclining to brown; those varieties that border on graphite, are of a steel-grey colour.

It occurs massive.

It

It is shining and glistening, and the lustre is intermediate between metallic and semi-metallic.

The principal fracture is more or less perfect slaty; the cross fracture small and flat conchoidal.

The fragments are pretty sharp-edged, and sometimes trapezoidal.

It is soft; in some rare varieties, viz. those passing into graphite, it is very soft.

It is easily frangible.

It is intermediate between sectile and brittle.

It is light.

Specific gravity, 1 530, *Klaproth.* 1.800, *Hauy.* 1.415, *Thomson.* 1.300, *La Metherie.* 1.468, *Groess.* 1.526, *Kirwan.*

Chemical Characters.

According to Dolomieu, when reduced to powder, and heated in a crucible, it does not give any sulphureous or bituminous odour, and, on distillation, it affords neither sulphur nor bitumen. By exposure to a considerable heat, it burns without flame, and at length is consumed, leaving a greater or lesser portion of ash, according to its purity.

Constituent Parts.

	Panzenberg.	*Dolomieu.*
Carbon,	90	72.05
Silica,	4 to 2	13.19
Alumina,	4 to 5	3.29
Oxide of iron,	2 to 3	3.47
Loss,		8.00
	100	100.00

Geognostic

Geognostic Situation.

It occurs in imbedded masses, beds and veins, in primitive, transition and flœtz rocks. It occurs in Spain, in gneiss; in Switzerland and Savoy, in mica-slate and clay-slate; at Lischwitz, near Gera in Saxony, in transition rocks; in trap rocks, as in the Calton Hill at Edinburgh; in the coal formation in the river district of the Forth; and in a similar formation near the village of Brandau, in the Saatzer circle in Bohemia.

Geographic Situation.

Europe.—It is found in several flœtz districts in Scotland, as near West Craigs in West Lothian, Dunfermline in Fifeshire, Cumnock, and Kilmarnock in Ayrshire, and in the island of Arran. In similar rocks in England, as in the southern parts of Brecknock, Caermarthenshire, Pembrokeshire, and Birch Hill, near Walsal in Staffordshire: also at Kilkenny in Ireland. On the Continent, it is met with at Kongsberg in Norway, where it is associated with native silver, in veins that traverse mica-slate; in the Hartz, in veins of red and brown ironstone, which traverse grey-wacke; in imbedded masses in grey-wacke in Dauphiny; in mineral veins at Schemnitz in Hungary.

America.—In beds in transition rocks, in the United States *.

Asia.—In the goverment of Katharinoslow in Siberia.

Observations.

In this country it is named *Blind Coal.*

Third

* Vid. Maclure's interesting Sketch of the Mineralogy of the United States, and Dr Bruce's Mineralogical Journal.

Third Subspecies.

Columnar Glance-Coal.

Stangenkohle, *Voigt.*

Stangenkohle, *Leonhard,* Tabel. s. 50. *Id. Karsten,* Tabel. s. 58. *Id. Lenz,* b. ii. s. 1067.

External Characters.

Its colour is iron-black, and sometimes passing into steel-grey. It occasionally exhibits a tempered-steel tarnish.

It occurs massive, and disseminated.

The lustre is shining and semi-metallic.

The fracture is more or less perfect, and small conchoidal.

The fragments are sharp-edged.

It occurs in prismatic concretions, which are sometimes straight, sometimes curved ; and vary in thickness from a few lines to upwards of an inch, and from an inch to four or five inches in length.

It is opaque.

It is soft.

It is brittle, and easily frangible.

It is light.

Chemical Characters.

It burns without flame or smoke.

Geognostic and Geographic Situations.

It forms a bed several feet thick, in the coal field of Sanquhar in Dumfriesshire; at Saltcoats in Ayrshire, it occurs,

occurs, not only in beds, along with strata of greenstone, slate-clay, clay-ironstone, and bituminous-shale, in the coal formation of that district, but also imbedded in the greenstone; about four miles from New Cumnock, also in Ayrshire, there is a bed of columnar glance-coal, from three to six feet thick, in which the columns are arranged in rows like basalt, and which is intermixed with compact, scaly, and columnar graphite. Both the graphite and columnar glance-coal are contained in the coal formation, and in some places, cotemporaneous masses of greenstone are imbedded in the coal *. It occurs also at the Meissner in Hessia, where it is associated with conchoidal glance-coal, pitch-coal, brown-coal, bituminous wood, and earth-coal, and covered with greenstone and basalt.

2. Graphite.

Graphit, *Werner*.

Ferrum molybdena, *Wall.* t. ii. p. 249.—Plombagine, *Romé de Lisle,* t. ii. p. 500. *Id. De Born,* t. ii. p. 295.—Graphites plumbago, *Lin.* Syst. Nat. edit. 13. cura Jo. Frid. Gmelin, t. iii. p. 284.—Plumbago, *Kirw.* vol. ii. p. 58.—Graphit, *Emm.* b. ii. s. 97. *Id. Wid.* s. 651.—Graphite, *Broch.* t. ii. p. 76.—Fer carburé, *Hauy,* t. iv. p. 98.—Graphite, *Reuss,* b. iii. 3. s. 176. *Id. Lud.* b. i. s. 196. *Id. Suck.* 2ter th. s. 73. *Id. Bert.* s. 335 *Id. Mohs,* b. ii. s. 327. *Id. Leonhard,* Tabel. s. 50.—Plumbago, *Kid,* vol. ii. p. 58. *Id. Aikin,* p. 6.

This species is divided into two subspecies, viz. Scaly Graphite, and Compact Graphite.

First

---

* Jameson's Mineralogical Description of Dumfriesshire, p. 160, 161, 162.

First Subspecies.

Scaly Graphite.

Schuppiger Graphit, *Werner.*

Graphite lamellaire, *Brong.* t. ii. p. 53.—Schuppiger Graphit, *Karsten,* Tabel. s. 58. *Id. Haus.* s. 115.—Graphite granulaire, *Hauy,* Tabl. p. 70.—Schuppiger Graphit, *Lenz,* b. ii. s. 1084.

External Characters.

Its colour is dark steel-grey, which approaches to light iron-black.

It occurs massive, disseminated, and crystallised in thin six-sided tables.

It is usually glistening, sometimes glimmering, and the lustre is metallic.

The fracture is scaly foliated, which passes sometimes into large conchoidal, and it is sometimes slaty and uneven at the same time.

The fragments are indeterminate angular, and sometimes trapezoidal.

It occurs in coarse, small, and fine granular distinct concretions.

The streak is shining, even splendent, and its lustre is metallic.

It is very soft.

It is perfectly sectile.

It is rather difficultly frangible.

It writes and soils.

It feels very greasy.

Specific

Specific gravity, (but uncertain whether of scaly of compact graphite),

Kirwan,	1.987	2.267
Brisson,	2.1500	2.456
Hauy,	2.0891	2.2456

Second Subspecies.

Compact Graphite.

Dichter Graphit, *Werner.*

Graphite granuleux, *Brong.* t. ii. p. 54.—Dichter Graphit, *Karsten,* Tabel. s. 58. *Id. Haus.* s. 115. *Id. Lenz,* b. ii. s. 1085.

External Characters.

The colour is nearly the same with the preceding, only rather blacker.

It occurs massive, and disseminated.

Internally it is glimmering, sometimes glistening, and the lustre is metallic.

The fracture is fine-grained uneven, which passes into even, and also into large and flat conchoidal.

Some varieties shew a slaty longitudinal fracture.

It occurs in columnar distinct concretions.

In other characters it agrees with the preceding subspecies.

Chemical Characters.

When heated in a furnace, it burns, and during combustion emits a great deal of carbonic acid, and leaves a residuum of red oxide of iron.

Constituent

Constituent Parts.

				Graphite of Pluffier.	
Carbon	90.9	Carbon,	81	Carbon,	23
Iron, -	9.1	Oxygen,	9	Iron, -	2
	——	Iron, -	10	Alumina,	37
	100.00		——	Silica,	38
	Berthollet.		100		——
			Scheele.		100
				Journal des Mines, N. 12. p. 16.	

According to John, it sometimes contains Chrome, Nickel, and Manganese; and Schrader mentions Oxide of Titanium as one of its ingredients.

Geognostic Situation.

It occurs usually in beds, sometimes disseminated, and in imbedded masses, in granite, gneiss, mica-slate, clay-slate, coal and trap formations, and probably also in transition-slate.

Geographic Situation.

Europe.—Graphite occurs in the coal formation near Cumnock in Ayrshire, where it is imbedded in greenstone, and in columnar glance-coal *. At Borrodale in Cumberland, where there is the most considerable mine of this mineral in the world; it occurs in large imbedded masses or kidneys in transition rocks. On the Continent, it is met with in the granite of Langsdorf in Bavaria; in mica-slate, near Monte-Rosso in Calabria; in gneiss in Piedmont; in serpentine in the mountain of Mora, near to Marbella in Andalusia; along with felspar at Krageroe in Norway; and in Iceland, in trap, along with green earth

* Jameson's Mineralogical Description of Dumfriesshire, p. 161.

earth and zeolite. It is also enumerated amongst the mineral productions of France, Savoy, Bohemia, Austria, Stiria, Salzburg, Hungary and Transylvania.

America.—It occurs in syenite, near New-York; in transition rocks on Rhode Island *; and in granite in Greenland.

Asia.—At Thutskoi Noss.

Africa.—It is said to occur in rocks near the Cape of Good Hope.

Uses.

The finer kinds are first boiled in oil, and then cut into tables or pencils: the coarser parts, and the refuse of the sawings, are melted with sulphur, and then cast into coarse pencils for carpenters; they are easily distinguished by their sulphureous smell. It is also used for brightening and preserving grates and ovens from rust; and, on account of its greasy quality, for diminishing the friction in machines. Crucibles are made with it, which resist great degrees of heat, and have more tenacity and expansibility than those manufactured with the usual clay mixtures.

3. Mineral

* Bruce's Mineralogical Journal, vol. i. p. 345.

Mineral Charcoal.

Mineralische Holzkohle, *Werner.*

Mineralische Holzkohle, *Leonhard,* Tabel. s. 50.—Fasriger Anthracit, *Karsten,* Tabel. s. 58. *Id. Haus.* s. 115. *Id. Lenz,* b. ii. s. 1082.

External Characters.

Its colour is dark greyish-black, which sometimes approaches to bluish-black.

It occurs massive, disseminated, in flattened roundish pieces, and crusts.

It is glimmering, bordering on glistening, and the lustre is silky.

The fracture is fibrous which sometimes shews the woody texture.

The fragments are indeterminate angular, blunt-edged, sometimes also splintery.

It soils strongly.

It is soft, passing into friable.

It is very easily frangible.

It is light.

Chemical Characters.

When exposed to a strong heat, it burns without flame or smoke.

Geognostic and Geographic Situations.

It occurs in thin layers in slate-coal, brown-coal, earth-coal and moor-coal, in countries where these coals abound.

IV. RESIN FAMILY.

This Family contains the following species, viz. Amber, Honeystone, Retin-Asphalt, and Fossil Copal.

1. Amber.

Bernstein, *Werner*.

This species contains two subspecies, 1. White Amber: 2. Yellow Amber.

First Subspecies.

White Amber.

Weisser Bernstein, *Werner*.

Id. Werner, Pabst. b. i. s. 367.—Le Succin blanc, *Broch.* t. ii. p. 69.—Weisser Bernstein, *Reuss,* b. iii. s. 166. *Id. Leonhard,* Tabel. s. 47. *Id. Karsten,* Tabel. s. 58. *Id. Haus.* s. 117. *Id. Lenz,* b. ii. s. 1093.

External Characters.

Its colour is straw-yellow, which sometimes inclines to yellowish-white.

It occurs massive, and sometimes inclosed in the yellow subspecies.

It is glistening, approaching to shining, and the lustre is resinous.

The fracture is conchoidal, but not so perfect as in the yellow subspecies.

The

The fragments are indeterminate angular, and sharp-edged.

It is only translucent.

In other characters, it resembles the following subspecies.

Second Subspecies.

Yellow Amber.

Gelber Bernstein, *Werner.*

Id. Werner, b. i. s. 367.—Le Succin jaune, *Broch.* t. ii. p. 70. —Gelber Bernstein, *Reuss,* b. iii. s. 169 *Id. Leonhard,* Tabel. s. 47. *Id. Karsten,* Tabel. s. 58. *Id. Haus.* s. 117. *Id. Lenz,* b. ii. s. 1095.

External Characters.

Its colour is wax-yellow, which passes, by a kind of lemon-yellow, into honey-yellow; from this it passes into dark yellowish-brown, and hyacinth-red *.

It generally occurs in indeterminate angular blunt-edged pieces, having a rough uneven surface.

Externally it is generaly dull; internally it is splendent, and the lustre is intermediate between vitreous and resinous.

The fracture is large and perfect conchoidal.

The fragments are indeterminate angular, and very sharp-edged.

It is transparent, semi-transparent, and translucent.

 It

* It is said to occur of an olive-green colour.

It is soft.
It is rather brittle.
It is easily frangible.
Specific gravity, 1.078 to 1.085.

Chemical Characters.

It burns with a yellow-coloured flame, and fragrant odour, at the same time intumescing, but scarcely melting.

Physical Characters.

When rubbed, it acquires a very strong resino-electric virtue. This property was known to the ancients, who termed amber *electrum;* from whence is derived the word *electricity* *.

Constituent

---

* The appearances and electrical property of amber are so often alluded to in ancient authors, that it is not necessary to shew by quotations that they were familiar with that substance ; and though the history of its origin is much involved in fable, yet they seem to have had some idea that it was found in the north of Europe :—

——————— Κελτοὶ δ' ἐπὶ βάξιν ἔθεντο
Ὡς ἄρ' Ἀπόλλωνος τάδε δάκρυα Λητοΐδαο,
Ἐμφέρεται δίναις, ἅτε μυρία χεῦε πάροιθεν,
Ἦμος Ὑπερβορέων ἱερὸν γένος εἰσαφίκανεν *.

Pliny says, in speaking of amber, " Certum est gigni in insulis Septentrionalis Oceani †:" and, in another place, " Ab adverso (Britanniarum) in Germanicum mare sparsæ Glessariæ (insulæ) ; quas Electridas Græci recentiores appellavere, quod ibi electrum nasceretur ‡." In another place, he says, that in the spring-time it was washed on a part of the coast of Germany, from an island in the North Sea ; concluding with these words: " Incolas pro ligne ad ignem uti eo, proximisque Teutonis vendere ||."

From

* Apoll. Rhod. lib. iv. lin 611,—614.
† Nat. Hist. t. vi. p. 266. ed. Brot.
‡ Hist. Nat. lib. iii.
|| Hist. Nat. lib. xxxvii.

Constituent Parts.

It is composed of Carbon, Hydrogen, and Oxygen. An acid named *Succinic* is obtained from it by distillation.

Geognostic Situation.

This mineral occurs in beds of bituminous-wood and moor-coal; also in a conglomerate formed by the aggregation of fragments on the sea shores, in sandy soil, and frequently floating on the sea. It is said to have been observed imbedded in flœtz limestone, bituminous marl-slate, slate-coal, and has been lately met with imbedded in flœtz gypsum.

Geographic Situation.

Europe,—It is thrown up by the sea on the coasts of Norfolk, Suffolk, and Essex, and imbedded in a gravel-pit at Kensington, near London. It occurs in greatest quantity in East Prussia; also on the coast of the Baltic, in Courland, Liefland, Russia, Swedish Pomerania, and West Prussia. It is found in a sandy soil in Poland, at a great distance from the sea, where it is intermixed with the fruit of the Pinus abies. It is imbedded in brown-coal in the department of Aisne in France; in slate-clay in the Lower Alps; imbedded in a bituminous marl-slate at Aarau in Switzerland; on the coasts of Sicily; in

 Spain,

From the foregoing passages, it seems very probable, that the opinion of Solinus respecting the origin of amber is correct: he says that it was originally brought from the northern sea, through Pannonia and Illyria, into the country bordering on the river Po; and hence Phaeton's sisters, or the poplars of that river, are fabled to have wept amber; this substance being easily mistaken for a vegetable gum.—*Kid's* Mineralogy, vol. ii. p. 37.

Spain, near Alicant, and in the Asturias, in one of which it is said to occur imbedded in limestone; in gypsum in the Segeberg, near Kiel in Holstein; at Wittenberg, in the kingdom of Saxony; and in Moravia, Austria, and the Bannat of Temeswar.

Asia.—It is found imbedded in slate-coal at the mouth of the Jenisei, in Siberia, and in a similar situation in the Bay of Penschincha, in the same country.

America.—The beds of brown-coal which occur in Greenland, occasionally contain imbedded grains and masses of amber.

Africa.—It is said to occur on the coast of Madagascar.

Uses.

On account of its beautiful colour, great transparency, and the fine polish it receives, it is considered as an ornamental stone, and is cut into necklaces, bracelets, snuff-boxes, and other articles of dress. Before the discovery of the diamond, and the other precious stones of India, it was considered to be the most precious of jewels, and was employed in all kinds of ornamental dress. Great quantities of it are annually exported from Dantzig to Constantinople, the Levant, Persia, and France. The most considerable purchasers of amber are the merchants of Armenia and Greece; but it is still uncertain how they dispose of it. It is conjectured by some, that it is purchased from them by pilgrims, previous to their journey to Mecca, and that on their arrival there, it is burnt in honour of the prophet Mahomet. It is also an important article of exchange in Africa. When dissolved in oil, it forms a species or varnish, named *Amber varnish*.

Observations.

Observations.

1. The only minerals with which it is likely to be confounded, are *Honeystone* and *Fossil Copal:* its strong electrical property, and single refracting power, distinguish it from Honeystone; and its colour and difficult fusibility from Fossil Copal.

2. It frequently includes bodies of different kinds, as grains of sand, pieces of iron-pyrites, and also insects, which, according to Jussieu, are not natives of Europe. Born mentions a specimen of amber, containing a species of gorgonia: another author describes a specimen containing the seed-vessels of the *Pinus abies:* in some cabinets, there are specimens including pinnated leaves, resembling those of ferns, and other specimens inclosing drops of transparent water.

3. Masses of considerable size have been met with in the amber-mines on the coasts of the Baltic. Thus, in the year 1576, a piece of amber weighing 11 pounds was found in Prussia, and sent to Prague, as a present to Rodolph II.; and a few years ago, a mass weighing upwards of 13 pounds, and whose contents amounted to 318¾ cubic inches, was dug up in the same country. Five thousand dollars are said to have been offered for this latter mass; and the Armenian merchants assert, that in Constantinople it would sell for thirty or forty thousand dollars *.

4. Various conjectures have been proposed in regard to its origin and formation. By some, it is held to be a vegetable gum or resin, altered by processes unknown to us: others consider it a variety of mineral oil, thickened

* Neues allgemeines Journal der Chemie, b. i. s. 224.

by absorption of oxygen; and it has also been alleged to be inspissated mineral oil.

5. The pitch-coal sometimes found along with it, is by the amber-diggers named *Black Amber*, and is sold at a great price.

6. This mineral is sometimes named *Succinum*, from the word *succus*, it having been conceived that amber was an inspissated juice. Thus, Pliny remarks, "Arboris succum esse prisci nostri credidere, ob id *succinum* appellantes *." It was also by Pliny, and other ancient writers, named *Electrum*, from its resemblance in colour to the metallic alloy of the ancients, which consisted of gold and silver, and was called by the same name; or from Ηλεκτωρ, one of the names of the sun *

7. When one part of the empyreumatic oil obtained by distilling mineral pitch, is boiled several times with one and a half parts of turpentine, a compound is formed, which bears a great resemblance to amber, and which is frequently cut into necklaces, and other ornaments, and sold as true amber.

2. Honeystone.

---

* Plin. Hist. Nat. t. vi. p. 266. ed. Brot.

† Kid's Mineralogy, vol. ii. p. 36.

2. Honeystone.

Honigstein, *Werner.*

Id. Wid. s. 639.—Succin transparent en Cristaux octaedres, *De Born,* t. ii. p. 90.—Mellilite, *Kirw.* t ii. p. 68.—Honigstein, *Emm.* t. ii. p. 86.—La Pierre de Miel, ou le Mellite, *Broch.* t. ii. p. 73. *Id. Hauy,* t. iii. p. 335.—Honigstein, *Reuss,* b. ii. s. 52. *Id. Leonhard,* Tabel. s 47. *Id Karsten,* Tabel. s. 58. —Mellite, *Brong.* t. ii. p. 52.—Mellilite, *Kid,* vol. ii. p. 39. Honigstein, *Lenz,* b. ii. s. 1100.—Mellite, *Aikin,* p. 7.

External Characters.

Its colour is honey-yellow, sometimes inclining to hyacinth-red, seldom to yellowish-brown, and wine-yellow.

It occurs massive, in angular pieces, grains; and crystallised:

1. Flat octahedron, which is sometimes truncated on the angles of the common basis. When these truncations increase, it passes into the
2. Four-sided prism, acuminated by four planes, which are set on the lateral edges.

Externally it is smooth and splendent.

The lustre is shining or splendent, and intermediate between vitreous and resinous.

The fracture is perfect and flat conchoidal.

The fragments are indeterminately angular, and rather sharp-edged.

It is semi-transparent, or translucent, and refracts double.

It is soft; softer than amber.

It is brittle.

It is easily frangible.

It

It is light.

Specific gravity, 1.5858 to 1.666.

Chemical Characters.

Before the blowpipe, it becomes white and opaque, with black spots, and is at length reduced to ashes: when heated in a close vessel, it becomes black.

Physical Character.

It becomes slightly resino-electric by friction.

Constituent Parts.

Alumina, - - -	16
Mellilitic Acid, - -	46
Water of crystallisation, -	38
	100

Klaproth, Beit. b. iii. s. 114.

Geognostic and Geographic Situations.

It occurs superimposed on bituminous wood and earth-coal, and is usually accompanied with sulphur. It has been hitherto found only at Aertern in Thuringia, and Langenbogen in the circle of Saal.

Observations.

1. Its name is borrowed from its honey-yellow colour.

2 It differs from *Amber*, in being crystallised, refracting double, and in being softer, heavier, and less powerfully electric.

3. It is chemically distinguished from *Amber:* on burning coal amber intumesces, and diffuses a fragrant odour; Honeystone, on the contrary, becomes white, without intumescence or fragrant odour.

3. Retin

3. Retin-Asphalt, *Hatchett.*

Retin-Asphalt, *German.*

Retin-asphalt, *Hatchett,* in Phil. Trans. for 1804. *Id. Kid,* vol. ii. p. 66.—Erdharze, *Wagner,* in Von Moll's Ephemeriden der Berg und Hüttenkunde, b. iv. s. 20.—Retin-asphalt, *Aikin,* p. 68.

External Characters.

Its colours are yellowish-brown, reddish-brown, and sometimes inclining to ochre-yellow.

It occurs massive, in angular pieces, and disseminated.

Its lustre is glistening, and vitreo-resinous.

The fracture is imperfect conchoidal.

It is opaque.

It is soft.

It is brittle.

Specific gravity, 1.13.

Chemical Characters.

When placed on a hot iron it melts, smokes, and burns with a bright flame, giving out a fragrant odour; it is soluble in potash, and partly so in spirit of wine.

Constituent Parts.

Resin, - - -	55
Asphalt, - - -	42
Earth, - - -	3
	100

Hatchett, Phil. Trans. for 1804.

Geognostic

Geognostic and Geographic Situations.

It is found at Bovey Tracey in Devonshire, adhering to brown-coal. A similar mineral has been met with near Wildshut in the Innviertel, and near Halle on the Saale, and in both these places along with brown-coal.

Observations.

This curious mineral was first discovered, described, and analysed by Mr Hatchett.

4. Fossil Copal.

Fossil Copal, or Highgate Resin, *Aikin.*

External Characters.

Its colour is pale muddy yellowish-brown.
It occurs in irregular roundish pieces.
The lustre is resinous.
It is semi-transparent.
It is brittle.
It yields easily to the knife.
Specific gravity, 1.046.

Chemical Characters.

It gives out a resinous aromatic odour when heated; melts into a limpid fluid; takes fire when applied to the flame of a candle; and burns away entirely before the blowpipe. Insoluble in potash ley.

Geognostic

Geognostic and Geographic Situations.

It is found in the bed of blue clay at Highgate, near London.

Observations.

The preceding description of this mineral is that of Mr Aikin, which I have extracted from his Manual of Mineralogy.

END OF VOLUME SECOND.

APPENDIX.

APPENDIX.

No. I.

ADDITIONAL SPECIES.

1. Pyreneite.

Pyrenit, *Werner.*

External Characters.

Its colour is greyish-black.

It occurs massive, and crystallised in rhomboidal dodecahedrons.

The crystals are small, all around crystallised, and imbedded.

Externally the crystals are glistening, inclining to shining.

Internally it is glistening, and the lustre is vitreous.

The fracture is small grained uneven.

The fragments are indeterminate angular, and pretty sharp-edged.

It is opaque.

It is hard.

Specific gravity, 2.500, *Klaproth.*

Chemical Characters.

It loses its colour before the blowpipe, and melts easily, and with intumescence, into a yellowish-green porous slag.

Constituent Parts.

Silica, - - - -	43
Alumina, - - -	16
Lime, - - - - -	20
Oxide of Iron, - -	16
Water, and other volatile matters, - - - -	4
	99

Vauquelin, Journ. des Mines, N. 44. p. 571.

Geognostic and Geographic Situations.

It occurs in small layers, or in imbedded crystals, in primitive limestone, in the Pic Eres-Lids, near Bareges in France.

Observations.

1. It was for some time considered as a variety of garnet, until Werner ascertained it to be a distinct substance, and placed it in the system between Leucite and Melanite, under its present name.

2. I am indebted to Mr Vivian for a description of this species.

2. Humite.

2. Humite.

Hümite, *Bournon.*

Id. Bournon, Cat. Min. p. 52.

External Characters.

Its colour is reddish-brown.

It occurs crystallised in octahedrons, which are always more or less truncated and bevelled.

The planes are frequently transversely streaked.

Its lustre is shining.

It is transparent.

It scratches quartz with difficulty.

Geognostic and Geographic Situations.

It occurs at Somma, near Naples, in a rock composed of grey-coloured granular topaz, mixed with grains of pale yellow and green topaz, which latter is sometimes crystallised in cavitie ; also with brown and olive-green mica, and white hauyne.

Observations.

The preceding account is from Bournon's *Catalogue Mineralogique*, and it contains all that is known of the species. It was named by Bournon in honour of Sir Abraham Hume, Baronet, a zealous cultivator of mineralogy, and possessor of one of the most valuable and splendid mineralogical cabinets in England.

3. Fibrolite.

Id. Bournon, Ph. Trans. 1802, p. 289. *Id. Hauy. Id. Delam. Id. Karsten. Id. Lucas.*

External Characters.

Its colours are white and grey.

It occurs crystallised in rhomboidal prisms, the angles of whose planes are 80° and 100°.

Internally it is glistening.

The principal fracture is fibrous, the cross fracture is uneven.

It is harder than quartz.

Specific gravity, 3.214.

Constituent Parts.

Alumina, - - -	58.25
Silica, - - -	38.00
Iron, and loss, - -	3.75
	100

Chenevix, Phil. Trans. 1802, p. 335.

Geographic Situation.

It is found in the Carnatic.

Observations.

It was first described and named by Bournon.

4. Lythrodes.

4. Lythrodes.

Lythrodes, *Karsten.*

Id. Karsten, in Gesel. Naturforsh. Freunde zu Berlin Magazin, 1810, p. 78.—*Thomson's* Annals, i. p. 111.

External Characters.

Its colour is aurora-red, passing into brownish-red; and sometimes through flesh-red into yellowish-brown; and it is occasionally marked with cream-yellow, and greenish spots.

It occurs massive and disseminated.

In the principal fracture, the lustre is resinous and glimmering; in the cross fracture dull.

The principal fracture is concealed foliated; the cross fracture splintery.

It is feebly translucent on the edges.

It is semi-hard in a high degree.

Specific gravity, 2.510, *John.*

Constituent Parts.

Silica, - - -	44.62
Alumina, - - -	37.36
Lime, - - -	2.75
Soda, - - -	8.00
Water, - - -	6.00
Oxide of Iron, - -	1.00
Loss, - - - -	0.27
	100

John, Chem. Unters. p. 171.

 Geographic

Geographic Situation.

It occurs at Friedrichswärn and Laurwig in Norway.

Observations.

1. It appears to be nearly allied to Elaolite.

2. It was first described by Karsten, who gave it the name *Lythrodes*, from το λυθρον, because, when fresh broken, it appears as if spotted with coagulated blood.

5. Rhaetizite.

Rhätizit, *Werner*.

External Characters.

Its colours are cream-yellow, and brick-red.

It occurs massive.

It is glistening, and the lustre is pearly.

The fracture is radiated, and is long and narrow, and either parallel, scopiform, or promiscuous.

It is feebly translucent on the edges.

It is soft.

It affords a white coloured streak.

It is difficultly frangible.

Specific gravity, 3.100.

Geographic Situation.

It occurs at Pfizsch in the Tyrol.

Observations

Observations.

The preceding description was given to me by Mr Vivian, and seems to apply to the mineral described in p. 35. vol. ii. under the name *Fibrous Cyanite.* The present name has been given to this mineral by Werner, who considers it as forming a distinct species.

6. Platiniferous Copper-Ore.

Cuivre gris platinifere, *Lucas.*

Id. Vauquelin, Ann. de Chim. lxx. p. 317. & Journ. de Phys. t. 73. p. 412.

External Characters.

This mineral is of a grey colour, and nearly resembles in other characters grey copper-ore.

Constituent Parts.

It contains Copper, Lead, Antimony, Iron, Silver, Platina, and sometimes Sulphur. Some specimens have afforded 10 *per cent.* of platina, while others have scarcely yielded any; and the proportion of silver is also very variable.

Geographic Situation.

It occurs at Guadalcanal, in Estremadura in Spain, where it is associated with arsenical silver-ore, calcareous-spar, heavy-spar, and quartz.

7. Crichtonite.

7. Crichtonite.

Craitonite, *Bournon.*

Id. Bournon, Cat. Min. p. 430.

External Characters.

Its colour is velvet-black.

It occurs crystallised in the following figures:

1. Very acute rhomboid, with angles of 18° and 162°; or it may be described as an acute double three-sided pyramid, in which the lateral planes of the one are set on the lateral edges of the other.
2. In which the summits of the pyramids are more or less deeply truncated.
3. The summits of the pyramids acuminated with three planes, which are set on the lateral planes, and the summits of the acumination truncated.
4. In which the angles of the summits of the pyramids, are bevelled.

The crystals are very small.

Externally and internally it is splendent, and the lustre is vitreous, inclining to metallic.

The fracture in one direction is foliated, in another conchoidal.

It is opaque.

It is harder than octahedrite; it scratches fluor-spar, but it does not affect glass.

It does not affect the magnet.

Chemical Characters.

It is infusible without addition before the blowpipe.

Geognostic

Geognostic Situation.

It occurs in primitive rocks, along with octahedrite, in the different countries where that mineral is found.

Observations.

1. It was named Crichtonite by Bournon, in honour of Dr Crichton, in the service of the Emperor of Russia.

2. It appears to have been confounded, sometimes with octahedrite, sometimes with micaceous iron-ore.

No. II.

ADDITIONAL British Localities of Minerals *.

Beryl.—In granite veins, which traverse quartz in Rannoch. M.

Rose-quartz.—Ben Gloe, Perthshire. M.

Heliotrope.—In the Hill of Kinnoul at Perth. M.

Hyalite, along with *Prehnite* and *Cubicite.*—Kilpatrick in Dunbartonshire, where it was first observed by my excellent friend *Dr Thomas Brown* of Glasgow.

Pitchstone.—Cumberhead, Lanarkshire.

Prehnite.—In greenstone in the island of Raasay. M.

Apophyllite.—Dunvegan, Skye. M.

Green Cubicite.—Dunvegan, Skye. M.

Chabasite.—At Storr and Talisker in Skye. M.

Cross-stone.—Kilpatrick Hills. *Dr Thomas Brown.*

Laumonite.—Frisky Hall, Dunbartonshire. M.

Andalusite.—Near Macduff in Banffshire, and Kinnaird Head in Aberdeenshire.

Lepidolite.—In primitive limestone at Dalmally; in primitive limestone in a quarry on the north side of Loch Fyne, opposite to the Inn at Cairndow, situated on the south side; in primitive limestone near Ballachulish slate-quarry; and in primitive limestone, from a quarry on the east

* Those marked *M.* were communicated by Dr Macculloch.

east side of Loch Leven, nearly opposite to the Inn at Ballachulish, situated on the west side. The credit of the first discovery of this rare mineral in Great Britain, is due to that ingenious chemist, *Mr Holme* of Peter House, Cambridge, who met with it in the places above mentioned. Dr Clarke of Cambridge, politely communicated to me from that gentleman the following analyses of the limestones in which the lepidolite occurs:

" 1. An analysis of 100 grains, from the most brilliant part of a specimen of primitive limestone, from a quarry near Dalmally, in the Highlands of Scotland:

	Grains.
Lime, - - -	$22\frac{2}{16}$
Carbonic Acid, - -	$17\frac{4}{16}$
Lepidolite and Hornblende,	$60\frac{10}{16}$
	100

Analysis of 100 grains from the least brilliant part of the same specimen:

	Grains.
Lime, - - -	$39\frac{6}{16}$
Carbonic Acid, - -	$30\frac{10}{16}$
Lepidolite and Hornblende,	30
	100

2. Primitive limestone, from a quarry on the north side of Loch-Fyne, opposite to the Inn at Cairndow, situated on the south side:

Lime,

	Grains.
Lime, - -	$42\frac{6}{16}$
Carbonic Acid, - -	$32\frac{15}{16}$
Lepidolite and Hornblende,	$24\frac{11}{16}$
	100

3. Primitive limestone from Ballachulish, near the slate-quarry:

	Grains.
Lime, - - - -	$42\frac{9}{16}$
Carbonic Acid, - -	$33\frac{1}{16}$
Lepidolite, and small quantities of powdery Schistus, and of Iron-pyrites, -	$24\frac{6}{16}$
	100

4. Primitive limestone of a whitish colour, slightly tinged with blue, from a quarry on the east side of Loch Leven, nearly opposite to the Inn at Ballachulish, situated on the west side:

	Grains.
Lime, - - - -	$51\frac{3}{16}$
Carbonic Acid, - -	$39\frac{13}{16}$
Lepidolite, and a small quantity of Iron-pyrites, -	$5\frac{10}{16}$
Oxide of Iron, - -	$3\frac{6}{16}$
	100

As this limestone receives no colour from the iron which it contains, we may infer, that the metallic body is neither in the state of the black oxide, nor of the red, but in that of the grey oxide, and is in all probability united of with carbonic acid. This last oxide, during the process of calcination, is converted into the red oxide, and being mixed with the lime, occasions it to assume a reddish-brown colour. See Dr Thomson's Chemistry, vol. i. p. 391.

The lepidolite in the limestones before mentioned, occurs in fine distinct granular concretions. These substances are of a white colour, and are either transparent or semi transparent. The hornblende, when present, has also a fine granular form, and is slightly attracted by the magnet. These limestones, on account of the hard foreign bodies which they contain, are capable of cutting glass, and of yielding a few feeble sparks to steel. The lepidolite mentioned by Klaproth, when exposed with soda to a red heat, and melted, exhibited a mass of a yellow colour, with red and blue spots, according to the oxidation of the manganese contained in the minerals. But the lepidolite which is present in these limestones being free from manganese, or any other metallic body, exhibits only a yellow globule when fused with soda."

Pinite.—In porphyry in Ben Gloe, and Blairgowrie. M.

Serpentine.—Bervie, and Cortachie.

Tremolite, fibrous, common, and *glassy.*—In limestone, Glen Tilt. M.

Diallage.—Island of Coll. M.

Hyperstene.—In syenite at Loch-Scavig, in Skye. M. Also near Portsoy.

Schiller-spar.—In serpentine at Cortachie.

Sahlite.—In limestone in Glen Tilt; and in primitive rocks in Glenelg and Rannoch. M. Also in the island of Unst.

Fibrous

Fibrous Gypsum.—Cumbray-More, and near Kelso.

Native Copper.—In the island of Unst,—not island of Yell, as stated in p. 98. vol. iii.

Pisiform Ciay Ironstone.—Galston, Ayrshire.

Molybdena.—In granite at Peterhead.

No. III.

No. III.

TABULAR VIEW

OF

SYSTEMS

OF

MINERALOGY,

Since the first Publication of the Arrangement of LINNÆUS, A. D. 1736.

It was my intention to have prefixed to this Volume, a History of Oryctognosy; but the Work having swelled much beyond what I had anticipated, the following Tabular View of some of the Mineral Systems, from the time of LINNÆUS, is all that I can spare room for at present.

LINNÆI (*Car.*) Systema Naturæ, *Lugdb.* 1736, 1748.

I. PETRÆ.

1. Vitrescentes.
   Cos.
   Quartzum.
   Silex.

2. Calcariæ.
   Marmor.
   Spatum.
   Schistus.

3. Apyræ.
   Mica.
   Talcum.
   Amiantus.
   Asbestus.

II. MINERÆ.

1. Salia.
   Natrum.
   Selenites.
   Nitrum.
   Muria.
   Alumen.
   Vitriolum.

2. Sulphura
   Electrum.
   Bitumen.
   Pyrites.
   Arsenicum.

3. Mercurialia.
   Hydrargyrum.
   Stibium.
   Zincum.
   Vismuthum.
   Ferrum.
   Stannum.
   Plumbum.
   Cuprum.
   Argentum.
   Aurum.

III. FOSSILIA.

1. Concreta.
   Saxum,
   Tophus.
   Stalactites.
   Pumex.
   Aëtites.
   Tartarus.
   Calculus.

2. Petrificata.
   Helmintholithus.
   Entomolithus.
   Ichthyolithus.
   Ornitholithus.
   Zoolithus.
   Phytolithus.
   Graptolithus.

3. Terræ.
   Marga.
   Ochra.
   Creta.
   Argilla.
   Arena.
   Humus.

LINNÆI (*Car.*) SYSTEMA NATURÆ, *Holm.* 1768.

I. PETRÆ.

I. HUMOSÆ.
1. Schistus.

II. CALCARIÆ.
2. Marmor.
3. Gypsum.
4. Stirium.
5. Spatum.

III. ARGILLACEÆ.
6. Talcum.
7. Amiantus.
8. Mica.

IV. ARENATÆ.
9. Cos.
10. Quartzum.
11. Silex.

V. AGGREGATA.
12. Saxum.

II. MINERÆ.

I. SALIA.
13. Nitrum.
14. Natrum.
15. Borax.
16. Muria.
17. Alumen.
18. Vitriolum.

II. SULPHURA.
19. Ambra.
20. Succinum.
21. Bitumen.
22. Pyrites.
23. Arsenicum.

III. METALLA.
24. Hydrargyrum.
25. Molybdænum.
26. Stibium.
27. Zincum.
28. Vismuthum.
29. Cobaltum.
30. Stannum.
31. Plumbum.
32. Ferrum.
33. Cuprum.
34. Argentum.
35. Aurum.

III. FOSSILIA.

I. PETRIFICATA.
36. Zoolithus.
37. Ornitholithus.
38. Amphibiolithus.
39. Ichthyolithus.
40. Entomolithus.
41. Helmintholithus.
42. Phytolithus.
43. Graptolithus.

II. CONCRETA.
44. Calculus.
45. Tartarus.
46. Aëtites.
47. Pumex.
48. Stalactites.
49. Tophus.

III. TERRÆ.
50. Ochra.
51. Arena.
52. Argilla.
53. Calx.
54. Humus.

WALLERII (*J. G.*) MINERALOGIA, *Stockh.* 1747, 8vo.

I. TERRÆ.

1. MACRÆ.
Humus.
Creta.

2. PINGUES.
Argilla.
Marga.

3. MINERALES.
Salinæ.
Sulphureæ.
Metallicæ.

4. ARENACEÆ.
Arena.
Glarea.
Metallicæ.
Animales.

II. LAPIDES.

1. CALCARII.
Calcareus.
Marmor.
Gypsum.
Spatum.

2. VIRTRESCENTES.
Fissilis,
Cos.
Silex.
Petrosilex.
Quartzum.
Crystallus.

3. APYRI.
Mica.
Talcum.
Ollaris
Corneus.
Amiantus.
Asbestus.

4. SAXA.
Simplicia.
Mixta.
Grisea.
Petrosa.

III. MINERÆ.

1. SALIA.
Vitriolum.
Alumen.
Nitrum.
Muria.
Alcalia.
Acida.
Neutra.
Ammoniacum.
Borax.

2. SULPHURA.
Bitumen.
Succinum.
Ambra.
Sulphura.

3. SEMIMETALLA.
Hydrargyrum.
Arsenicum.
Cobaltum.
Antimonium.
Wismuthum.
Zincum.

4. METALLA.
Ferrum.
Cuprum.
Plumbum.
Stannum.
Argentum.
Aurum.

IV. CONCRETA.

1. PORI.
Ignei.
Aquei.

2. PETRIFICATA.
Vegetabilia.
Corrallia.
Animalia.
Testacea.

3. FIGURATA.
Lithomorphi.
Lithoglyphi.
Lithotomi.

4. CALCULI.
Vegetabilium.
Animalium.

WALLERII (*J. G.*) Systema Mineralogicum, *Holm.*, 1772. 8vo.

I. TERRÆ.

1. Macræ.
Humus.
Calcareæ.
Gypseæ.
Manganenses.

2. Tenaces.
Argilla.
Marga.

3. Minerales.

4. Duræ.
Glarea.
Tripela.
Carmentum.
Arena.
Ar. metallica.
Ar. animalis.

II. LAPIDES.

1. Calcarei.
Calcareus.
Marmor.
Spatum.
2. Gypsum.
Fluor mineralis.

2. Vitrescentes.
Cos.
Spat. scintillans.
Quartzum.
Gemma.
Granatus.
Silex.
Petrosilex.
Achates.
Jaspis.

3. Fusibiles.
Zeolithus.
Basaltes.
Magnesia.
Schistus.
Margodes.
Corneus.

4. Apyri.
Mica.
Talcum.
Steatites.
Serpentinus.
Ollaris.
Asbestus.
Amiantus.

5. Saxa.
Mixta.
Aggregata.

III. MINERÆ.

1. Salia.
Acida.
Vitriolum.
Alumen.
Nitrum.
Muria.
Natron.
Alc. volatile.
Neutra.
Ammoniacum.
Borax.

2. Sulphura.
Bitumen.
Succinum.
Ambra.
Sulphur.

3. Semimetalla.
Mercurius.
Arsenicum.
Cobaltum.
Niccolum.
Antimonium.
Wismuthum.
Zincum.

4. Metalla.
Ferrum.
Cuprum.
Plumbum.
Stannum.
Argentum.
Aurum.
Platina.

IV. CONCRETA.

1. Pori.
Ignei.
Aquei.

2. Petrificata.
Vegetabilia.
Corallia.
Zooli.
Helmintholithi.
— testaceorum
Entomolothi.
Amphibiolithi.
Ichthyolithi.
Ornitholithi.
Zoolithi.
Anthropolithi.

3. Figurata.
Lithomorphi.
Lithoglyphi.
Lithotomi.

4. Calculi.
Vegetabilium.
Animalium.

CRONSTEDT

CRONSTEDT MINERALOGIA,

Stockh. 1758, 8vo.

I. TERRÆ.

1. *Calcareæ.*
Puræ.
Vitriolaceæ.
Phlogisticæ.
Argillaceæ.

2. *Siliciæ.*
Adamas.
Sapphirus.
Topazius.
Smaragdus.
Quartzum.
Silex.
Jaspis.

3. *Granatina.*
Granatus.
Basaltes.

4. *Argillaceæ.*
Porcellana.
Lithomarga.
Bolus.
Tripolitana.
Argilla.

5. *Micaceæ.*
Mica pura.
Mica martialis.

6. *Fluores.*
Indurati.

7. *Asbestinæ.*
Asbestus.
Amiantus.

8. *Zeolithicæ.*
Zeol. purus.
Zeol. metallicus.

9. *Magnesiæ.*
Magnesia terrea.
Magnesia indurata.

II. SALIA.

1. *Acida.*
Vitriolum.
Muria.

2. *Alcalina.*
Fixa.
Volatilia.

III. PHLOGISTICA.
Ambra.
Succinum.
Petroleum.
Sulphur.
Phlogistic. terreum
—— metallicum.

IV. METALLA.

1. *Perfecta.*
Aurum.
Argentum.
Platina.
Stannum.
Plumbum.
Cuprum.
Ferrum.

2. *Semimetalla.*
Hydrargyrum.
Wismuthum.
Zincum.
Antimonium.
Arsenicum.
Cobaltum.
Niccolum.

WALKER,

WALKER,

CLASSIS FOSSILIUM, Edin. 1789.

Classis I.—*Terræ.*

Ord. 1. Figulinæ. Ord. 2. Fimosæ. Ord. 3. Calcareæ. Ord. 4. Absorbentes. Ord. 5. Gypsea. Ord. 6. Siliceæ. Ord. 7. Asperæ. Ord. 8. Steatiticæ. Ord. 9. Apyræ. Ord. 10. Lapideæ. Ord. 11. Inflammabiles. Ord. 12. Ochræ.

Classis II.—*Calcarea.*

Ord. 1. Cementaria. Ord. 2. Dædalea. Ord. 3. Stiriacea. Ord. 4. Spata.

Classis III.—*Gypsea.*

Ord. 1. Plastica. Ord. 2. Selenitica.

Classis IV.—*Phosphorea.*

Classis V.—*Zeolitica.*

Classis VI.—*Ponderosa.*

Classis VII.—*Amandina.*

Ord. 1. Shorlacea. Ord. 2. Garamantica. Ord. 3. Ignigena.

Classis VIII.—*Silicea.*

Ord. 1. Quartzosa. Ord. 2. Jaspidea. Ord. 3. Lithidea. Ord. 4. Gemmæ.

Classis IX.—*Steatitica.*

Ord. 1. Saponacea. Ord. 2. Ollaria.

Classis X.—*Apyra.*

Ord. 1. Amiantina. Ord. 2. Asbestina.

Classis XI.—*Micacea.*

Classis XII.—*Petræ.*

Ord. 1. Quadrinæ. Ord. 2. Cotaceæ. Ord. 3. Schistosæ, Ord. 4. Siliciæ.

Classis XIII.—*Saxa.*

Ord. 1. Calcarea. Ord. 2. Arenaria. Ord. 3. Porphyria. Ord. 4. Granitæ. Ord. 5. Schistosa. Ord. 6. Amandina. Ord. 7. Steatitica. Ord. 8. Amiantina. Ord. 9. Micacea. Ord. 10. Metallica.

Classis XIV.—*Concreta.*

Ord. 1. Terrestria. Ord. 2. Aquea. Ord. 3. Ignea. Ord. 4. Metallica.

Classis XV.—*Salia.*

Ord. 1. Acida. Ord. 2. Alcalina. Ord. 3. Acido-alcalina. Ord. 4. Acido-terrea. Ord. 5. Alcalino-terrea. Ord. 6. Vitriola.

Classis XVI.—*Inflammabilia.*

Ord. 1. Acria. Ord. 2. Sulphurea. Ord. 3. Bitumina. Ord. 4. Carbonaria. Ord. 5. Electrica.

Classis XVII.—*Pyritæ.*

Ord. 1. Sulphureæ. Ord. 2. Arsenicales. Ord. 3. Ferreæ. Ord. 4. Amandinæ.

Classis XVIII.—*Semi-metalla.*

Ord. 1. Arsenicalia. Ord. 2. Sulphurea. Ord. 3. Fluida. Ord. 4. Dubia.

Classis XIX.—*Metalla.*

Ord. 1. Dura. Ord. 2. Flexilia. Ord. 3. Fixa.

KIRWAN's

KIRWAN's

MINERAL SYSTEM IN 1794.

I. EARTHY SUBSTANCES.

I. *Calcareous Genus.*

1. Native Lime
2. Carbonat de Chaux
3. Agaric Mineral
4. Chalk
5. Ganil
6. Testaceous Tufa
7. Limestone, Compact
    - Splintery fracture
    - Conchoidal fracture
    - Earthy fracture
    - Slaty fracture
    - Foliated and granular
    - Sparry
    - Arragon-spar
    - Striated or fibrous
8. Swinestone
9. Oviform
10. Baryto-calcite
11. Muri-calcite
12. Argillo-calcite
13. Marl, semi-indurated
    - Indurated
    - Siliciferous
14. Marlites
    - Argilliferous
    - Siliciferous
    - Bituminous
15. Pyritaceous Limestone
16. Argentine
17. Sidero-calcite
18. Ferri-calcite
19. Dolomite
20. Elastic Marble
21. Gypsum
    - Farinaceous
    - Compact
    - Fibrous or striated
    - Foliated
    - Specular

32. Fluor
    - Sandy
    - Compact
    - Foliated or Sparry

23. Phosphorate
24. Tungsten
    - Grey
    - Brown

2. *Barytic*

2. *Barytic Genus.*

1. Barolite, or Aerated Barytes
2. Baroselenite
   Earthy
   Compact
   Baroselenite
   Foliated
   Striated or Fibrous
   Acicular
3. Liver-stone

3. *Muriatic Genus.*

1. Kiffekill
2. Martial Muriatic-spar
3. Calci-muriate
4. Argillo-muriate
5. Chlorite
   Indurated
   Slaty
6. Talcite
7. Talc
   Shistose
8. Steatites, Common
   Indurated
   Foliated
9. Potstone
10. Serpentine
11. Asbestus
    Ligniform
12. Amianthus
13. Suber montanum
14. Amianthinite
15. Asbestinite
16. Asbestoid Common
    Metalliform
17. Actinolyte
    Lamellar
    Shorlaceous
    Glassy
18. Jade
19. Boracite

4. *Argillaceous Genus.*

1. Native Argil
2. Porcelain Clay
3. Potter's Clay
4. Indurated Clay
5. Shistose Clay
6. Shale, Bituminous
7. Fuller's Earth
8. Lithomarga
   Friable
   Indurated
10. Bole
11. Argillaceous Marl
12. Chalk
    Red

Chalk,

Chalk, Yellow
Black
13. Green Earth
14. Umber
15. Tripoli
16. Phospholite
17. Lepidolite
18. Sappare
19. Mica
20. Micarelle
21. Hornblende
22. Basaltine
23. Labrador Hornblende
24. Schiller-spar
25. Shistose Hornblende
26. Wacken
27. Mullen
28. Kragg
89. Trap
30. Basalt
31. Calp
32. Argillite
33. Novaculite

5. Siliceous Genus.

1. Mountain Crystal & Quartz
2. Amethyst
3. Emerald
4. Beryll
5. Prasium
6. Oriental Ruby
Topaz
Sapphire
7. Spinel and Balas Rubies
8. Occidental Ruby
Topaz or Brazilian
Sapphire
9. Hyacinth
10. Garnet
Oriental
Common
Amorphous
11. Chrysoberyl
12. Chrysolite
13. Oliven
14. Obsidian
15. Shorl
16. Tourmaline
17. Thumerstone
18. Prehnite
19. Ædilite, or Siliceous Zeolite
20. Zeolite
21. Staurolite, or Cross-stone of St Andreasberg in the Hartz
22. Lapis Lazuli
23. Chrysoprasium.
24. Vesuvian, or White Garnet of Vesuvius.
25. Shorlite
26. Rubellite
27. Opal
28. Semi-opal
29. Pitchstone
30. Hydrophane
31. Hyalite, Muller's Glass of the Germans
32. Calcedony

32. Calecdony
    Common
    Carnelian
33. Cat's-Eye
34. Flint
35. Hornstone
36. Schistose Hornstone
37. Siliceous Shistus
38. Basanite
39. Hornslate
40. Jasper
    Common
    Striped
41. Egyptian Pebble
42. Sinople
43. Porcellanite
44. Heliotropium
45. Woodstone
46. Elastic Quartz.
47. Felspar
    Common
    Moonstone
    Continuous
48. Labradore Stone
49. Petrilite
50. Argentine Felspar
51. Red stone of Rawenstein
52. Siliceous Spar
53. Agates

6. *Strontian Genus.*

1. Strontianite

7. *Jargon Genus.*

1. Jargon or Zircon.

8. *Sydneia Genus.*

9. *Adamantine Genus.*

1. Adamantine Spar.

II. SALINE SUBSTANCES.

1. Acids
2. Alkalies
3. Neutral Salts
   Spec. 1. Tartar Vitriolate
   2. Glauber's Salt
   3. Vitriolic Ammoniac
4. Epsom Salt
5. Alum
6. Aluminous Ores
   Stony
   Earthy

1st Family

1st Family, Slaty
2d. Compact
3d. Ligneous
7. Vitriol of Iron
8. —— of Copper
9. —— of Zinc
10. Mixed Iron, Copper and Zinc
11. Nitre
12. Nitrated Soda
13. Nitrous Ammoniac
14. Nitrated Lime
15. Nitrated Magnesia
16. Muriated Tartarin
17. Common Salt
Var. 1. Lamellar
2. Fibrous
18. Sal Ammoniac
19. Muriated Barytes
20. —— Lime
21. —— Magnesia
22. —— Argil
23. —— Iron
24. —— Copper
25. —— Manganese.
26. Tincal

III. INFLAMMABLES.

Genus 1. *Inflammable Air*
2. *Bituminous*
Species 1. Naphtha
2. Petrol
3. Mineral Tar
4. —— Pitch
Var. 1. Cohesive
2. Maltha
3. Asphalt
5. Mineral Tallow
—— Cahoutchou
Genus 3. *Carbonaceous Substances*
Spec. 1. Native Carbon
2. Bituminous Carbon
Fam. 1, Cannel Coal
Var. 1. Compact
2. Slaty
Fam. 2. Impregnated with Asphalt and Maltha
Var. 1. From Whitehaven
2. — Wigan
3. — Swansea
4. — Leitrim
5. — Irvine
Fam. 3. Spurious Coal
Spec. 3. Antracolite
4. Plumbago
Genus 4. *Vegeto-Carbonated Substances*
Spec. 1. Carbonated Wood
Var. 1. Ligniform
2. Scaly or Earthy
3. Compact

Spec. 2.

Spec. 2. Turf and Peat
Genus 5. *Vegeto-Bituminous*
Spec. 1. Jet
2. Amber
Ambergris and Copal
Appendix. Mellilite
Genus 6. *Sulphur and its Ores*
Spec. 1. Native Sulphur
2. Hepatic Air
3. Sulphur combined with Argil
4. — mixed with Argil
5. United with Calcareous Earth
6. United to Fixed Alkalies
7. — to Metallic Substances
8. Martial Pyrites
Fam. 1. United to Iron in its metallic state
Var. 1. Common Sulphur Pyrites
2. Striated
3. Capillary
4. Magnetic
Fam. 2. United to Calx of Iron
Hepatic Pyrites

IV. METALLIC SUBSTANCES.

1. *Gold.*

Species 1. Native gold
Auriferous ores

2. *Platina.*

3. *Silver and its Ores.*

Spec. 1. Native silver
Fam. 1. Pure
2. Auriferous
3. Cupriferous
4. Antimoniated
5. Arsenicated
Spec. 2. Calciform
3. Mineralised by Acids
Fam. 1. Corneous
2. Argillo-Muriated
Spec. 4. Sulphurated Silver-ore
5. Light Lamellar Silver-ore
6. Sooty Silver Ore
Spec. 7. Antimoniated Silver-ore
8. Plumbiferous Antimoniated Silver-ore
Fam. 1. Light Grey
2. Dark Grey
9. Cupriferous Sulphurated Silver Ore
10. Red Silver Ore
Fam. 1. Light Red
2. Dark Red
11. Scoriaceous
12. Bismuthic
13. Greenish and Reddish black Silver-Ore
Subsidiary

4.

4. *Copper and its Ores.*

Spec. 1. Native Copper
Var. 1. —— ——
2. Cement Copper
Spec. 2. Calciform Copper-Ores
Tribe i. Blue
Var. 1. Mountain Blue
2. Striated
Tribe ii. Green
Fam. 1. Malachite
Var. 1 Fibrous
2. Compact
Fam. 2. Mountain Green
Tribe iii. Red
Fam. 1. Cochineal Red
Var. 1. Compact
2. Foliated
3. Fibrous
Fam. 2. Brick Red
Var. 1. Earthy
2. Indurated
Spec. 3. Mineralized by Sulphur
Fam. 1. Copper-Pyrites
2. Purple
3. Black
4. Vitreous
Var. 1. Compact
2. Foliated
Fam. 5. Grey
Spec. 4. Mineralized by Muriatic Acid
5. Mineralized by Arsenical Acid
Olive Copper-Ore
Earthy Iron shot
Mountain Green
Glassy Iron shot
Mountain Green
6. Mineralized by Arsenic
White Copper
Various Cupriferous Compounds

5. *Iron and its Ores.*

Spec. 1. Native Iron
2. Mineralized by pure Air
Tribe i.
Fam. 1. Common Magnetic Ironstone
2. Fibrous
3. Magnetic Sand
Tribe ii.
Fam. 1. Specular Iron Ore
2. Brown Hæmatites
3. Compact Brown Ironstone
4. Brown scaly Ore
5. Brown Iron Ochre
6. Black Ironstone
Tribe iii.
Fam. 1. Red Hæmatites
2. Compact Red Ironstone
3. Red Ochre
4 Red scaly Iron Ore
Tribe iv. Argillaceous Iron Ores
Fam. 1. Upland Argillaceous

Var. 1. Common Argillaceous Iron stone
Var. 2. Columnar Iron Ore
3. Acinose
4. Nodular
5. Pisiform
Fam. 2. Lowland Argillaceous Iron Ore
Siderite
Sideritic Calx
Var. 1. Lowland Ores
Meadow
Var. 2. Swampy
3. Morassy
Spec. 3. Mineralized by Carbon
Plumbaginous
4. Blue Martial Earth
5. Blue Iron-Ore of Vorau
6. Green Martial Earth
7. Mineralized by Sulphur
8. Mineralised by Arsenic
9. Mineralized by the Arsenical Acids
10. Sparry Iron-Ore
11. Emery
12. Tungstenic Iron-Ore

6. *Tin and its Ores.*

Spec. 1. Native Tin
2. Mineralized by Oxygen
Fam. 1. Common Tin stone
Fam. 2. Fibrous, or Wood Tin Ore
3. Tin Pyrites

7. *Lead and its Ores.*

Spec. 1. Native Lead
2. Mineralized by Oxygen and Fixed Air
Fam. 1. White Lead Ore
2. Earthy, Yellowish, Greenish, &c. Lead Ore
3. Earthy red Lead Ore
3. Phosphorated Lead Ore
4. Arsenicated Lead Ore
5. Arsenico-Phosphorated
Spec. 6. Vitriolated
7. Yellow Molybdenated Lead Ore
8. Red Lead-Spar
9. Mineralized by Sulphur
Fam. 1. Common Galena
2. Compact
3. Blue Lead Ore
4. Black Lead Ore
Brown Lead Ore of Werner

8. *Mercury*

8. *Mercury and its Ores.*

Spec. 1. Native
2. Natural Amalgama
3. Mineralized by Oxygen
Fam. 1. Compact
2. Slaty
4. Corneous Mercurial Ore
5. Mineralized by Sulphur
Fam. 1. Native Æthiops
2. Native Cinnabar
Dark Red
Bright Red
3. Greyish Black

9. *Zinc and its Ores.*

Spec. 1. Calamine
Fam. 1. Loose
2. Compact
3. Striated
Spec. 2. Blende
Fam. 1. Yellow
2. Brown
3. Black

10. *Antimony and its Ores.*

Spec. 1. Native
2. Sulphurated
Fam. 1. Compact
2. Foliated
3. Striated
Spec. 3. Sulphurated and Arsenicated Plumose
4. Red Antimonial Ore
5. Muriated
Antimonial Ochre
Supposed Phosphorated Antimony

11. *Arsenic and its Ores.*

Spec. 1. Native Arsenic
2. Do. Alloyed with Iron
3. Do. with Sulphurated Iron
Spec. 4. Mineralised by Oxygen
Loose
Indurated
5. Mineralised by Sulphur
Fam. 1. Orpiment
2. Realgar

12. *Bismuth and its Ores.*

Spec. 1. Native Bismuth
2. Bismuth Ochre
Earthy
Crystallised
3. Mineralised by Sulphur

13. *Cobalt and its Ores.*

Spec. 1. Dull grey Cobalt Ore
2. Bright white Cobalt Ore
3. Mineralized by Oxygen
Fam. 1. Black Cobalt Ore
Loose
Indurated
2. Brown
Fam. 3. Yellow
Spec. 4. Red Cobalt Ore
Fam. 1. Cobaltic Germinations
2. Cobaltic Incrustations
Green and Violet Cobalt-Ores

14. *Nickel and its Ores.*

Spec. 1. Native Nickel alloyed by Iron
2. Nickel Ochre, and Vitriol, Loose
Indurated
Spec. 3. Arsenicated Nickel
4. Sulphurated Nickel

15. *Manganese and its Ores.*

Spec. 1. Native
2. Mineralized by Oxygen
Fam. 1. Grey
2. Black
Earthy
Indurated
Spec. 3. Mineralized by Oxygen and fixed air
Fam. 1. White
2. Red
4. Vitriolated Manganese

16. *Uranite and its Ores*

Spec. 1. Mineralized by Acids
Fam. 1. Uranitic Ochre
2. Micaceous
Spec. 2. Sulphurated

17. *Tungstenite and its Ores.*

Spec. 1. Tungsten
Fam. 1. White or Grey
2. Brown
Spec. 2. Wolfram

18. *Molybdenite.*

Molybdena

19. *Sylvanite.*

20. *Menachanite.*

21. *Titanite.*

Calcareo Siliceous-Ore.

MOHS'

MINERAL SYSTEM IN 1804.

CLASS I.—EARTHY MINERALS.

1. Diamond family
2. Zircon family
3. Chrysoberyl family
4. Augite family
5. Garnet family
6. Spinel family
7. Hardstone (hartstein) family
8. Schorl family
9. Quartz family
10. Opal family
11. Obsidian
12. Zeolite
13. Felspar family
14. Clay family
15. Clay-slate family
16. Mica family
17. Trap family
18. Lithomarge family
19. Bole family
20. Talc family
21. Actynolite family
22. Limestone family
23. Brown-spar family
24. Marl family
25. Apatite family
26. Fluor family
27. Gypsum family
28. Baryte family
29. Saltstone family

CLASS II.—SALINE MINERALS.

30. Family of carbonats
31. Family of nitrats
32. Family of muriats
33. Family of sulphats

CLASS III.—INFLAMMABLE MINERALS.

34. Sulphur family
35. Amber family
36. Coal family
37. Graphite family

CLASS IV.—METALLIC MINERALS.

38. Family, Native Gold
39. Family, Mercurial-ores
40. Family, Native Silver
41. Family, Silver-ores
42. Family, Native Copper
43. Family of Copper-pyrites
44. Family, Malachite
45. Family, Copper Emerald
46. Family, Native Iron
47. Family, Iron-pyrites
48. Family, Ironstone
49. Family, Iron Earth
50. Family, Manganese
51. Family, Manakan
52. Family, Lead-ore
53. Family, Tinstone
54. Family, Speiss-Cobalt
55. Family, Cobalt-ochre
56. Family, Earthy Cobalt-ores
57. Family, Native Arsenic
58. Family, Antimony-ores
59. Family, Uranium Ores

APPENDIX.

262. Needle-ore
263. Chrome-ochre

BRONG-

BRONGNIART's

MINERAL SYSTEM IN 1807.

Classe I.—*Les Oxigénés non Metalliques.*

L'oxigêne combiné avec des bases non metalliques.

Ord. 1.—Les Oxigénés non Acides.

L'oxigène formant avec ces bases des corps non acides.
Generes, Air, Eau.

Ord. 2.—Les Oxigénés non acides.

L'oxygêne formant avec ces bases des corps acides.
Acides Sulphurique, Muriatique, Carbonique et Boracique.

Classe II.—*Les Sels non Metalliques.*

Une base non metallique combinée avec un acide.

Ord. 1.—Les Sels Alcalins.

Une base alcaline avec un acide.

Ord. 2.—Les Sels Terreux.

Une base terreuse avec un acide.

Ord. 3.—Les Sels Terreux.

Une base terreuse combinée avec un acide.

Classe III.—*Les Pierres.*

Les terres combinées entr'elles, et quelquefois avec des principes accessoires alcalins, acides ou metalliques.

Ord. 1.—Les Pierres dures.

Seches et apres au toucher, une dureté assez considerable pour rayer le verre à vitre blanc.

 Ord.

Ord. 2.—Les Pierres Onctueuses.

Ne rayant point le verre, le plus tendre, douces, et même onctueuses au toucher.

Ord. 3.—Les Pierres Argilloides.

Aspect argilleux, odeur argilleuse, souvent douces au toucher.

Classe IV.—*Les Combustibles.*

Mineraux qui peuvent se combiner immediatement avec l'oxigêne.

Ord. 1. Les Combustibles Composés.

Donnant de la fumée huileuse en brulant.

Ord. 2.—Les Combustibles Simples.

Ne donnant point de fumée huileuse dans leur combustion.

Classe V.—*Les Metaux.*

Mineraux ayant pour base une substance metallique.

Ord 1.—Les Metaux Fragiles.

N'etant susceptibles de s'alonger ni sous le marteu ni sous le laminoir.

Ord. 2.—Les Metaux Ductiles.

Susceptibles de s'étendre sous le laminoir ou sous le martean.

KARSTEN's

KARSTEN'S

MINERAL SYSTEM IN 1808.

I. Class. *Earthy Minerals.*

1. Order, Zirconia.
2. ——— Yttria.
3. ——— Glucina.
4. ——— Silica.
    - *a.* Silica and glucina.
    - *b.* Silica, with a very slight intermixture of other substances.
    - *c.* Silica with water.
    - *d.* Silica with water and alumina.
    - *e.* Silica, alumina, and lime, or an alkali.
    - *f.* Silica and magnesia.
    - *g.* Silica and lime.
    - *h.* Silica, lime, alumina, and gypsum.
5. Order, Alumina.
6. —— Magnesia.
7. —— Lime.
    - *a.* Lime with carbonic acid.
    - *b.* Lime with phosphoric acid.
    - *c.* ——— fluoric acid.
    - *d.* ——— sulphuric acid,
    - *e.* ——— boracic acid.
8. Order, Strontian.
9. —— Barytes.

II. Class. *Saline Minerals.*

1. Order, Carbonates.
2. —— Borates.
3. —— Nitrates.
4. —— Muriates.
5. —— Sulphates.

III. Class. *Inflammable Minerals.*

IV. Class. *Metallic Minerals.*

1. Order, Platina
2. —— Gold
3. —— Mercury
4. —— Silver
5. —— Copper
6. —— Iron
7. —— Lead
8. —— Molybdena
9. —— Tin
10. —— Zinc
11. —— Bismuth
12. —— Tellurium
13. Order, Antimony
14. —— Manganese
15. —— Nickel
16. —— Cobalt
17. —— Arsenic
18. —— Uranium
19. —— Titanium
20. —— Scheel
21. —— Chrome
22. —— Tantalum
23. —— Cerereum
24. —— Columbium

THOMSON's

THOMSON'S

MINERAL SYSTEM IN 1810.

CLASS I.—STONES.

Order 1.—*Earthy Stones.*

Families.—Diamond, Zircon, Chrysolite, Garnet, Ruby, Topaz, Schorl, Quartz, Pitchstone, Zeolite, Felspar, Clayslate, Mica, Trap, Lithomarge, Soapstone, Talc, Actynolite, and Gadolinite.

Order 2.—*Saline Stones.*

i. Genus, Calcareous Salts.

1. Family of Carbonates
2. Family of Phosphates
3. Family of Fluates
4. Family of Sulphates
5. Family of Borates

ii. Genus, Barytic Salts.

Carbonate
Sulphate

iii. Genus, Strontian Salts.

Carbonate
Sulphate

iv. Genus, Magnesian Salts.

Sulphate
Carbonate
Borate

v. Genus, Aluminous Salts.

CLASS II.—SALTS.

Genus i. Potash
ii. Soda
Genus iii. Ammonia

CLASS III.—COMBUSTIBLES.

Genus i. Sulphur
ii. Resin
Genus iii. Bitumen
iv. Graphite

CLASS

Class IV.—ORES.

Order i. *Gold*
1. Alloys

Order ii. *Platinum*
1. Alloys

Order iii. *Iridium*
1. Alloys

Order iv. *Silver*
1. Alloys. 2. Sulphurets. 3. Oxides. 4. Salts

Order v. *Mercury*
1. Alloys. 2. Sulphurets. 3. Salts

Order vi. *Copper*
1. Alloys. 2. Sulphurets. 3. Oxides. 4. Salts

Order vii. *Iron*
1. Alloys. 2. Sulphurets. 3. Oxides. 4. Salts

Order viii. *Nickel*
1. Alloys. 2. Oxides

Order ix. *Tin*
1. Sulphurets. 2. Oxides

Order x. *Lead*
1. Sulphurets. 2. Oxides. 3. Salts

Order xi. *Zinc*
1. Sulphurets 2. Oxides. 3. Salts

Order xii. *Bismuth*
1. Alloys. 2. Sulphurets. 3. Oxides

Order xiii. *Antimony*
1. Alloys 2. Sulphurets. 3. Oxides 4. Salts

Order xiv. *Arsenic*
1. Alloys. 2. Sulphurets. 3. Oxides. 4. Salts.

Order xv. *Cobalt*
1. Alloys. 2. Oxides. 3. Salts

Order xvi. *Manganese*
1. Oxides. 2. Salts

Order xvii. *Chromium*
1. Alloys. 2. Oxides. 3. Salts

Order xviii. *Uranium*
1. Oxides

Order xix. *Molybdenum*
1. Sulphurets

Order xx. *Tungsten*
1. Salts

Order xxi. *Titanium*
1. Oxides

Order xxii. *Columbium*
1. Oxides

Order xxiii. *Cerium*
1. Oxides.

MURRAY's

MURRAY'S

MINERAL SYSTEM IN 1812.

I. *Saline Minerals.*

1. Native Salts, with a base of Ammonia.
2. Native Salts, with a base of Potash.
3. Native Salts, with a base of Soda.

II. *Earthy Minerals.*

1. Barytic Fossils.
2. Strontitic Fossils.
3. Calcareous Fossils.
4. Magnesian fossils.
5. Argillaceous Fossils.
6. Glucine Fossils.
7. Siliceous Fossils.
8. Zircon Fossils.
9. Gadolinite.

III. *Metallic Minerals.*

1. Native Gold.
2. Native Platina.
3. Ores of Silver.
4. Ores of Quicksilver.
5. Ores of Copper.
6. Ores of Iron.
7. Ores of Lead.
8. Ores of Tin.
9. Ores of Zinc.
10. Ores of Nickel.
11. Ores of Cobalt.
12. Ores of Manganese.
13. Ores of Arsenic.
14. Ores of Bismuth.
15. Ores of Antimony.
16. Ores of Tellurium.
17. Ores of Chrome.
18. Ores of Molybdena.
19. Ores of Tungsten.
20. Ores of Titanium.
21. Ores of Uranium.
22. Ores of Tantalum.
23. Ores of Cerium.

IV. *Inflammable*

IV. *Inflammable Minerals.*

1. Native Sulphur.
2. Carbonaceous Minerals.
3. Inflammable Minerals, in which Hydrogen predominates.

HAUSMANN's

HAUSMANN's

MINERAL SYSTEM, IN 1813.

CLASS I.—COMBUSTIBLES.

I. Order. INFLAMMABILES.

Non-Metallic Combustibles,

1. Sub-Order,—*Simple*,—Ex. *Diamond*, &c.
2. Sub-Order,- -*Compound*.

Combinations of two or more non-Metalliic Combustibles, —Ex. *Graphite*, &c.

2. Order. METALS.

Native Metals and Alloys.—Ex. *Native Silver* and *Antimonial Silver*.

3. Order. ORES.

Combinations of Metals and Sulphur.—Ex. *Copper-Pyrites*.

CLASS II.

Class II.—INCOMBUSTIBLES.

Oxidised Minerals, and Combinations of these.

1. Order. Oxides.

1. Sub-Order. *Metallic Oxides.* Oxidised Metals, either simple, or in combination with each other, and sometimes also combined with earths or with water.—Ex. *Magnetic Ironstone,* and *Brown Ironstone.*

2. Sub-Order. *Earths,* variously combined with each other, and with metallic oxides and water.

1. Series. Simple, *Quartz.*

Indeterminate combinations of earths with each other, or with other matters.

2. Series. Compound, *Opal.*

Determinate combinations of earths with each other, or with other substances.

2. Order. Oxydoide.

Combinations of combustible bodies with oxygen, which possess neither the properties of bases nor of acids. —Ex. *Water.*

3. Order. Acids.

4. Order. Salts.

Combinations of Bases with Acids.

1. Sub-Order. *Earthy.*

With Earthy Bases,

1. Series. Aluminous Salts.

2. —— Magnesian Salts.

2. Sub-

2. Sub-Order. *Alkaline.*

With Alkaline Bases,

1.	Series.	Salts of Soda.
2.	——	Salts of Potash,
3.	——	Salts of Ammonia.
4.	——	Salts of Lime.
5.	——	Salts of Strontian.
6.	——	Salts of Barytes.

3. Sub-Order. *Metallic.*

With Metallic Oxide Bases,

1.	Series.	Salts of Silver,
2.	——	Salts of Mercury.
3.	——	Salts of Copper.
4.	——	Salts of Iron.
5.	——	Salts of Manganese,
6.	——	Salts of Lead.
7.	——	Salts of Zinc.
8.	——	Salts of Cobalt.
9.	——	Salts of Nickel,

AIKIN's

AIKIN's

MINERAL SYSTEM IN 1815.

Class I.—*Non-Metallic Combustible Substances.*

1. Combustible with flame. Mineral Oil.
2. Combustible without flame. Graphite.

Class II.—*Native Metals, and Metalliferous Minerals.*

Order I.—Volatilisable, wholly or in part, by the blowpipe on charcoal, into a vapour, which condenses in a pulverulent form on a piece of charcoal held over it.

1. Entirely, or almost entirely volatilisable.
   Lustre metallic. Native Arsenic.
   Lustre non-metallic. Cinnabar.

2. Partly volatilisable; the residue affording metallic grains with borax, on charcoal.
   Lustre metallic. Silver-white Cobalt-ore.
   Lustre non-metallic. Red Silver-ore.

3. Partly volatilisable; the residue not reducible to the metallic state.
   Lustre metallic. Common Iron-pyrites.
   Lustre non-metallic. Red Cobalt-ochre.

Order II.

Order II.—Fixed; not volatilisable except at a white heat.

1. Assume or preserve the metallic form, after roasting on charcoal while any thing is dissipated, and subsequent fusion with borax.

   Lustre metallic. Native Copper.
   Lustre non-metallic. Malachite.

2. Not reducible to the metallic state before the blow-pipe on charcoal, either with or without borax.

   Magnetic after roasting. Common Pyrites.
   Not magnetic after roasting. Blende.

Class III.—*Earthy Minerals.*

Order I.—Soluble with effervescence, either wholly, or in considerable proportion, in cold and moderately dilute muriatic acid; yield to the knife.

1. Effervesce vigorously. Marl.
2. Effervesce very feebly in cold, but more vigorously in warm, muriatic acid. Carbonate of Magnesia.

Order II. *Fusible before the blowpipe.*

1. Hardness equal or superior to that of quartz. Garnet.
2. Hardness superior to that of common window-glass; generally yield in some degree to the knife. Felspar.
3. Yield to the knife; and sometimes feebly scratch glass. Tremolite.
4. Yield easily to the knife, and sometimes to the nail. Heavy-spar.
5. Very soft; yield to the nail. Gypsum.

Order III.—*Infusible before the blowpipe.*

1. Hardness equal or superior to that of quartz. Flint.
2. Scratch glass; sometimes yield to the knife. Opal.
3. Yield to the knife. Serpentine.
4. Yield to the nail. Mountain-cork.

Class IV.—*Saline Minerals.*

Soluble in Water; Sapid.

Order I.—Afford a precipitate with carbonated alkali. Blue Vitriol.

Order II.—Do not afford a precipitate with carbonated alkali. Natron.

HAUY's

HAUY'S

SYSTEM IN 1813.

PREMIERE CLASSE.
SUBSTANCES ACIDIFERES.

PREMIERE ORDRE.
Substances acidifères libres.

1. I. Acide sulfurique

2. II. Acide boracique

SECOND ORDRE.
Substances acidifères terreuse.

† A BASE SIMPLE.

I. Genre.—*Chaux.*

3. 1. Chaux carbonatée
  i. Chaux carb. *ferrifère*
  ii. Chaux carb. *manganèsifère rose*
  iii. Chaux carb. *ferro-manganèsifère*
  iv. Chaux carb. *quarzifère*
  v. Chaux carb. *magnésifère*
  vi. Chaux carb. *nacrée*
  vii. Chaux carb. *fétide*
  viii. Chaux carb. *bituminifère*

4. 2. Arragonite

5. 3. Chaux phosphatée
  Chaux phosphatée *quarzifère*

6. 4. Chaux fluatée
  Chaux fl. *aluminifère*

7. 5. Chaux sulfatée
  Chaux sul. *calcarifère*

8. 6. Chaux anhydro-sulfatée
  i. Chaux an.-sul. *muriatifère*
  ii. Chaux an.-sul. *quarzifère*
  iii. Chaux an.-sul. *épigène*

9. 7. Chaux nitratée

10. 8. Chaux arseniatée

II. Genre.

II. Genre.—*Baryte.*

11. 1. Baryte sulfatée
Baryte sulfatée *fétide*

12. 2. Baryte carbonatée

III. Genre.—*Strontiane*

13. 1. Strontiane sulfatée
Strontiane sul. *calcarifère*

14. 2. Strontiane carbonatée

IV. Genre.—*Magnesie.*

15. 1. Magnésie sulphatée
i. Magnésie sul. *ferrifère*
ii. Magnésie sul. *cobaltifère*

16. 2. Magnésie boratée
Magnésie bor. *calcarifère*

17. 3. Magnésie carbonatée
Magnésie carb. *silicifère*

†† A BASE DOUBLE.

V. Genre.—*Chaux et Silice.*

18. Chaux boratée siliceuse.

VI. Genre.—*Silice et Alumine.*

19. Silice fluatée alumineuse *ou* Topaze.

TROISIEME ORDRE.

Substances acidifères alkalines.

I. Genre.—*Potasse.*

20. Potasse nitratée

II. Genre.—*Soude.*

21. 1. Soude sulfatée
22. 2. Soude muriatée
23. 3. Soude boratée
24. 4. Soude carbonatée

III. Genre.—*Ammoniaque.*

25. 1. Ammoniaque sulfatée
26. 2. Ammoniaque muriatée

QUATRIEME ORDRE.

Substances acidifères alkalino-terreuses.

Genre unique.—*Alumine.*

27. 1. Alumine sulfatée alkaline
28. 2. Alumine fluatée alkaline

Appendice.—Glauberite

SECONDE

SECONDE CLASSE.

SUBSTANCES TERREUSES.

30. 1. Quarz
    i. Quarz-*hyalin*
    ii. Quarz-*agathe*
    iii. Quarz-*resinite*
    iv. Quarz-*jaspe*
    v. Quarz-*pseudomorphique*
31. 2. Zircon
32. 3. Corindon
    i. Corindon-*hyalin*
    ii. Corindon-*harmophane*
    iii. Corindon-*granulaire*
33. 4. Cymophane
34. 5. Spinelle
35. 6. Emeraude
36. 7. Euclase
37. 8. Grenat
    Grenat *ferrifère*
38. 9. Amphigène
39. 10. Idocrase
40. 11. Meïonite
41. 12. Feld-spath
    i. Feld-spath *tenace*
    ii. Feld-spath *décomposé*
42. 13. Apophyllite
43. 14. Triphane
44. 15. Axinite
45. 16. Tourmaline
    Tourmaline *apyre*
46. 17. Amphibole
47. 18. Pyroxène
48. 19. Yenite
49. 20. Staurotide
50. 21. Epidote
    Epidote *manganesifère*
51. 22. Hypersthène
52. 23. Wernerite
53. 24. Paranthine
54. 25. Diallage
55. 26. Gadolinite
56. 27. Lazulite
57. 28. Mésotype
    Mesotype *altérée*
58. 29. Stilbite
59. 30. Laumonite
60. 31. Prehnite
61. 32. Chabasie
62. 33. Analcime
    Analcime *cubo-octaèdre ?*
63. 34. Népheline
64. 35. Harmotome
65. 36. Péridot
    Péridot *décomposé*
66. 37. Mica
67. 38. Pinite
68. 39. Disthène
69. 40. Dipyre
70. 41. Asbeste
71. 42. Talc
    Talc *pseudomorphique*
72. 43. Macle

TROISIEME CLASSE.

SUBSTANCES COMBUSTIBLES.

PREMIERE ORDRE.

Substances combustibles simples.

73. 1. Soufre
74. 2. Diamant
75. 3. Anthracite

SECOND ORDRE.

Substances combustibles composées.

76. 1. Graphite
77. 2. Bitume
78. 3. Houille
79. 4. Jayet
80. 5. Succin
81. 6. Mellite

QUATRIEME CLASSE.

SUBSTANCES METALLIQUES.

PREMIERE ORDRE.

Non oxydables immédiatement, si ce n'est à un feu très violent, et réductibles immédiatement.

I. Genre.—*Platine.*

82. Platine natif *ferrifère*

II. Genre.—*Or.*

83. Or natif

III. Genre.—*Argent.*

84. 1. Argent natif
85. 2. Argent antimonial
    Argent antimonial *ferro-arsenifère*
86. 3. Argent sulfuré
87. 4. Argent antimonié sulfuré
    Argent antimonié sulfuré *noir*
88. 5. Argent carbonaté
89. 6. Argent muriaté

SECOND

SECOND ORDRF.

Oxydables et réductibles immédiatement.

Genre Unique.—*Mercure.*

90. 1. Mercure natif
91. 2. Mercure argental
92. 3. Mercure sulfuré
    Mercure sulfuré *bitu-minifère*
93. 4. Mercure muriaté

TROISIEME ORDRE.

Oxydables, mais non réductibles immédiatement.

SENSIBLEMENT DUCTILES.

I. Genre.—*Plomb.*

94. 1. Plomb natif *volcanique*
95. 2. Plomb sulfuré
    i. Plomb sulfuré *antimo-nifère*
    ii. Plomb sulfuré *antimo-nio-arsenifère*
96. 3. Plomb oxydé rouge
97. 4. Plomb arsenié
98. 5. Plomb chromaté
99. 6. Plomb carbonaté
    i. Plomb carbonaté *noir*
    ii. Plomb carbonaté *cu-prifère*
100. 7. Plomb phosphaté
    i. Plomb phosphaté *ar-senifère*
    ii. Plomb sulfuré *épigène*
101. 8. Plomb molybdaté
102. 9. Plomb sulfaté

II. Genre.—*Nickel.*

103. 1. Nickel natif
104. 2. Nickel arsenical
    Nickel arsenical *ar-gentifère*
105. 3. Nickel oxydé

III. Genre.—*Cuivre.*

106. 1. Cuivre natif
107. 2. Cuivre pyriteux
    Cuivre pyriteux *hé-patique*
108. 3. Cuivre gris
    i. Cuivre gris *arsenifère*
    ii. Cuivre gris *antimoni-fère*
    iii. Cuivre gris *platinifère*

109. 4. Cuivre sulfuré
Cuivre sulfuré *hépatique*
110. 5. Cuivre oxydulé
Cuivre oxydulé *arsenifère*
111. 6. Cuivre muriaté
112. 7. Cuivre carbonaté bleu
Cuivre carbonaté vert *épigène*
113. 8. Cuivre carbonaté vert
114. 9. Cuivre arseniaté
i. Cuivre arseniaté *altéré*
ii. Cuivre arseniaté *ferrifere*
115. 10. Cuivre dioptase
116. 11. Cuivre phosphaté
117. 12. Cuivre sulfaté

IV Genre.—*Fer.*

118. 1. Fer natif
i. Fer natif *volcanique*
ii. Acier natif *pseudo-volcanique*
iii. Fer natif *météorique*
119. 2. Fer oxydulé
Fer oxydulé *titanifère*
120. 3. Fer oligiste
121. 4. Fer arsenical
Fer arsenical *argentifère*
122. 5. Fer sulfuré
i. Fer oxydé *épigène*
ii. Fer sulfuré *ferrifère*
iii. Fer sulfuré *aurifère*
iv. Fer sulfuré *titanifère*
123. 6. Fer oxydé
i. Fer oxydé noir *vitreux*
ii. Fer oxydé *résinite*
iii. Fer oxydé *carbonaté*

V. Genre.—*Etain.*

128. 1. Etain oxydé
129. 2. Etain sulfuré

VI. Genre.—*Zinc.*

130. 1. Zinc oxydé
131. 2. Zinc carbonaté
Zinc carbonaté *pseudomorphique*
132. 3. Zinc sulfuré
133. 4. Zinc sulfaté

NON

NON DUCTILES.

VII. Genre.—*Bismuth.*

134. 1. Bismuth natif
135. 2. Bismuth sulfuré
Bismuth sulfuré *plumbo-cuprifère*
136. 3. Bismuth oxydé

VIII. Genre.—*Cobalt.*

137. 1. Cobalt arsenical
138. 2. Cobalt gris
139. 3. Cobalt oxydé noir
140. 4. Cobalt arseniaté
Cobalt arseniaté *terreux argentifère*

IX. Genre.—*Arsenic.*

141. 1. Arsenic natif
142. 2. Arsenic oxydé
143. 3. Arsenic sulfuré
Arsenic sulfuré *rouge*
Arsenic sulfuré *jaune*

X. Genre.—*Manganese.*

144. 1. Manganèse oxydé
i. Manganèse oxydé *noirâtre barytifère*
ii. Manganèse oxydé *carbonaté*
145. 2. Manganèse sulfuré
146. 3. Manganèse phosphaté *ferrifere*

XI. Genre.—*Antimoine.*

147. 1. Antimoine natif
Antimoine natif *arsenifère*
148. 2. Antimoine sulfuré
i. Antimoine sulfuré *argentifère*
ii. Antimoine oxydé *épigène*
iii. Antimoine oxydé sulfuré *épigène*
149 3. Antimoine oxydé
150. 4. Antimoine oxydé sulfuré

XII. Genre.—*Urane.*

151. 1. Urane oxydulé
152. 2. Urane oxydé

XIII. Genre.

XIII. Genre.—*Molybdène.*

153. Molybdène sulfuré

XIV. Genre.—*Titane.*

154. 1. Titane oxydé
i. Titane oxydé *chromifère*
ii. Titane oxydé *ferrifère*

155. 2. Titane anatase

156. 3. Titane siliceo-calcaire

XV. Genre.—*Schéelin.*

157. 1. Schéelin ferruginé

158. 2. Schéelin calcaire

XVI. Genre.—*Tellure.*

159. Tellure natif
i. Tellure natif *auro-ferrifère*
ii. Tellure natif *auro-argentifère*
iii. Tellure natif *auro-plumbifère*

XVII. Genre.—*Tantale.*

160. Tantale oxydé
i. Tantale oxydé *ferro-manganesifère*
ii. Tantale oxydé *yttrifère*

XVIII. Genre.—*Cerium.*

161. Cerium oxydé *silicifère*

XIX. Genre.—*Chrome.*

WERNER's

WERNER's

MINERAL SYSTEM IN 1815.

[Communicated to me by M. VIVIAN.]

CLASS I.—EARTHY FOSSILS.

1. DIAMOND GENUS.

1. Diamond.

2. ZIRCON GENUS.

Zircon Family.

2. Zircon.
3. Hyacinth.
4. Cinnamon-stone.

3. FLINT GENUS.

Augite Family.

5. Chrysoberyl.
6. Chrysolite.
7. Olivine.
8. Coccolite.
9. Augite.
    a. granular.
    b. foliated.
    c. conchoidal.
    d. common.
10. *Baikalite.*
11. Sahlite.
12. Diopside.
13. *Fassaite.*

Garnet

Garnet Family.

14. Vesuvian.
15. Grossulare.
16. Leucite.
17. *Pyreneite.*
18. Melanite.
19. Allochroite.
20. *Colophonite.*
21. Garnet.
    a. Precious.
    b. Common.
22. Staurolite or Grenatite.
23. Pyrope.

Ruby Family.

24. Automalite
25. Ceylanite.
26. Spinel.
27. Sapphire.
28. Emery.
29. Corundum.
30. Diamond-spar.

———

31. Topaz.

Beryl Family.

32. Iolite.
33. Euclase.
34. Emerald.
35. Beryl.
    a. Precious beryl.
    b. Common.
36. Schorlous Beryl.
37. Tourmaline.

Pistacite Family.

38. *Lievrite.*
39. Pistacite.
40. Diaspore.
41. Zoisite.
42. Anthophylite.
    a. Radiated.
    b. Foliated.
43. Axinite.

Quartz Family.

44. Quartz.
    a. Amethyst.
        α. Common.
        β. Thick fibrous.
    b. Rock crystal.
    c. Milk quartz.
    d. Common quartz.
    e. Prase.
45. Iron-flint.
46. Hornstone.
    a. Splintery.
    b. Conchoidal.
    c. Woodstone.
47. Flinty-slate.
    a. Common.
    b. Lydian-stone.
48. Flint.

48. Flint.
49. Chalcedony.
    - *a.* Common.
    - *b.* Carnelian.
        - *α.* Common.
        - *β.* Fibrous.
50. Hyalite.
51. Opal.
    - *a.* Precious.
    - *b.* Common opal.
    - *c.* Semi-opal.
    - *d.* Wood-opal.
52. Menilite.
    - *a.* Brown menilite.
    - *b.* Grey menilite.
53. Jasper.
    - *a.* Egyptian jasper.
        - *α.* Red.
        - *β.* Brown.
    - *b.* Striped jasper.
    - *c.* Porcelain jasper.
    - *d.* Common jasper.
        - *α.* Conchoidal.
        - *β.* Earthy.
    - *e.* Opal jasper.
    - *f.* Agate jasper.
54. Heliotrope.
55. Chrysoprase.
56. Plasma.
57. Cat's-eye.
58. *Faser Kiesel.*
59. Elaolite.

Pitchstone Family.

60. Obsidian.
61. Pitchstone.
62. Pearlstone.
63. Pumice.

Zeolite Family.

64. Prehnite.
    - *a.* Fibrous.
    - *b.* Foliated.
65. Natrolite.
66. Zeolite.
    - *a.* Mealy Zeolite.
    - *b.* Fibrous do.
    - *c.* Radiated Zeolite.
    - *d.* Foliated do.
67. Ichthyophthalm.
68. Cubicite.
69. Cross-stone or Crucite.
70. Laumonite.
71. Schmelzstein.

Azurestone Family.

72. Azure stone.
73. Azurite.
74. Blue-spar.

Felspar

Felspar Family.

75. Andalusite.
76. Felspar.
    a. Adularia.
    b. Labrador.
    c. Glassy.
    d. Common felspar.
        α. Fresh.
        β. Disintegrated.
    e. Hollow spar.
    f. Compact felspar.
        α. Common.
        β. Variolite.
77. Spodumene.
78. Scapolite.
    a. Red scapolite.
    b. Grey scapolite.
        α. Radiated.
        β. Foliated.
79. Meionite.
80. Nepheline.
81. Ice-spar.

4. Clay Genus.

Clay Family.

82. Pure clay.
83. Porcelain earth.
84. Common clay.
    a. Loam.
    b. Potter's clay.
        α. Earthy.
        β. Slaty.
    c. Variegated clay.
    d. Slate clay.
85. Claystone.
86. Adhesive slate.
87. Polishing or polier slate.
88. Tripoli.
89. Floatstone.
90. Alum-stone.

Clay-Slate Family.

91. Alum-slate.
    a. Common.
    b. Glossy.
92. Bituminous shale.
93. Drawing-slate.
94. Whet-slate.
95. Clay-slate.

Mica Family.

96. Lepidolite.
97. Mica.
98. Pinite.
99. Potstone.
100. Chlorite.

100. Chlorite.
   a. Chlorite earth.
   b. Common chlorite.
   c. Chlorite-slate.
   d. Foliated chlorite.

Trap Family.

101. *Paulite.*
102. Hornblende.
   a. Common.
   b. Basaltic.
   c. Hornblende-slate.
103. Basalt.
104. Wacke.
105. Clinkstone.
106. Iron-clay.

———

107. Lava.

Lithomarge Family.

103. Green earth.
104. Lithomarge.
   a. Friable.
   b. Indurated.
105. Rock-soap.
106. Umber.
107. Yellow earth.

5. Talc Genus.

Soapstone Family.

108. Native magnesia, or talc-earth.
109. Meerschaum.
110. Bole.
111. Fuller's-earth.
112. Steatite.
113. Figurestone.

Talc Family.

114. Nephrite.
   a. Common nephrite.
   b. Axe-stone.
115. Serpentine.
   a. Common.
   b. Precious.
      α. Conchoidal.
      β. Splintery.
116. Schillerstone.
117. Talc.
   a. Earthy.
   b. Common.
   e. Indurated.
118. Asbestus.
   a. Rock-cork.
   b. Amianthus.
   c. Common asbestus.
   d. Rock-wood.

Actynolite

Actynolite Family.

119. Kyanite.
120. Actynolite.
    - *a.* Asbestous.
    - *b.* Common.
    - *c.* Glassy,
    - *d.* Granular.
121. *Spreustein or Chaff-stone.*
122. Tremolite.
    - *a.* Asbestous.
    - *b.* Common,
    - *c.* Glassy.
123. Sahlite.
124. *Rhætizite.*

6, Calcareous Genus.

A. *Carbonates.*

125. Rock-milk.
126. Chalk.
127. Limestone.
    - *a.* Compact.
        - *α.* Common.
        - *β.* Roestone.
    - *b.* Foliated.
        - *α.* Granular.
        - *β.* Calcareous-spar.
    - *c.* Fibrous.
        - *α.* Common.
        - *β.* Calc-sinter.
    - *d.* Pea-stone.
128. Calc-tuff.
129. Schaum-earth, or foam-earth.
130. Slate-spar.
131. Brown-spar.
    - *a.* Foliated.
    - *b.* Fibrous.
132. Schaalstone.
133. Dolomite.
134. Rhomb-spar.
135. *Anthracolite.*
136. Stinkstone.
137. Marl.
    - *a.* Marl earth.
    - *b.* Indurated marl.
138. Bituminous marl-slate.
139. Arragon.
    - *a.* Common.
    - *b.* Prismatic.

B. *Phosphates.*

140. Appatite.
141. Asparagus-stone.

C. *Fluates.*

142. Fluor.
    - *a.* Compact.
    - *b.* Fluor-spar.

D. *Sulphates.*

143. Gypsum.
    - *a. Spumous gypsum.*
    - *b.* Earthy gypsum.
    - *c.* Compact gypsum.
    - *d.* Foliated gypsum.
    - *e.* Fibrous gypsum.
144. Selenite.
145. Muriacite.

145. Muriacite.
    a. Anhydrite.
    b. Gekrösstein.
    c. Conchoidal Mur.
    d. Fibrous Mur.
    e. Compact Mur.

E. *Borates.*

146. Datolite.
147. Boracite.
148. Botryolite.

7. Baryte Genus.

149. Witherite.
150. Heavy-spar.
    a. Earthy heavy-spar.
    b. Compact heavy-spar.
    c. Granular heavy-spar.
    d. Curved lamellar heavy-spar.
    e. Straight lamellar heavy-spar.
        α. Fresh.
        β. Disintegrated.
    f. Columnar spar.
    g. Prismatic spar.
    h. Bolognese, or Bolognian spar.

8. Strontian Genus.

151. Strontian.
    a. Compact.
    b. Radiated.
152. Celestine.
    a. Fibrous.
    b. Radiated.
    c. Lamellar.
    d. Prismatic.

9. Hallite Genus.

153. Cryolite.

Class II.—FOSSIL SALTS.

1. *Carbonates.*
154. Natural soda or natron.

2. *Nitrates.*
155. Natural nitre.

3. *Muriates.*
156. Natural rock-salt.
   a. Stone-salt.
      α. Foliated.
      β. Fibrous.
   b. Lake-salt.
157. Natural sal-ammoniac.

4. *Sulphates.*
158. Natural vitriol.
159. Hair-salt.
160. Rock-butter.
161. Natural Epsom-salt.
162. Natural Glauber-salt.

Class III.—INFLAMMABLE FOSSILS.

1. Sulphur Genus.

163. Natural sulphur.
   a. Crystallised.
   b. Common.
      α. Earthy.
      β. Conchoidal.
   c. Mealy.
   d. Volcanic.

2. Bituminous Genus.

164. Mineral or fossil oil.
165. Mineral pitch.
   a. Elastic.
   b. Earthy.
   c. Slaggy.
166. Brown coal.
   a. Bituminous wood.
   b. Earth coal.
   c. Alum earth.
   d. *Paper coal.*
   e. Common brown coal.
   f. Moor coal.
167. Black coal.

a. Pitch

a. Pitch coal.
b. Columnar coal.
c. Slate coal.
d. Cannel coal.
e. Foliated coal.
f. Coarse coal.

3. Graphite Genus.

168. Glance-coal.
a. Conchoidal.
b. Slaty.

169. Graphite.
a. Scaly.
b. Compact.

170. Mineral charcoal.

4. Resin Genus.

171. Amber.
a. White.
b. Yellow.

172. Honey stone.

Class IV.—METALLIC FOSSILS.

1. Platina Genus.

173. Native Platina.

3. Gold Genus.

174. Native gold.
a. Gold yellow.
b. Brass yellow.
c. Greyish yellow.

3. Mercury Genus.

175. Native mercury.

176. Natural amalgam.
a. Semi-fluid.
b. Solid.

177. Mercurial horn-ore.

178. Mercurial liver-ore.
a. Compact.
b. Slaty.

179. Cinnabar.
a. Dark-red.
b. Light-red.

4. Silver

4. Silver Genus.

180. Native silver.
    a. Common.
    b. Auriferous.
181. Antimonial silver.
182. Arsenical silver.
183. *Molybdena-silver.*
184. Corneous silver-ore, or horn-ore.
185. Silver-black.
186. Silver-glance.
187. Brittle silver-glance,
188. Red silver-ore.
    a. Dark.
    b. Light.
189. White silver-ore.

5. Copper Genus.

190. Native copper.

Family of Copper Sulphurets.
191. Copper-glance.
    a. Compact.
    b. Foliated.
192. Variegated copper-ore.
193. Copper-pyrites.
194. White copper-ore.
195. Grey copper-ore.
196. Black copper-ore.

197. Red copper-ore.
    a. Compact.
    b. Foliated.
    c. Capillary.
198. Tile-ore.
    a. Earthy.
    b. Indurated.
199. Azure copper-ore.
    a. Earthy.
    b. Indurated or radiated.
200. *Velvet copper-ore.*
201. Malachite.
    a. Fibrous.
    b. Compact.
202. Copper-green.
203. Ironshot copper-green.
    a. Earthy.
    b. Slaggy.
204. Emerald copper-ore.
205. Copper mica.
206. Lenticular-ore.
207. Oliven-ore.
208. Muriat of copper.
209. Phosphat of copper.

6. Iron Genus.

210. Native iron.
211. Iron-pyrites.
    a. Common pyrites.
    b. Radiated pyrites.
    c. Liver or hepatic pyrites.
    d. Cock's-comb pyrites.
    e. Cellular pyrites.

212. Capillary

212. Capillary pyrites.
213. Magnetic pyrites.
214. Magnetic ironstone.
  a. Common.
  b. Iron-sand.
215. *Chrome-ironstone.*
216. *Menac-ironstone.*
217. Iron-glance.
  a. Common.
    α. Compact.
    β. Foliated.
  b. Iron-mica.
218. Red ironstone,
  a. Red iron-froth.
  b. Ochry red ironstone.
  c. Compact.
  d. Red hematite.
219. Brown ironstone.
  a. Brown iron-froth.
  b. Ochry brown ironstone.
  c. Compact.
  d. Brown hematite,
220. Sparry ironstone.
221. Black ironstone.
  a. Compact.
  b. Black hematite,
222. Clay-ironstone.
  a. Reddle.
  b. Columnar clay-ironstone.
  c. Lenticular clay-ironstone.
  d. Jaspery clay-ironstone.
  e. Common clay ironstone,
  f. Reniform clay ironstone.
  g. Pea-ore, or pisiform ironstone.
223. Bog iron-ore.
  a. Morass-ore.
  b. Swamp-ore.
  c. Meadow-ore.
224. Blue iron-earth.
225. Pitchy iron-ore.
226. Green iron-earth.
227. Cube-ore.
228. Gadolinite.

7. Lead Genus.

229. Galena or Lead-glance.
  a. Common.
  b. *Disintegrated.*
  c. Compact,
230. Blue lead-ore.
231. Brown lead-ore.
232. Black lead-ore.
233. White lead-ore.
234. Green lead-ore.
235. Red lead-ore.
236. Yellow lead-ore.
237. Lead-vitriol.
238. Earthy lead-ore, or Lead-earth.
  a. Coherent.
  b. Friable.

8. Tin

8. Tin Genu .

239. Tin pyrites.
240. Tinstone.
241. Cornish tin-ore.

9. Bismuth Genus.

242. Native bismuth.
244. Bismuth-glance.
245. Bismuth-ochre.
246. *Arsenical bismuth-ore.*

10. Zinc Genus.

246. Blende.
    - *a.* Yellow.
    - *b.* Brown.
        - *α.* Foliated.
        - *β.* Fibrous.
        - *γ.* Radiated.
    - *c.* Black.

247. Calamine.

11. Antimony Genus.

248. Native antimony.
249. Grey antimony-ore.
    - *a.* Compact.
    - *b.* Foliated.
    - *c.* Radiated.
    - *d.* Plumose.

250. Black antimony-ore.
251. Red antimony-ore.
252. White antimony-ore.
253. Antimony-ochre.

12. Sylvan Genus.

254. Native sylvan.
255. Graphic-ore.
256. White sylvan-ore.
257. Nagyag-ore.

13. Manganese Genus.

258. Grey manganese-ore.
    - *a.* Radiated.
    - *b.* Foliated.
    - *c.* Compact.
    - *d.* Earthy.

259. Black manganese-ore.
260. *Piedmontese manganese-ore.*
261. Red manganese-ore.
262. *Manganese-spar.*

14. Nickel Genus.

263. Copper-nickel.
264. *Capillary-pyrites.*
265. Nickel-ochre.

15. Cobalt

15. Cobalt Genus.

Family of Speiss-Cobalt.

266. White cobalt-ore.
267. Grey cobalt-ore.
268. Glance-cobalt.

Family of Cobalt-Ochre.

269. Black cobalt-ochre.
    a. Earthy.
    b. Indurated.
270. Brown cobalt-ochre.
271. Yellow cobalt-ochre.
272. Red cobalt-ochre.
    a. Cobalt-crust.
    b. Cobalt-bloom.

16. Arsenic Genus.

273. Native arsenic.
274. Arsenic pyrites.
    a. Common.
    b. Argentiferous.
275. Orpiment.
    a. Yellow.
    b. Red.
276. Arsenic bloom.

17. Molybdena Genus.

27. Molybdena.

18. Scheele Genus.

278. Tungsten.
279. Wolfram.

19. Menachine Genus.

280. Menachan.
281. Octahedrite.
282. Rutile.
283. Nigrine.
284. Iserine.
285. Brown menachine-ore.
286. Yellow menachine-ore.

20. Uran Genus.

287. Pitch-ore.
288. Uran-mica.
289. Uran-ochre.

21. Chrome Genus.

290. Acicular-ore.
291. Chrome-ochre.

22. Cerium Genus.

292. Cerium-stone.

ARRANGEMENT

ARRANGEMENT

OF

SIMPLE MINERALS

IN THIS

WORK, 1816.

CLASS I.—EARTHY MINERALS.

1. Diamond Family.
2. Zircon Family.
3. Ruby Family.
4. Schorl Family.
5. Garnet Family.
6. Quartz Family.
7. Pitchstone Family.
8. Zeolite Family.
9. Azurestone Family.
10. Felspar Family.
11. Clay Family.
12. Clay-slate Family.
13. Mica Family.
14. Lithomarge Family.
15. Soapstone Family.
16. Talc Family.
17. Hornblende Family.
18. Chrysolite Family.
19. Basalt Family.
20. Dolomite Family.
21. Limestone Family.
22. Apatite Family.
23. Fluor Family.
24. Gypsum Family.
25. Boracite Family.
26. Baryte Family.
27. Hallite Family.

CLASS II.

CLASS II.—SALINE MINERALS.

ORDER I.—EARTHY SALTS.

Genus 1. Salts of Alumina.
Genus 2. Salts of Magnesia.

ORDER II.—ALKALINE SALTS.

Genus 1. Salts of Soda.
2. Salts of Potash.
Genus 3. Salts of Ammonia.

ORDER III.—METALLIC SALTS.

Genus 1. Salts of Iron.
2. Salts of Copper.
Genus 3. Salts of Zinc.
4. Salts of Cobalt.

CLASS III.—INFLAMMABLE MINERALS.

1. Sulphur Family.
2. Bituminous Family.
3. Graphite Family.
4. Resin Family.

CLASS IV.—METALLIC MINERALS.

Order 1. Platina.
2. Gold.
3. Mercury.
4. Silver.
5. Copper.
6. Iron.
7. Manganese.
8. Titanium.
9. Lead.
10. Zinc.
11. Tin.
Order 12. Bismuth.
13. Tellurium.
14. Antimony.
15. Molybdena.
16. Cobalt.
17. Nickel.
18. Arsenic.
19. Tungsten, or Scheelium.
20. Uranium.
21. Tantalum.

END OF VOLUME SECOND.

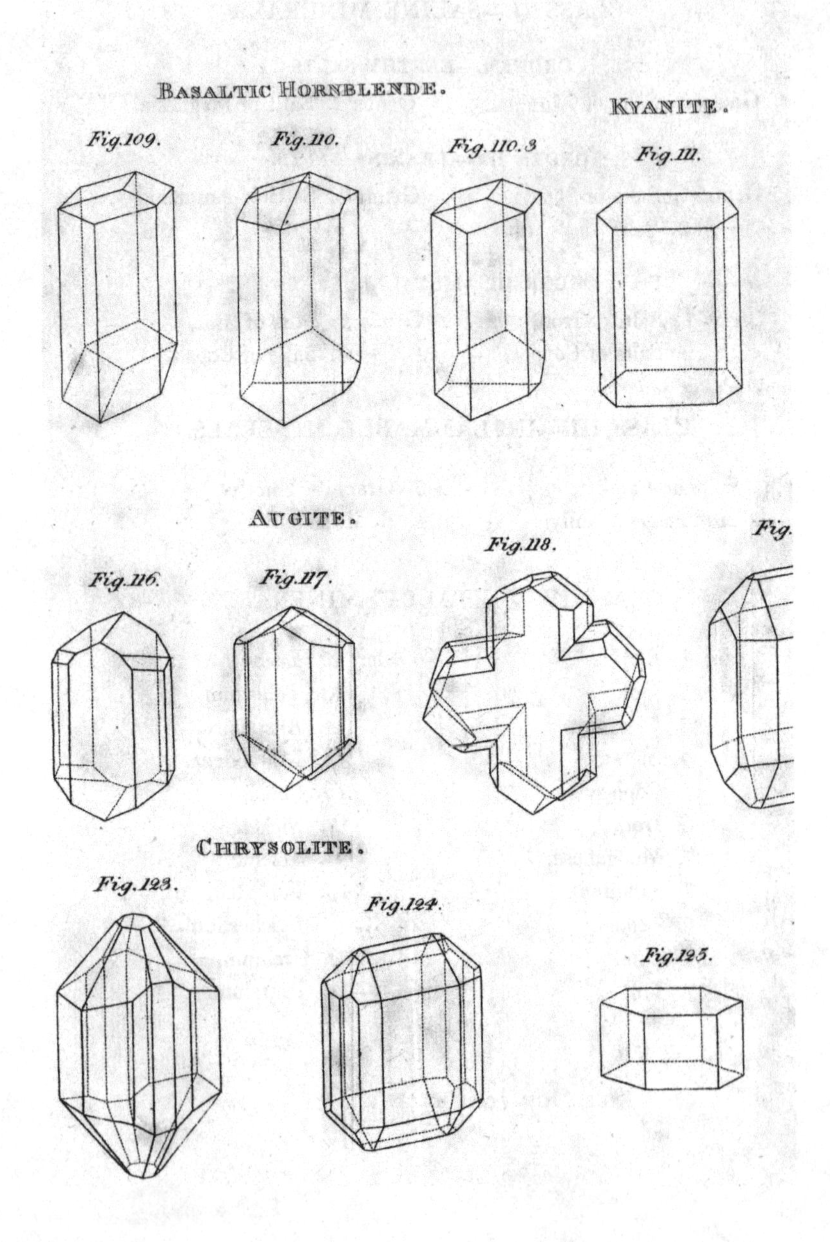

Basaltic Hornblende.
Kyanite.
Fig.109.
Fig.110.
Fig.110.3
Fig.111.
Augite.
Fig.118.
Fig.116.
Fig.117.
Chrysolite.
Fig.123.
Fig.124.
Fig.125.

AUGITE.

Fig. 112.

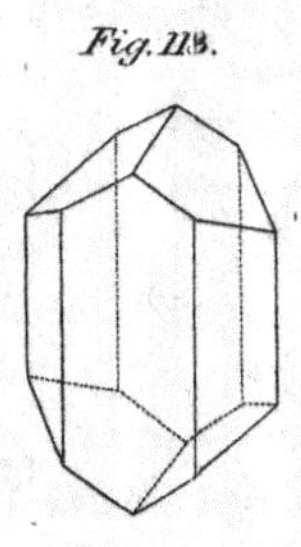

Fig. 113.

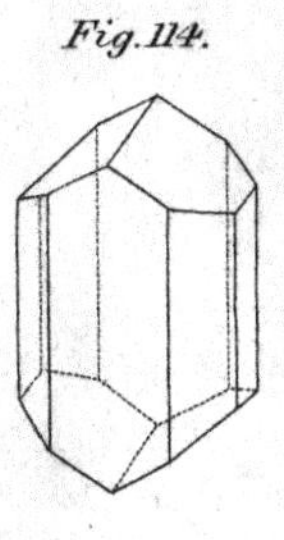

Fig. 114.

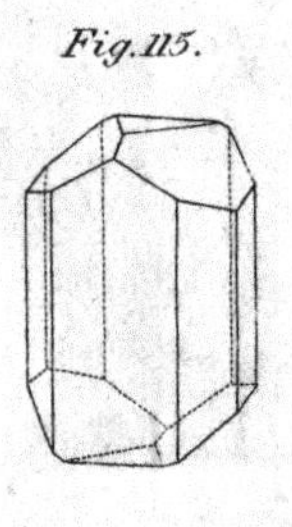

Fig. 115.

CHRYSOLITE.

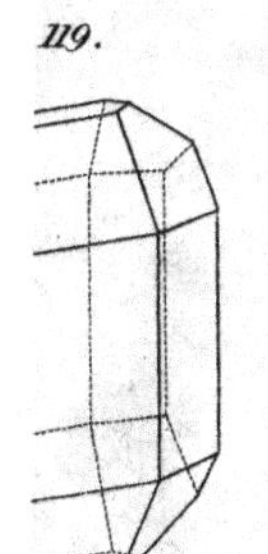

119.

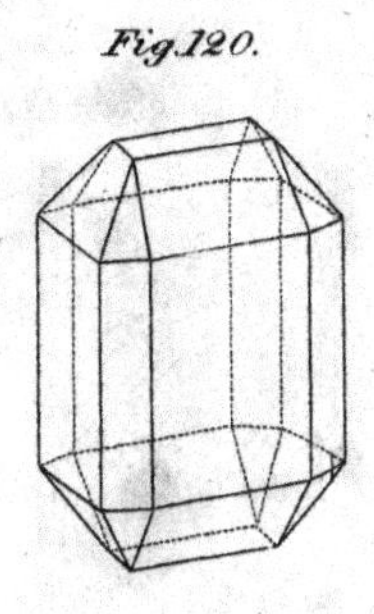

Fig. 120.

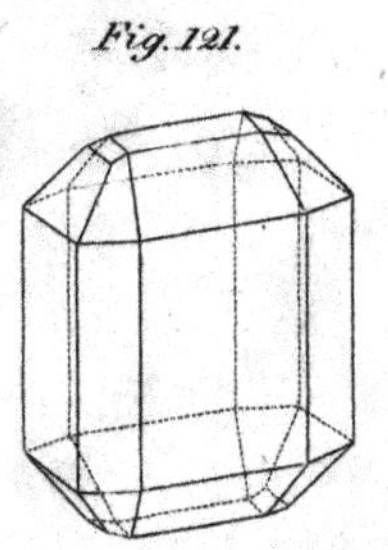

Fig. 121.

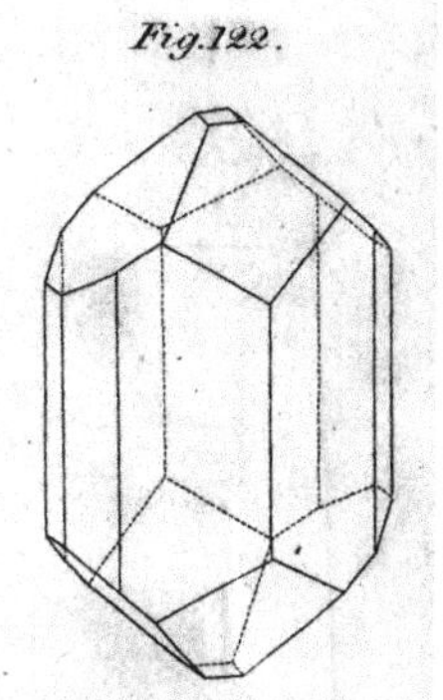

Fig. 122.

COMMON APATITE.

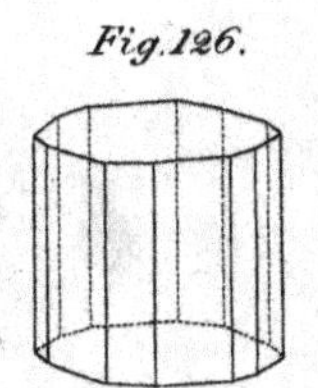

Fig. 126.

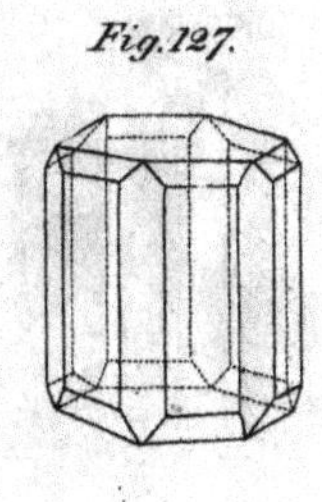

Fig. 127.

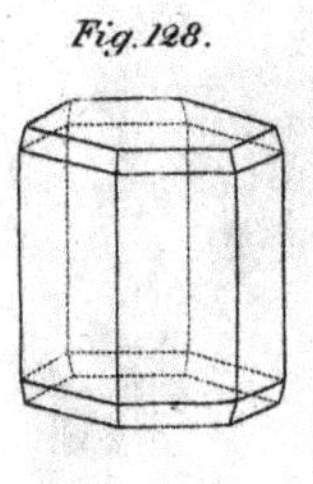

Fig. 128.

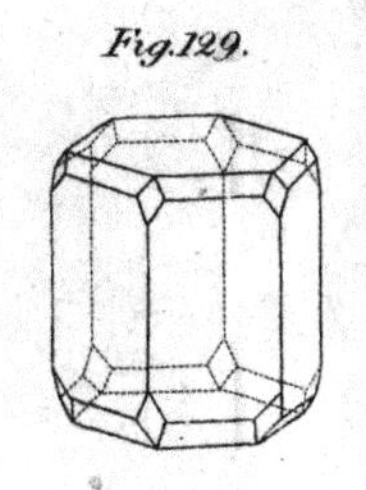

Fig. 129.

E. Mitchell sculp^t.

Conchoidal Apatite.

Fig. 130.

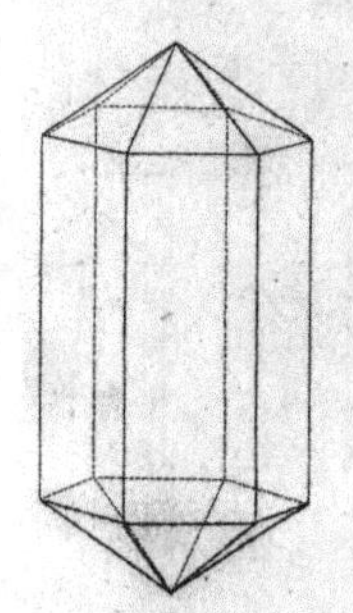

Fig. 131.

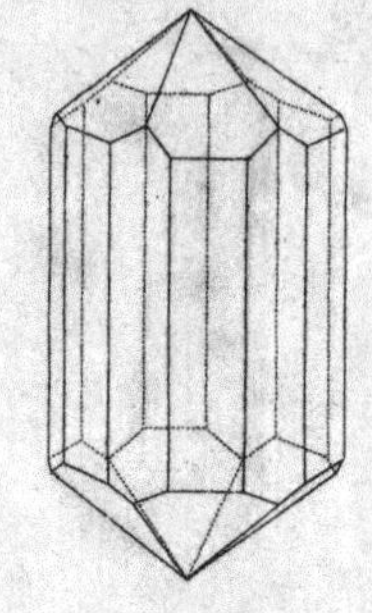

Fig. 132.

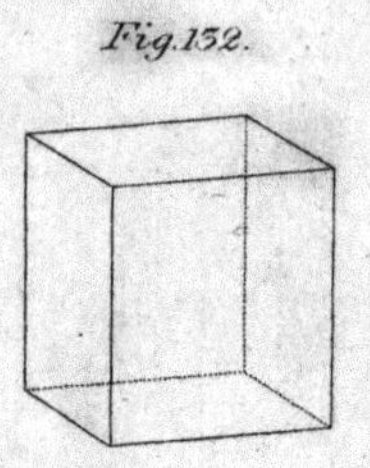

Fig. 133.

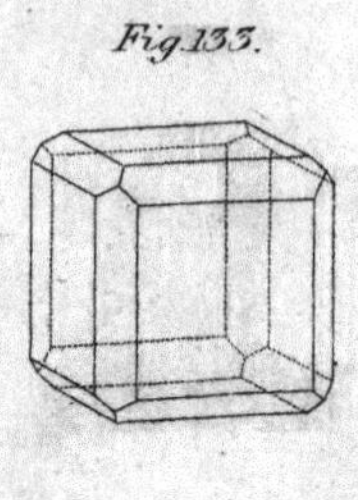

Fluor Spar.

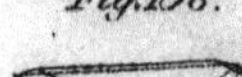

Fig. 138.

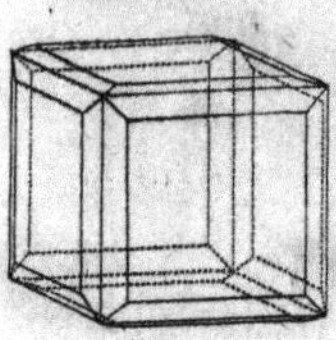

Fig. 139.

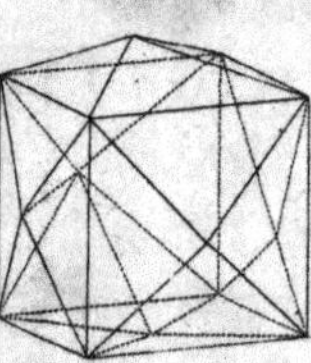

Fig. 140.

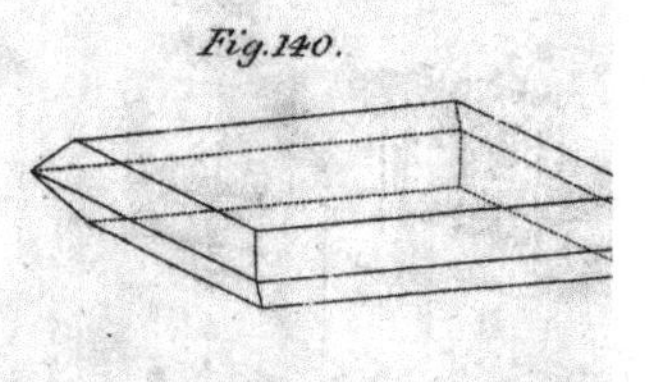

Prismatic Heavy Spar.

Fig. 143.

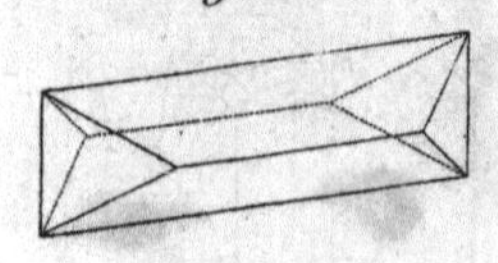

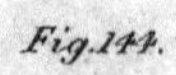

Fig. 144.

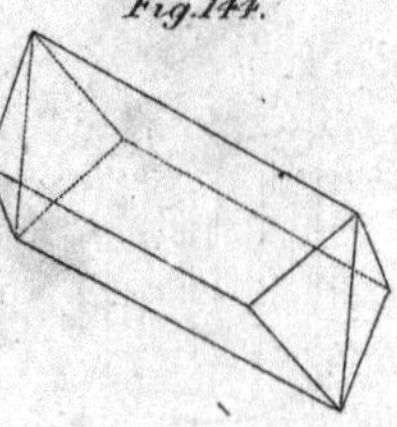

FLUOR SPAR.

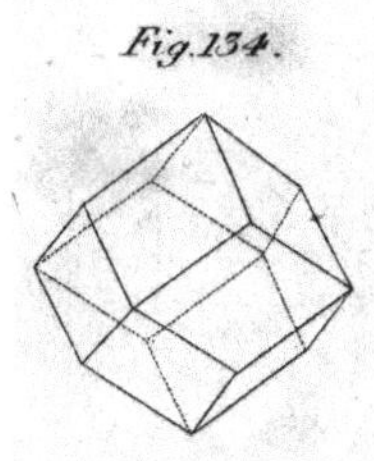
Fig. 134.

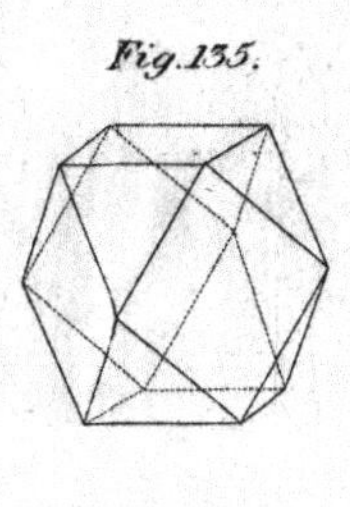
Fig. 135.

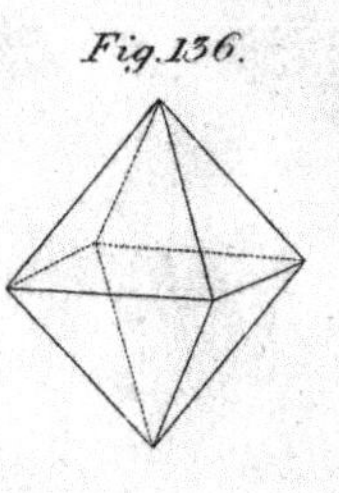
Fig. 136.

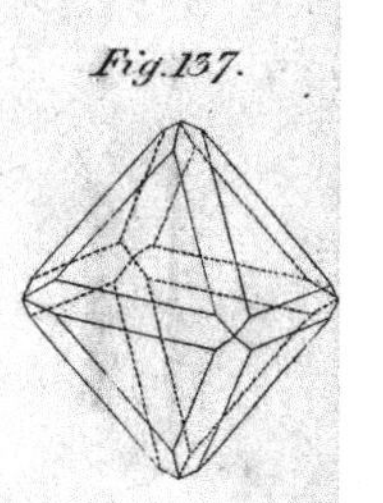
Fig. 137.

LAMELLAR HEAVY SPAR.

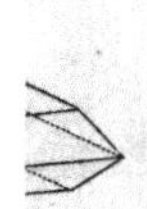

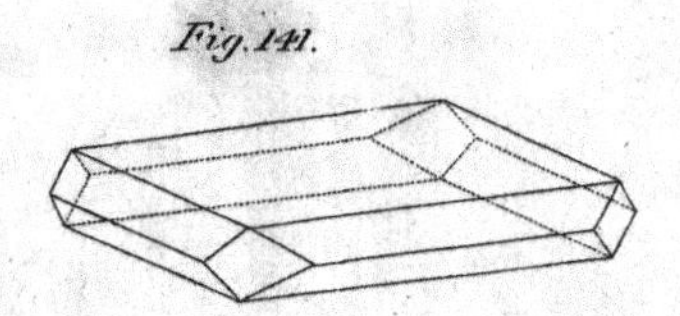
Fig. 141.

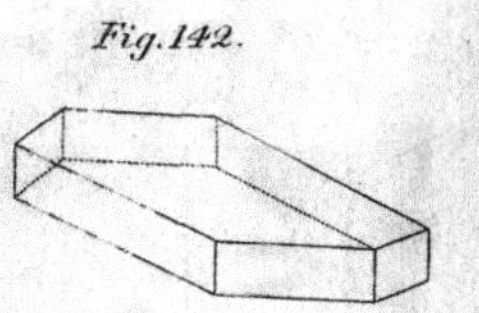
Fig. 142.

RADIATED CELESTINE.

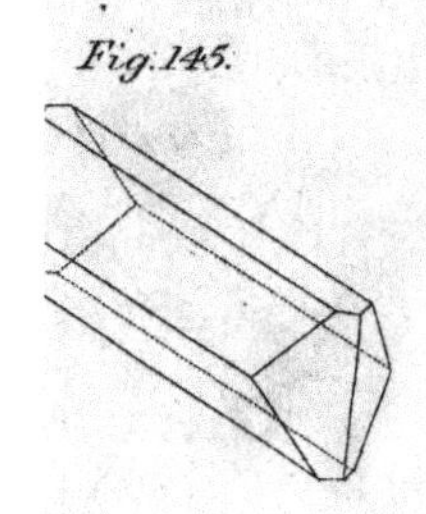
Fig. 145.

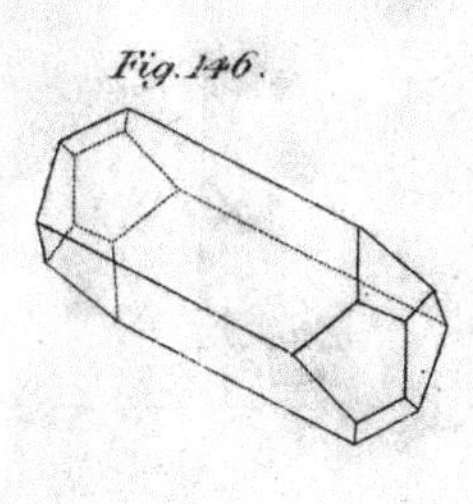
Fig. 146.

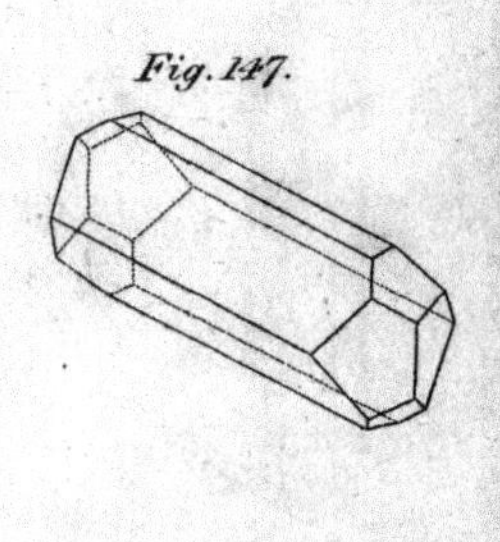
Fig. 147.

E. Mitchell sculp.t

NATIVE SUL

Fig. 148.

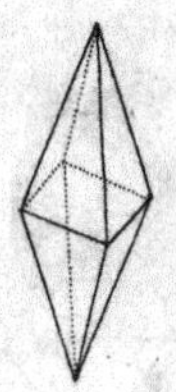

Fig. 149.

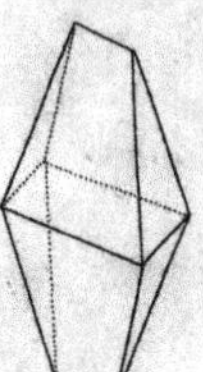

Fig. 150.

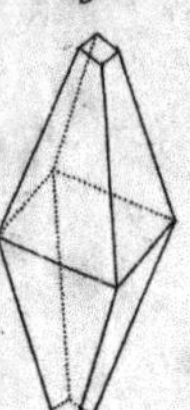

Fig. 151.

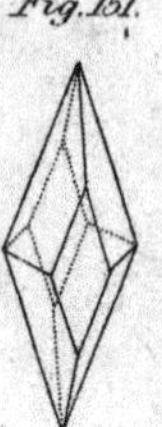

Fig. 152.

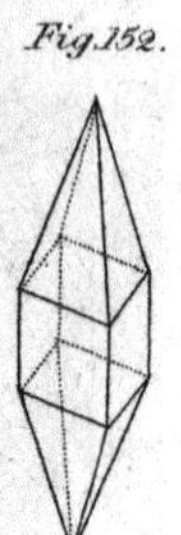

NATIVE GOLD, SILVER & COPPER.

Fig. 157.

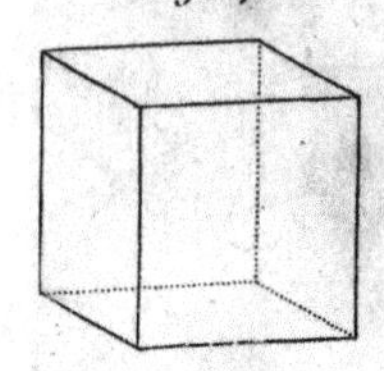

Fig. 158.

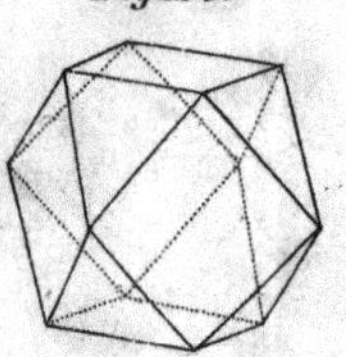

Fig. 159.

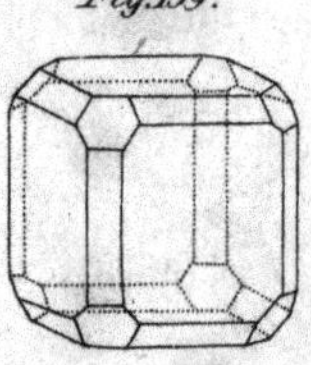

Fig. 160.

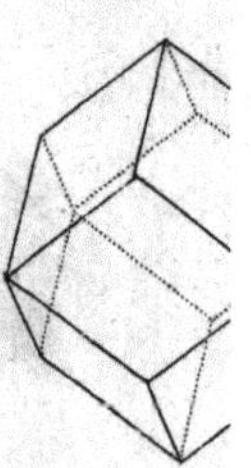

GREY COPP

Fig. 164.

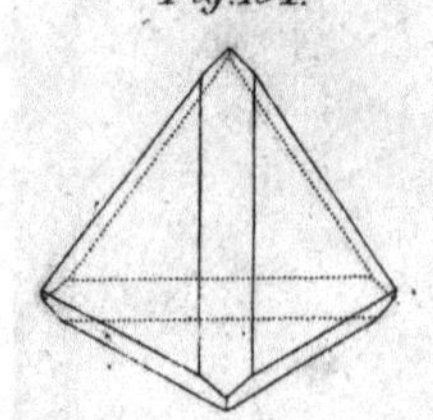

Fig. 165.

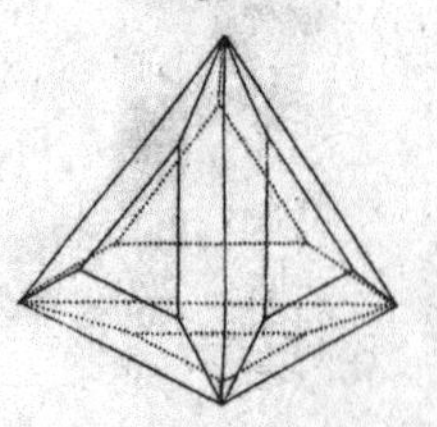

Fig. 1

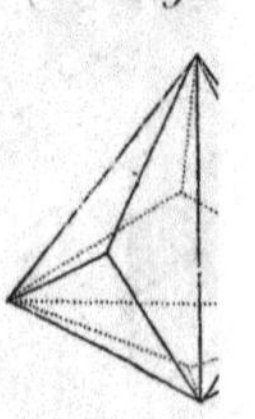

.PHUR.

Fig.153.

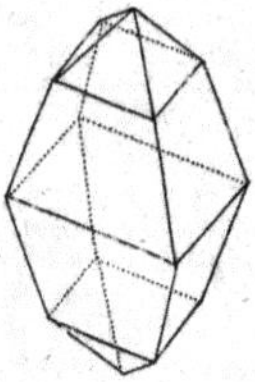

Fig.154.

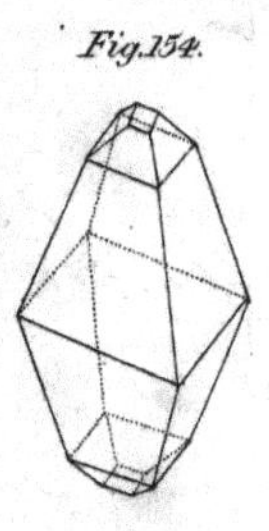

Fig.155.

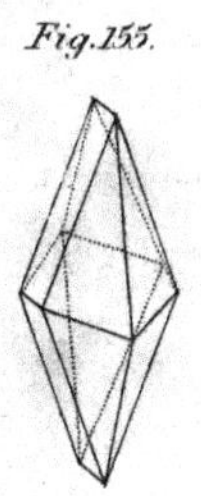

Fig.156.

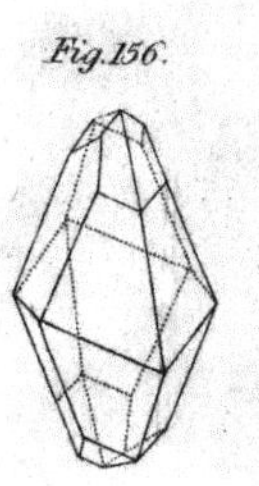

GREY COPPER ORE.

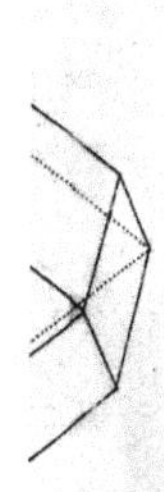

Fig.161.

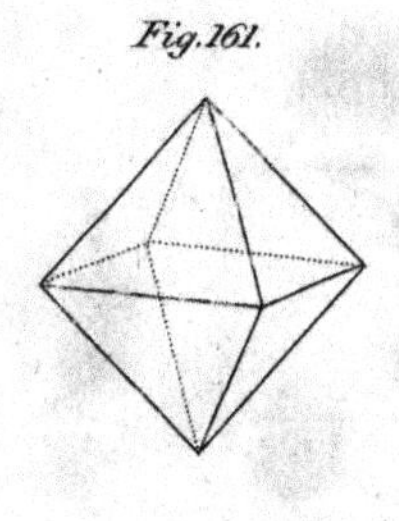

Fig.162.

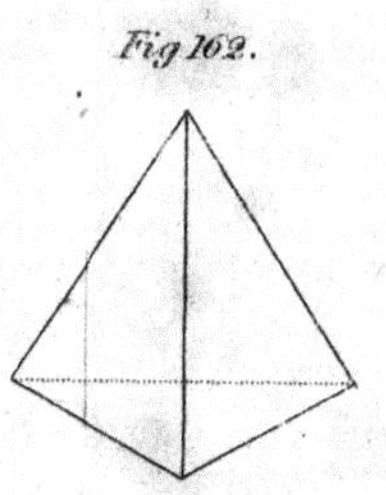

Fig.163.

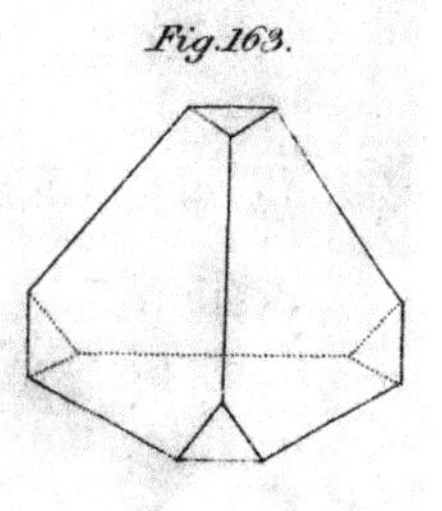

PER ORE.

166.

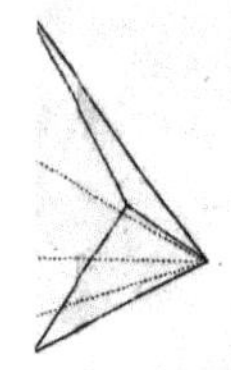

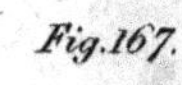

Fig.167.

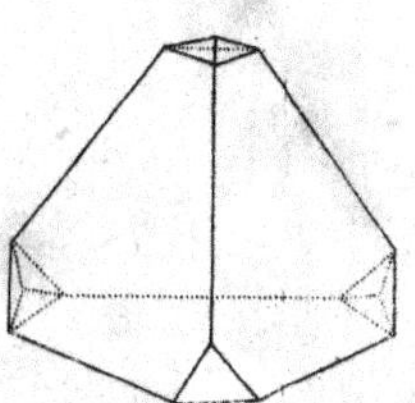

Fig.168.

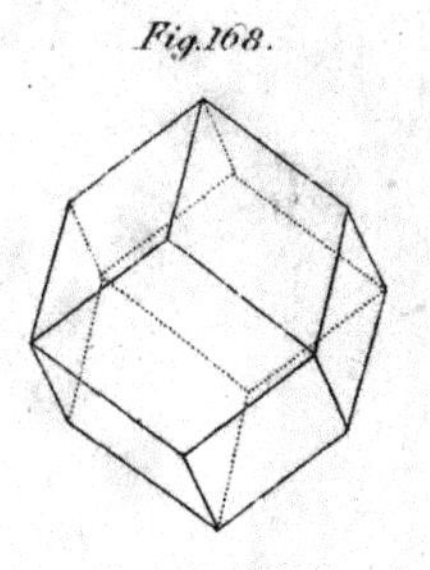

E. Mitchell, sculp^t.

Printed in the United States
By Bookmasters